OPTICAL PROPERTIES
AND SPECTROSCOPY
OF NANOMATERIALS

OPTICAL PROPERTIES AND SPECTROSCOPY OF NANOMATERIALS

Jin Zhong Zhang

University of California, Santa Cruz, USA

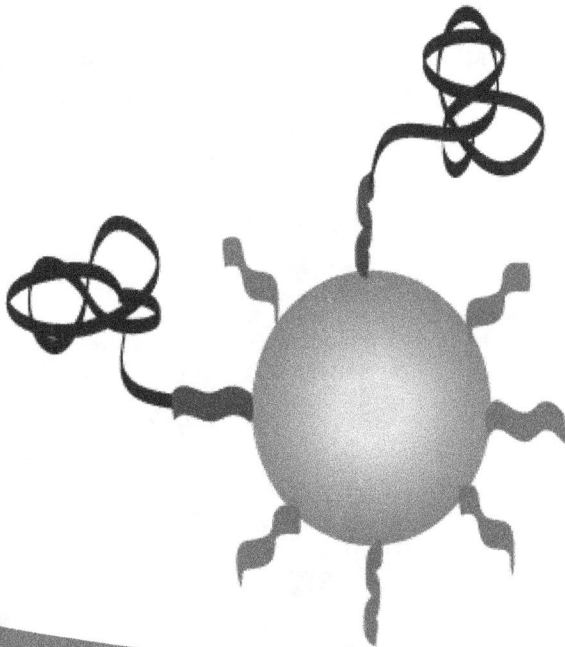

World Scientific

NEW JERSEY · LONDON · SINGAPORE · BEIJING · SHANGHAI · HONG KONG · TAIPEI · CHENNAI

Published by

World Scientific Publishing Co. Pte. Ltd.

5 Toh Tuck Link, Singapore 596224

USA office: 27 Warren Street, Suite 401-402, Hackensack, NJ 07601

UK office: 57 Shelton Street, Covent Garden, London WC2H 9HE

British Library Cataloguing-in-Publication Data
A catalogue record for this book is available from the British Library.

ISBN-13 978-981-283-664-9
ISBN-10 981-283-664-0
ISBN-13 978-981-283-665-6 (pbk)
ISBN-10 981-283-665-9 (pbk)

Typeset by Stallion Press
Email: enquiries@stallionpress.com

I wish to dedicate this book to all my teachers and mentors over the years, especially Mr. Wang Xue Yan, who first taught me how to read, count, and reason during the initial four years of my education.

Preface

This book is motivated by the need for an introductory level material focusing on the optical properties of nanomaterials and related spectroscopic techniques for upper level undergraduate and beginning graduate students who are interested in learning about the subject matter. While there are a number of excellent books on the market covering different aspects of nanomaterials, to date, there has not been a single monograph that specifically covers optical properties, optical spectroscopy and applications of nanomaterials.

Since optical properties are a major aspect of nanomaterials for both fundamental and technological reasons, I believe an introductory book specifically devoted to this subject is necessary and should be useful to both beginners and practitioners in the fields of nanoscience and nanotechnology. The objective is not to provide a comprehensive coverage or a review of all the nanomaterials studied to date, but rather to cover the very basics and illustrate the important fundamental principles and useful techniques with examples from recent literature. Given the fast pace of growth in nanoscience and nanotechnology, it is impossible to be comprehensive or inclusive. However, effort has been made to include as many significant and current examples as possible.

While nanomaterials are sometimes quite broadly defined to include inorganic, organic, biological, and various composite materials that involve a combination of these materials, this book will focus primarily on inorganic nanomaterials of semiconductor, metal and insulator. Some examples of more complex structures, including composites, will be briefly covered.

This book can be used as a textbook or used by students on their own as long as they have some basic knowledge of quantum chemistry and optical spectroscopy. I have strived to provide a balanced coverage of both the basic principles as well as related experimental optical techniques so that students can gain knowledge and skills that are practical and directly useful to them in their learning and research. I have also made a special effort to ensure that this book is relatively easy to read and follow and, if used for teaching, can be taught in roughly one quarter or semester. I have used many figures and illustrations to help the readability. A fair number of references, again not meant to be complete or comprehensive, are given wherever appropriate.

I welcome feedback from readers and will attempt to incorporate them in future editions of this book, if such an opportunity arises.

Jin Zhong Zhang
Santa Cruz, CA, USA
December 2008
zhang@chemistry.ucsc.edu

Acknowledgments

I would like to thank my mentors and many colleagues, collaborators, postdoctors, and students who have helped directly or indirectly with the writing of this book, through discussion, collaboration, and research work. An incomplete list of people to whom I wish to express my gratitude include: Ilan Benjamin, Rebecca Braslau, Mike Brelle, Frank Bridges, Guozhong Cao, Sue Carter, Ed Castner, Bin Chen, Jun Chen, Shaowei Chen, Wei Chen, Nerine Cherepy, Carley Corrado, Hai-Lung Dai, Elder De La Rosa, Hongmei Deng, Mostafa El-Sayed, Daniel Gamelin, Sarah Gerhardt, Daniel Gerion, Chris Grant, Claire Gu, Charles B. Harris, Greg Hartland, Eric J. Heller, Jennifer Hensel, Jianhua Hu, Thomas Huser, Dan Imre, Bo Jiang, Prashant Kamat, Alex Katz, Tav Kuykendall, Dongling Ma, Shuit-Tong Lee, Steve Leone, Chun Li, Can Li, Jinghong Li, Yadong Li, Yat Li, Tim Lian, Gang-yu Liu, Jun Liu, Xiaogang Liu, Tzarara Lopez-Luke, Glenn Millhauser, Rebecca Newhouse, Thaddeus Norman, Jr., Tammy Olson, Sergei Ostapenko, Umapada Pal, Cathy Phelan, Trevor Roberti, Lewis Rothberg, Nadya Rozanova, Archita Sengupta, George Schatz, Holger Schmidt, Adam Schwartzberg, Leo Seballos, Archita Sengupta, Chao Shi, Greg Smestad, Brian Smith, Jia Sun, Rebecca Sutphen, Chad Talley, Tony van Buuren, Changchun Wang, Zhong Lin Wang, Abe Wolcott, Fanxin Wu, Younan Xia, Xueming Yang, Kui Yu, Shuhong Yu, Yi Zhang, Zhongping Zhang, Yiping Zhao, Yingjie Zhu.

I wish to thank UCSC and a number of funding agencies for the partial financial support to do research in my lab over the years, and for the time I have spent on writing this book, including PRF-ACS, US NSF, US DOE and NSF of China.

I am grateful to my family, especially my parents, wife and daughters, for their unconditional support, love and understanding.

I wish to thank the book editor, Ms. Lakshmi Narayanan, for her wonderful and professional assistance.

Contents

Chapter 1

Introduction

Nanomaterials are cornerstones of nanoscience and nanotechnology. Many modern technologies have advanced to the point that the relevant feature size is on the order of a few to a few hundreds of nm. A classic example are computer chips with key features that now reach the length scale of <50 nm. The demand for smaller feature sizes is increasing at a faster and faster pace. At the fundamental level, there is an urgent need to better understand the properties of materials on the nanoscale level. At the technological front, there is a strong demand to develop new techniques to fabricate and measure the properties of nanomaterials and related devices. Fortunately, significant advancement has been made over the last decade in both fronts. It has been demonstrated that materials at the nanoscale have unique physical and chemical properties compared to their bulk counterparts and these properties are highly promising for a variety of technological applications.

One of the most fascinating and useful aspects of nanomaterials is their *optical properties*. Applications based on optical properties of nanomaterials include optical detector, laser, sensor, imaging, phosphor, display, solar cell, photocatalysis, photoelectrochemistry and biomedicine. Many of the underlying principles are similar in these different technological applications that span a variety of traditional disciplines including chemistry, physics, biology, medicine, materials science and engineering, electrical and computer science and engineering. It is essential for students in these various science and engineering fields to have a basic understanding of the fundamental optical properties and related spectroscopic techniques.

The optical properties of nanomaterials depend on parameters such as feature size, shape, surface characteristics, and other variables including doping and interaction with the surrounding environment or other nano-structures. The simplest example is the well-known blue-shift of absorption and photoluminescence spectra of semiconductor nanoparticles with decreasing particle size, particularly when the size is small enough. Figure 1.1 shows optical absorption spectra and colors of different sized CdTe nanoparticles or quantum dots (QDs) [1]. For semiconductors, size is a critical parameter affecting optical properties.

Likewise, shape can have dramatic influence on optical properties of metal nanostructures. Figure 1.2 shows optical absorption spectra and colors of hollow gold nanospheres with various outer diameter and wall thickness [2]. As can be seen, by simply controlling the physical dimensions, one can generate gold nanostructures with absorption covering the entire visible and near IR regions of the optical spectrum.

Doping is an effective and useful way to alter the optical, electronic, and magnetic properties of nanomaterials, particularly semiconductors and

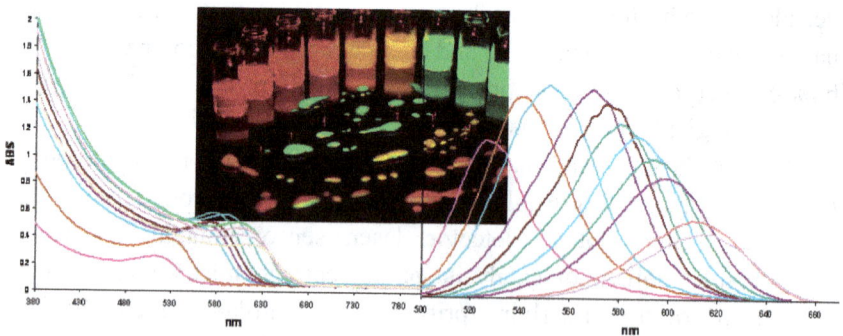

Fig. 1.1. UV-vis (left) and PL (right) spectra of CdTe QDs capped with thiolglycolic acid (TGA) in aqueous solution. The spectra were from aliquots taken from 1 to 42 h of reflux-ing. The exitonic peak position and emission spectra allow for the monitoring of particle growth. The image in the middle shows the different colors of the CdTe QDs with differ-ent average sizes, increasing in size from left to right or blue to red. Reproduced with permission from Ref. 1.

	Shell Diameter	Wall Thickness
—○—	24.5 ± 1.7 nm	5.7 ± 0.6 nm
—□—	28 ± 2.4 nm	6 ± 0.6 nm
—△—	32 ± 4.5 nm	5.8 ± 0.7 nm
—▽—	41 ± 3.7 nm	6.8 ± 0.5 nm
—◁—	41 ± 3 nm	6.3 ± 0.65 nm
—▷—	37 ± 2 nm	5.6 ± 0.6 nm
—○—	36 ± 2 nm	3.7 ± 0.6 nm
—◇—	44 ± 6 nm	5.4 ± 0.7 nm
—◇—	44 ± 5 nm	3 ± 0.6 nm

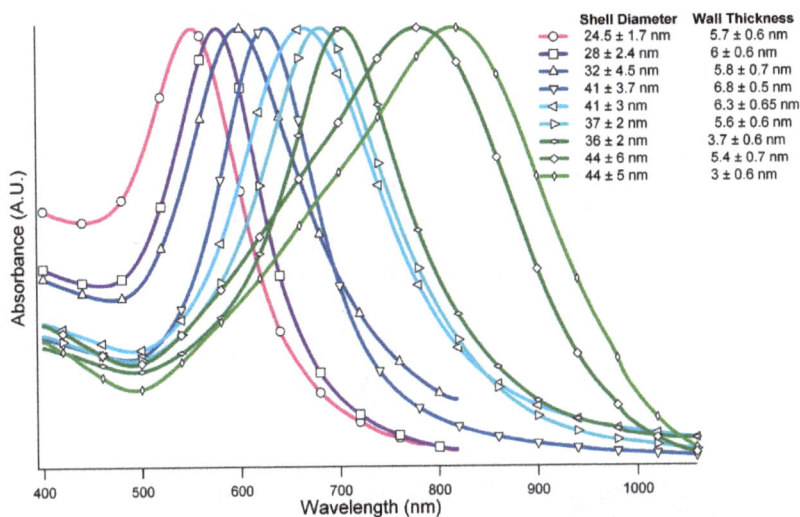

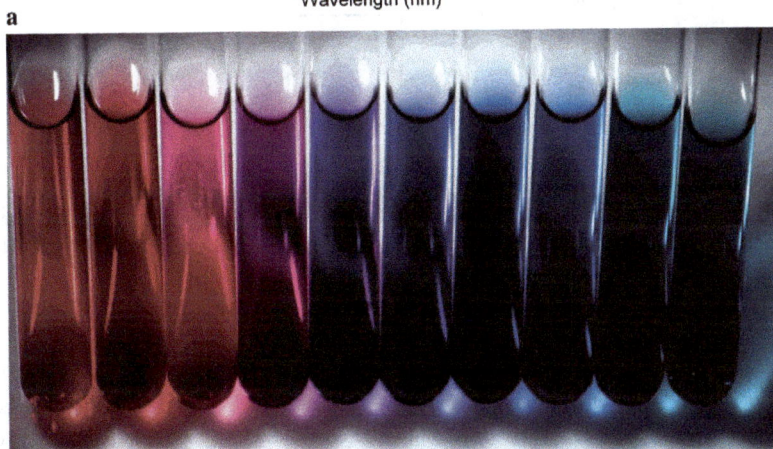

Fig. 1.2. (a) UV-visible absorption spectra of nine hollow gold nanosphere (HGN or HAuNS) samples with varying diameters and wall thicknesses. (b) Photograph showing the different colors of HGN solutions in different vials. The vial on the far left contains solid gold nanoparticles, the rest are HNGs with varying diameters and wall thicknesses. Reproduced with permission from Ref. 2.

insulators. Figure 1.3 shows up-conversion photoluminescence spectra of lanthanide-doped $NaYF_4$ nanoparticles [3]. This clearly illustrates the usefulness of doping for altering optical properties by controlling the chemical natures of the dopants, level of doping, and ratio of co-dopants

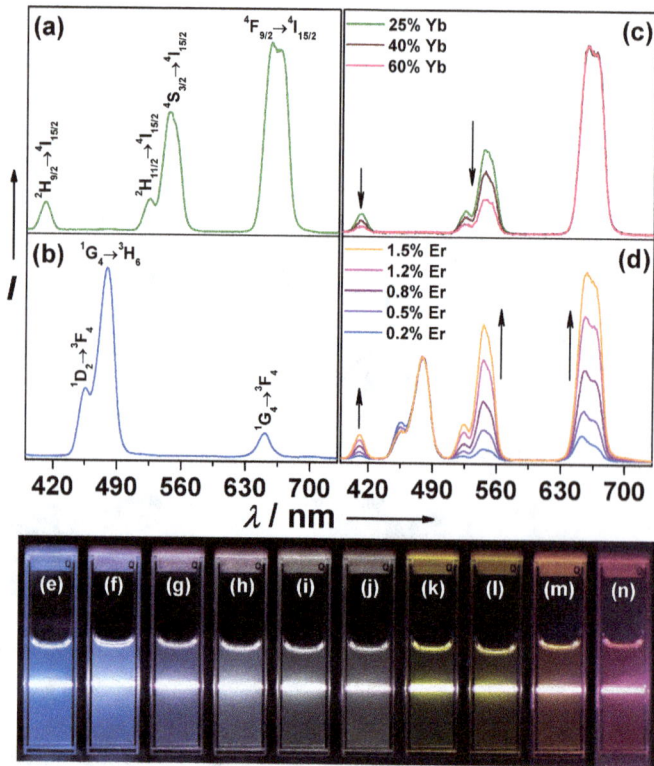

Fig. 1.3. Room temperature up-conversion emission spectra of (a) NaYF$_4$:Yb/Er (18/2 mol %), (b) NaYF$_4$:Yb/Tm (20/0.2 mol %), (c) NaYF$_4$:Yb/Er (25–60/2 mol %), and (d) NaYF$_4$:Yb/Tm/Er (20/0.2/0.2–1.5 mol %) particles in ethanol solutions (10 mM). The spectra in (c) and (d) were normalized to Er^{3+} 650 nm and Tm^{3+} 480 nm emissions, respectively. Compiled luminescent photos showing corresponding colloidal solutions of (e) NaYF$_4$:Yb/Tm (20/0.2 mol %), (f–j) NaYF$_4$:Yb/Tm/Er (20/0.2/0.2–1.5 mol %), and (k–n) NaYF$_4$:Yb/Er (18–60/2 mol %). The samples were excited at 980 nm with a 600 mW diode laser. The photographs were taken with exposure times of 3.2 s for e-l and 10 s for (m) and (n). Reproduced with permission from Ref. 3.

if used. Up-conversion is a nonlinear optical phenomenon that will be discussed in Chapter 5 later.

The above examples show the rich and fascinating optical properties that nanomaterials can exhibit. Indeed, novel optical properties manifest on the nanoscale for metal, semiconductor, and insulator. Besides fundamental interests, optical properties of nanomaterials are among the most exploited and useful properties for technological applications, ranging from sensing

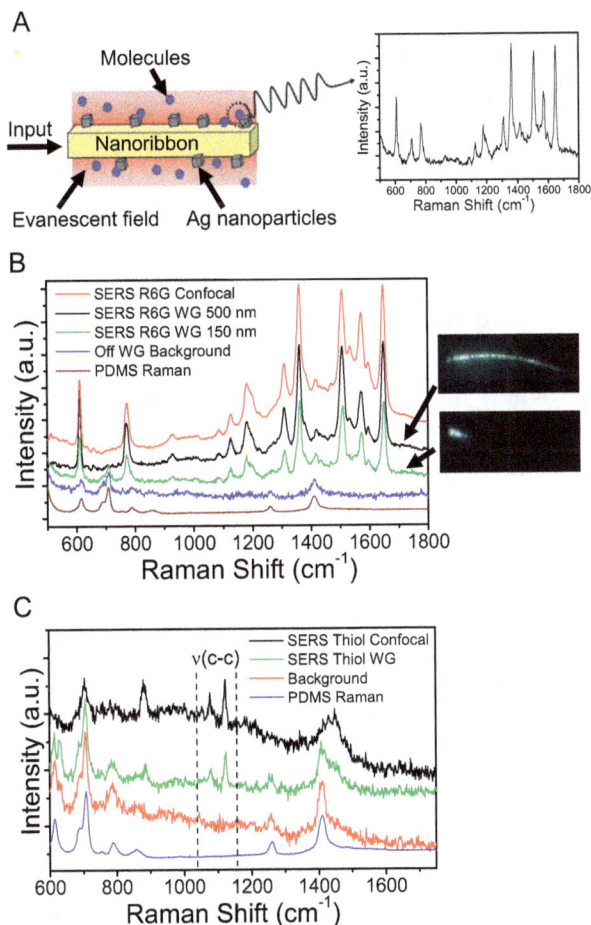

Fig. 1.4. Nanoribbon evanescent wave SERS sensor. (a) Cartoon of the sensing scheme. Analyte molecules in close proximity to a metal-decorated nanoribbon show amplified Raman scattering, which is then detected with the microscope objective. (b) Resonant SERS spectra of 100 μM Rhodamine 6G. Light is delivered to the particles by direct excitation (red line) or by the evanescent field of the waveguide (SERS R6G WG traces). Large ($d = 500$ nm) and small ($d = 150$ nm) waveguides yielded identical spectra. The background (blue trace) was acquired with the beam positioned off the end facet of the waveguide. A Raman spectrum of PDMS (brown trace) verifies the background results from PDMS. Insets: scattering images taken of the large (top) and small (bottom) waveguides. (c) Nonresonant SERS of bound dodecanethiol. Direct (black) and waveguide-excited (green) SERS spectra both show the distinct C–C stretching modes of the thiol ligands at 1080 cm^{-1} and 1122 cm^{-1}. Background (red) and PDMS Raman (blue) spectra are provided for clarity. Reproduced with permission from Ref. 7.

and detection, optical imaging, light energy conversion, environmental
protection, biomedicine, food safety, security and optoelectronics. For
example, many chemical and biochemical sensors and detection
systems take advantage of the unique optical properties of nanomaterials,
e.g. based on surface enhanced Raman scattering (SERS) [4–7] or fluorescence [8–11]. Figure 1.4 shows an example of a nanoribbon-based
optical SERS sensor that takes advantage of the waveguiding property of
the nanoribbon for improving the signal [7]. The principle of SERS will
be discussed in more detail in Chapter 7.

An area of growing interest is solar energy conversion in an environmentally friendly manner. Nanomaterials offer some unique features
including large surface-to-volume ratio, fast charge transport to surface, and controllable morphology, ease for large area processing, and
potential for inexpensive fabrication on flexible substrates. An example
is the so-called Gratzel cell making use of dye-sensitized TiO_2
nanocrystals [12]. Figure 1.5 shows a schematic of the idea behind the

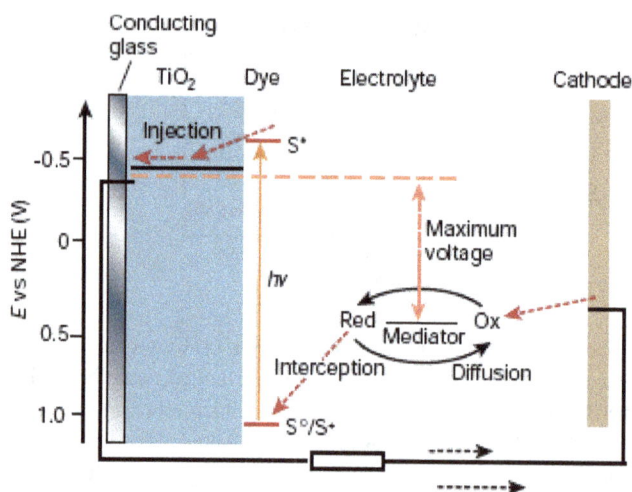

Fig. 1.5. Schematic of the operation of dye-sensitized electrochemical photovoltaic cell.
The photoanode, made of a mesoporous dye-sensitized semiconductor, receives electrons
from the photo-excited dye which is thereby oxidized, and which in turn oxidizes the
mediator, a redox species dissolved in the electrolyte. The mediator is regenerated by
reduction at the cathode by the electrons circulated through the external circuit. Adapted
with permission from Ref. 13.

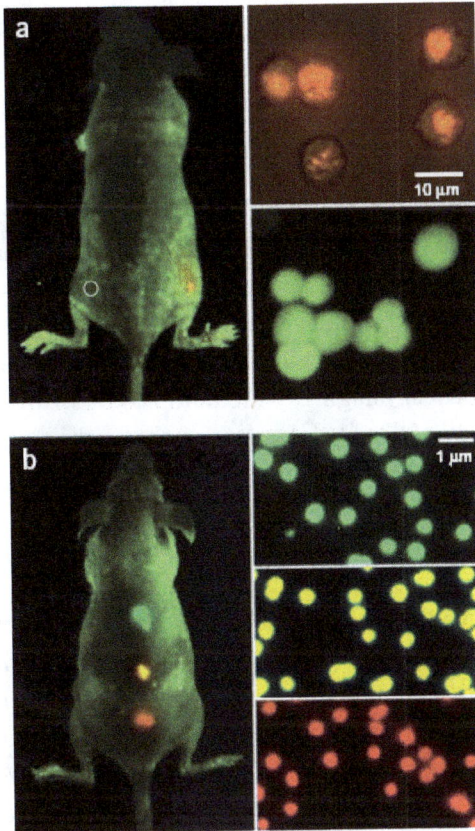

Fig. 1.6. Sensitivity and multicolor capability of QD imaging in live animals. (a, b) Sensitivity and spectral comparison between QD-tagged and GFP transfected cancer cells (a), and simultaneous *in vivo* imaging of multicolor QD-encoded microbeads (b). The right-hand images in (a) show QD-tagged cancer cells (orange, upper) and GFP-labeled cells (green, lower). Approximately 1,000 of the QD-labeled cells were injected on the right flank of a mouse, while the same number of GFP-labeled cells were injected on the left flank (circle) of the same animal. Similarly, the right-hand images in (b) show QD-encoded microbeads (0.5 μm diameter) emitting green, yellow or red light. Approximately 1–2 million beads in each color were injected subcutaneously at three adjacent locations on a host animal. In both (a) and (b), cell and animal imaging data were acquired with tungsten or mercury lamp excitation, a filter set designed for GFP fluorescence and true color digital cameras. Transfected cancer cell lines for high level expression of GFP were developed by using retroviral vectors, but the exact copy numbers of GFP per cell have not been determined quantitatively. Reproduced with permission from Ref. 16.

Gratzel solar cell that has reached solar to electrical conversion efficiency of over 10% [13].

Biomedical detection and treatment based on optical properties of nanomaterials is another area of fast growth. For example, in a process called photothermal ablation (PTA) for cancer imaging and treatment, hollow gold nanospheres with small size (30–50 nm), spherical shape, and narrow, strong as well as tunable surface plasmon resonance absorption have been demonstrated to be effective for both *in vitro* and *in vivo* imaging and destruction of human carcinoma and mice melanoma cancer tissues [14, 15]. Likewise, semiconductor QDs have been successfully used for *in vivo* cancer imaging, with multiple color capabilities based on different sized CdSe QDs [16]. Figure 1.6 shows the sensitivity and multicolor capability of QD imaging in live animals.

While we are at a historically high-tech era, there are still many challenges facing human beings. Examples include environmental change (global warming), increasing energy demand, limited water and other natural resources, health care and medicine, and food stocks for the ever growing population. While nanomaterials and nanotechnology should not be expected to solve all the problems we face, they are anticipated to play a critical role in addressing many scientific and technological issues, when hopefully used properly. Like any new technology, potential side-effects or adverse effects of nanomaterials and nanotechnology need to be addressed at the same time and are indeed receiving increasing attention in the scientific as well as non-scientific community.

References

1. A. Wolcott, D. Gerion, M. Visconte, J. Sun, A. Schwartzberg, S.W. Chen and J.Z. Zhang, *J. Phys. Chem. B* **110**, 5779 (2006).
2. A.M. Schwartzberg, T.Y. Olson, C.E. Talley and J.Z. Zhang, *J. Phys. Chem. B* **110**, 19935 (2006).
3. F. Wang and X.G. Liu, *J. Am. Chem. Soc.* **130**, 5642 (2008).
4. C.R. Yonzon, C.L. Haynes, X.Y. Zhang, J.T. Walsh and R.P. Van Duyne, *Anal. Chem.* **76**, 78 (2004).
5. X.Y. Zhang, M.A. Young, O. Lyandres and R.P. Van Duyne, *J. Am. Chem. Soc.* **127**, 4484 (2005).

6. Y. Zhang, C. Gu, A.M. Schwartzberg and J.Z. Zhang, *Appl. Phys. Lett.* **87**, 123105 (2005).
7. D.J. Sirbuly, A. Tao, M. Law, R. Fan and P.D. Yang, *Adv. Mater.* **19**, 61 (2007).
8. C.R. Kagan, C.B. Murray, M. Nirmal and M.G. Bawendi, *Phys. Rev. Lett.* **76**, 1517 (1996).
9. C.R. Kagan, C.B. Murray and M.G. Bawendi, *Phys. Rev. B-Condensed Matter* **54**, 8633 (1996).
10. D.M. Willard, L.L. Carillo, J. Jung and A. Van Orden, *Nano Lett.* **1**, 469 (2001).
11. S.P. Wang, N. Mamedova, N.A. Kotov, W. Chen and J. Studer, *Nano Lett.* **2**, 817 (2002).
12. B. Oregan and M. Gratzel, *Nature* **353**, 737 (1991).
13. M. Gratzel, *Nature* **414**, 338 (2001).
14. M.P. Melancon, W. Lu, Z. Yang, R. Zhang, Z. Cheng, A.M. Elliot, J. Stafford, T. Olsen, J.Z. Zhang and C. Li, Mole. *Cancer Therapeutics* 7, 1730 (2008).
15. W. Lu, C. Xiong, G. Zhang, Q. Huang, R. Zhang, J.Z. Zhang and C. Li, *Clinical Cancer Research*, in press, 1111 (2008).
16. X.H. Gao, Y.Y. Cui, R.M. Levenson, L.W.K. Chung and S.M. Nie, *Nat. Biotechnol.* **22**, 969 (2004).

Chapter 2

Spectroscopic Techniques for Studying Optical Properties of Nanomaterials

Optical spectroscopic techniques are widely used in the study of optical properties of different materials including nanomaterials. The different techniques are usually based on measuring absorption, scattering or emission of light that contains information about properties of the materials. Commonly used techniques include electronic absorption (UV-vis), photoluminescence (PL), infrared red (IR) absorption, Raman scattering, dynamic light scattering, as well as time-resolved techniques, such as transient absorption and time-resolved luminescence. Other more specialized techniques include single molecular spectroscopy and nonlinear optical techniques such as second harmonic or sum frequency generation and luminescence up-conversion. These different techniques can provide different information about the molecular properties of interest. In this chapter, several common spectroscopic techniques are reviewed with emphasis on their principle of operation as well as spectral interpretation. The main objective is to explain how one can get useful physical information about the nanomaterials under study from the optical spectrum measured experimentally.

2.1. UV-visible electronic absorption spectroscopy

2.1.1. *Operating principle: Beer's law*

The basic operating principle of electronic absorption spectroscopy is based on the measurement of light absorption due to electronic transitions in a sample. Since the wavelength of light required for electronic transitions is typically in the UV and visible region of the electromagnetic radiation spectrum, electronic absorption spectroscopy is often called UV-visible or UV-vis spectroscopy. It is named electronic absorption spectroscopy because the absorption in the UV-visible regions involves mostly electronic transitions.

Based on Beer's law, the absorbance, A, is related to the incident light intensity, I_0, and transmitted light intensity, I, concentration of a solution sample, c, path length of the sample, l, absorption coefficient, α, and molar absorptivity, ε, by the following equation:

$$A = \log I_0/I = \varepsilon l c = \alpha c \qquad (2.1)$$

In an experiment, both I_0 and I can be measured and thus A can be determined experimentally. If l and c are known, the absorption coefficient can be determined by Eq. (2.1).

The absorption coefficient is wavelength dependent and a plot of α as a function of wavelength λ is the spectrum of interest. The spectrum is often characteristic of a given sample and reflects the fundamental electronic properties of the sample.

For solid samples, concentration, c, has no meaning and similar expression can be derived in terms of the thickness of the sample. The solid sample needs to be thin enough to avoid saturation of absorption.

2.1.2. Instrument: UV-visible spectrometer

Electronic absorption or UV-visible spectroscopy is one of the simplest and yet most useful optical techniques for studying optical and electronic properties of nanomaterials. This technique is based on the measurement of light absorption by a sample, typically using commercially available spectrometers at reasonable cost. As illustrated in Fig. 2.1, the intensity of light from a light source, e.g. a lamp, is measured by a light detector, e.g. photodiode, photomultiplier tube (PMT) PMT or charge coupled device (CCD) detector, without (blank) and with a sample between the light source and detector. If the sample absorbs light at some wavelength, the transmitted light will be reduced. The intensity of the transmitted light

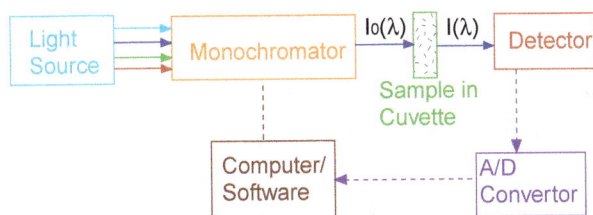

Fig. 2.1. Schematic diagram of key components of a typical UV-vis spectrometer that includes a light source, e.g. a lamp, monochromator to disperse the incident light, sample cuvette and holder, detector, e.g. PMT, photodiodes or CCD, analog-to-digital (A/D) convertors, and computer with software to control the scan of the monochromator and data acquisition.

plotted as a function of light wavelength will give a spectrum of the sample absorption. Most spectrometers cover the wavelength range from about 200 nm to 800 nm. Extending the measurement beyond 800 nm is possible but usually requires different light source, optics, and detector.

UV-visible spectrum is relatively easy to measure. However, there are still some common mistakes that beginners tend to make. First, the sample concentration cannot be too high. An optical density (OD) or absorbance (A), defined as $-\log_{10}(T)$ with T being the transmittance I/I_0, should usually be around 1 and not more than 3. Too high a concentration or OD leads to saturated absorption and distorted, unreal spectrum. Second, proper blank or background needs to be taken before the sample spectrum is measured. Ideally, one should use the same cuvette and solvent. If there are other species in the sample that could have absorption in the region measured but is of no interest, they should be part of the "blank" or background measurement if at all possible. Third, one needs to make sure that the sample is clear and has no floaters in a solution sample that are visible to the eye. Visible floaters will cause significant scattering and distort the measured spectrum. One common cause for floaters in nanomaterial solutions is agglomeration or aggregation of particles to form large structures with dimension on the scale of a few hundreds of nm or larger, thereby being visible to eye. If the sample contains only a small amount of aggregates, they can usually be filtered away or precipitated out by centrifugation before measurement is made.

2.1.3. *Spectrum and interpretation*

The electronic absorption spectrum contains information about the fundamental electronic properties of the sample, e.g. density of states, energy
levels and electronic dipole moment. In-depth discussion of electronic
absorption theory is beyond the scope of this book but can be found in
most quantum chemistry or spectroscopy books. Suffice to say here is that
at the most basic level it is the initial state electronic wavefunction, final
state wavefunction, and electrical dipole moment that determine an electronic transition. Density-of-states (DOS) plays an important role for
samples that involve multiple electronic transitions due to multiple electronic states involved for the initial states and/or final states, which often
result in broad spectral features.

Before discussing electronic absorption spectrum of nanomaterials, it
is useful to briefly review electronic spectrum of atoms and molecules.
Atomic spectrum results from electronic transition from a lower atomic
energy level, usually the ground electronic state, to a higher energy level
through interaction of the atom with an electromagnetic field such as
light. The transitions are governed by spectroscopic selection rules origi-
nating from electrical dipole moment. Other spectroscopic techniques can
be based on other moments such as magnetic dipole, electrical
quadrupoles or other higher order moments.

Likewise, for molecules electronic absorption spectrum is a result of
electronic transition from a lower electronic energy level or state to a
higher energy level. In most cases, the initial state is the ground electronic
state. Based on quantum mechanical perturbation theory for spectroscopy,
the electronic transition probability and intensity is determined by the
matrix element:

$$\langle \psi_f | \boldsymbol{\mu} \cdot \boldsymbol{E} | \psi_i \rangle \tag{2.2}$$

where ψ_i and ψ_f are the wavefunctions of the initial and final states, $\boldsymbol{\mu}$ is
the electrical dipole moment operator (a vector), and $\boldsymbol{E}$ is the applied
external electrical field (also a vector), e.g. light. The braket notation
implies complete integration over the entire space in all dimensions
involved by the wavefunctions and dipole moment operator. Since E is

usually considered as space independent on the scale of a typical molecule, it can be taken out of the integration or braket. The evaluation of Eq. (2.2) determines not only if there can be an electrically allowed transition (if the integration does not result in zero in value) but also how strong the transition is, i.e. the spectral intensity (proportional to the square of the value calculated by Eq. (2.2)).

Figure 2.2 illustrates schematically two typical situations involved in molecular electronic transitions. The left panel shows a transition from a bound stable ground electronic state to a bound, meta-stable excited electronic state, while the right panel shows a transition from a similar ground state to an unbound, unstable excited electronic state that is repulsive or dissociative. In this figure, $E(R)$, the nuclear potential energy is plotted as a function of the nuclear coordinate or the distance between two nuclei or atoms, R, which is usually called a *potential energy curve*. It is so called because $E(R)$ represents the potential energy in the Hamiltonian for the nuclear Schrödinger equation and it is the total energy for the electronic Schrödinger equation. The electronic energy levels have different functional dependence on the nuclear coordinate. The illustration is for a diatomic molecule. For molecules containing N atoms ($N > 2$), there are $3N-6$ (for nonlinear molecules) or $3N-5$

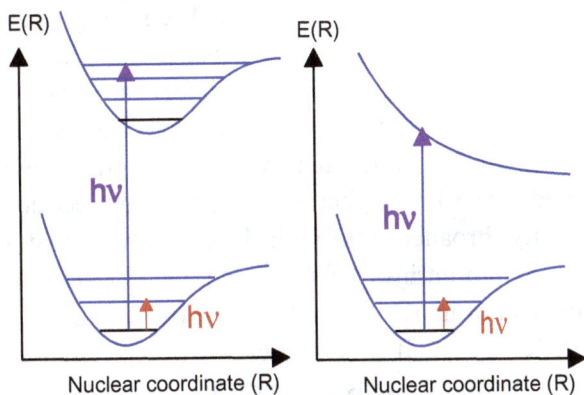

Fig. 2.2. Schematic illustration of electronic transitions in a diatomic molecule with two different excited state potential energy curves: bound (left) and repulsive (right). The long vertical arrows indicate electronic transition while the shorter red arrows indicate vibrational transition or IR absorption in the ground electronic state.

(for linear molecules) nuclear coordinates or modes. Thus, the electronic energy is a function of these multiple nuclear coordinates. A plot of the electronic energy as a function of these nuclear coordinates in more than one coordinate is often called a *potential energy surface or hypersurface*. Such potential energy surfaces are essential in describing chemical reactions that necessarily involve nuclear motion. In principle, such potential energy surfaces can be obtained from quantum mechanical electronic structure calculations by solving the electronic Schrödinger equation. In practice, this can be done usually only for small and light molecules. For large and heavy molecules, major approximations need to be introduced or models need to be developed to make the calculation practical. Nonetheless, Fig. 2.2 should serve as a convenient picture to consider electronic transitions in molecules.

The examples shown in Fig. 2.2 are the simplest possible scenarios, two states involved in a simple diatomic molecule with one vibrational mode and no interaction with other molecules. The electronic spectrum can be predicted or accurately calculated once the ground state and excited state potential energy curves and the operating dipole moment are known or obtained from electronic structure calculation. In reality, there are many possible factors that will make the spectrum as well as interpretation more complicated. First, all polyatomic molecules have more than one vibrational mode, as mentioned earlier. The nuclear potential energy is thus a function of all these modes. Some modes can be coupled or strongly interacting with each other. Second, there are many excited electronic states and some are coupled. Third, molecules can interact with each other, especially in liquids and solids, or with their environment or embedding medium. All these factors will make the spectrum more complicated, usually broader and with less resolved features due to homogeneous (due to intrinsic lifetime) and inhomogeneous (due to different environment of individual molecules) broadening.

Nanoparticles are small solid particles with typically a few tens to a few hundred or thousand atoms per particle. They can be considered as large molecules with $3N-6$ modes, with N being the number of atoms per particle. Their spectral features usually resemble those of large molecules in a condensed matter environment, e.g. solid or liquid. Figure 2.3 shows representative UV-vis absorption spectrum of CdS nanoparticles with

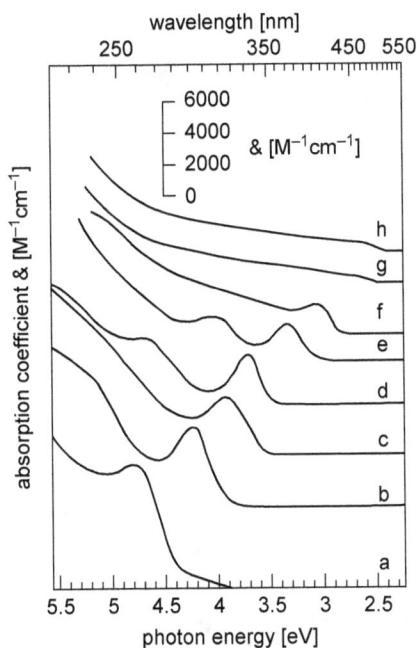

Fig. 2.3. UV-vis spectra of the isolated CdS nanoparticles. With decreasing nanoparticle size (from h to a), the excitonic transition is shifted toward higher energies and the molar absorption coefficient, which refers to the concentration of Cd, increases. The nanoparticles sizes are (all in Å): (a) 6.4; (b) 7.2; (c) 8.0, (d) 9.3; (e) 11.6; (f) 19.4; (g) 28.0; (h) 48.0. In this figure, the absorption coefficients refer to the analytical concentration of cadmium and not to the respective whole nanoparticle, i.e. the molarity refers to that of cadmium ions, not individual nanoparticles. This was done since the exact agglomeration number of the nanoparticles cannot be given, or the number of nanoparticles cannot be determined. Reproduced with permission from Ref. 1.

different sizes. As can be seen, the excitonic absorption peak blue shifts with decreasing particle size while the molar absorption coefficient, increases with decreasing size [1]. The first excitonic peaks are all blue shifted compared to that of bulk CdS.

Optical absorption is a fundamental property of materials including nanomaterials. Thus, the measurement of electronic absorption spectrum is essential to understanding the optical properties and applications of nanomaterials. The absorption will determine not only the color of the materials but also other important optical properties such as photoluminescence, lasing, electroluminescence, photovoltaic, photocatalytic, and photoelec-trochemical properties. Some of these properties will be elaborated further next or in later chapters.

2.2. Photoluminescence and electroluminescence spectroscopy

2.2.1. *Operating principle*

At the fundamental level, the principle underlying photoluminescence (PL) spectroscopy is very similar to that of electronic absorption spectroscopy. They both involve electronic transition of initial and final states coupled by the electrical dipole operator. The main difference is that the transition involved in PL is from a higher energy level or state to a lower energy level. There is also an important practical difference between the two techniques in that PL is a zero background experiment, i.e. no signal detected when there is no PL, which is in contrast to absorption spectroscopy that is a nonzero background experiment. Zero-background experiments are intrinsically more sensitive than nonzero background experiments. Therefore, PL is typically more sensitive than electronic absorption measurement.

A typical PL spectrum is just a plot of the PL intensity as a function of wavelength for a fixed excitation wavelength. A photoluminescence excitation (PLE) spectrum, however, is a measure of PL at a fixed emission wavelength as a function of excitation wavelength. To a good approximation, PLE is similar to the electronic absorption spectrum as long as no complications are involved, e.g. involvement of multiple over-lapping excited states or formation of excimers (excited dimers). PLE is useful for studying samples for which electronic absorption spectrum is challenging to obtain, e.g. due to low transmission as a result of thickness or high concentration of the sample.

2.2.2. *Instrumentation: spectrofluorometer*

Figure 2.4 show a schematic of the key components of a typical spectro-fluorometer used for PL measurement. While several components are similar to that in a UV-vis spectrometer, including light source, sample

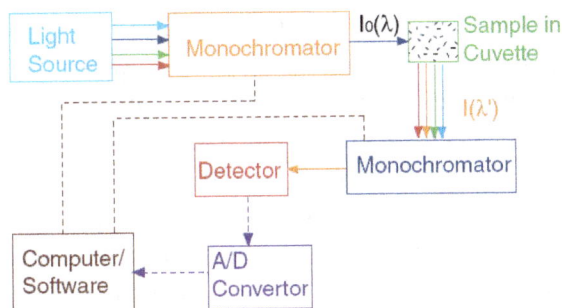

Fig. 2.4. Schematic illustration of key components of a typical spectrofluorometer that include a light source, e.g. a lamp, monochromator to disperse the incident light, sample cuvette and holder, a second monochromator to disperse the emitted light, detector, e.g. PMT, photodiodes or CCD, analog-to-digital (A/D) convertors, and computer with software to control the scan of the monochromators and data acquisition.

cuvette and detector, the detection scheme is different in that emitted light from the sample is detected, rather than transmitted light in UV-vis spectroscopy. Rayleigh scattering, which has the same wavelength as the excitation light, should be avoided in the detection of PL. Raman scattering can show up in a PL spectrum, especially when the Raman signal is strong while PL is relatively weak. The nature of Raman scattering will be discussed in Sec. 2.3 later.

Basically, in a typical PL measurement, a specific wavelength of light is selected from a light source by a monochromator, and directed at the sample of interest. Light emitted from the sample is collected through lenses, dispersed by another monochromator, and detected by a photo detector. The analog electrical signal generated by the photodetector is converted into a digital signal by an A/D (analog to digital) convertor and processed by software on a computer. The spectrum is displayed in terms of intensity of emitted PL light (proportional to the electrical signal generated) as a function of the wavelength of emitted light.

PL is usually red-shifted with respect to the incident excitation light, i.e. appearing at longer wavelength. Unwanted Rayleigh scattering is at the same wavelength of the incident light, so PL can be easily distinguished from Rayleigh scattering that is usually blocked by optical filters and excluded in the spectral range scanned. Raman scattering is usually much weaker than PL so it does not present a problem. However, when PL is weak and/or Raman is strong, e.g. from solvent molecules, Raman can be readily observed. For Stokes Raman scattering, to be explained later, the signal is also red-shifted with respect to the incident light, similar to PL. Thus, it is sometimes not easy to tell if the signal is Raman or true PL from the sample of interest. There are a couple of practical ways to determine PL versus Raman. First, Raman spectral features are usually much narrower than that of PL, especially for samples in liquid or solid forms. Second, and more reliably, Raman peaks should shift with changes in the excitation wavelength while PL usually does not, especially when the change in the excitation wavelength is small. The reason that the Raman peaks shift with excitation wavelength is that the energy difference between the Raman scattered light and the incident light is a constant, equal to the vibrational frequency of molecules, and therefore, shift in the excitation wavelength will result in shift in the Raman frequencies, in the same direction and with the same amount of energy or frequency.

2.2.3. *Spectrum and interpretation*

In order to extract useful physical information from the measured PL or Raman spectrum, it is necessary to understand the basic principles behind PL and Raman and their connection to the properties of the sample. Figure 2.5 shows a simple illustration of the PL and Raman scattering processes in a simple diatomic molecule with two different excited state potential energy curves. Interpretation of Raman scattering will be given in Sec. 2.3 later. We mention both Raman and PL here because of some common features they share. The focus is primarily on PL in this section. The data interpretation of PL (and Raman) is more complex for large molecules because of the many more nuclear degrees of freedom. In principle, one needs to consider all the possible nuclear degrees of freedom. In practice, however, only some modes are active and need to be considered.

The situation is even more complicated for liquids and solid due to strong inter-molecule interactions that have effects on the PL (and Raman) spectra. Therefore, PL and Raman can be used, in turn, to probe properties of individual molecules as well as interaction between molecules

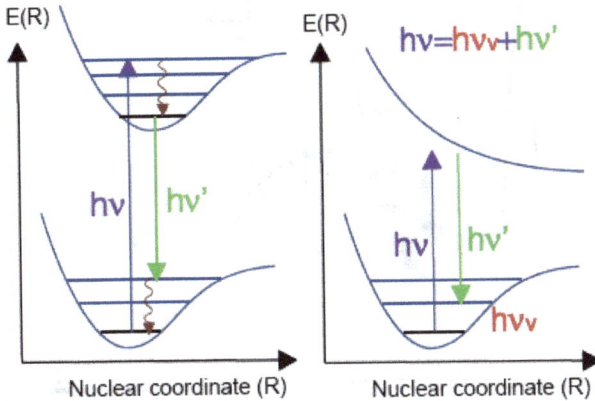

Fig. 2.5. Illustration of photoluminescence (left) from a bound excited state and (Stokes) Raman scattering involving a dissociative excited state (right).

in a liquid or solid. For example, in nanomaterials, there are unavoidably trap states due to defects or surface states in the bandgap for a semiconductor or insulator nanomaterial. PL measurement is usually much more sensitive to the presence of these trap states than electronic absorption spectroscopy, as to be discussed in more detail in Chapters 3 and 4. Therefore, PL is a useful tool for probing trap states besides the intrinsic electronic band structure. For instance, the PL spectrum from trap states provides useful information about the distribution and density of trap states within the bandgap. This is important for understanding the structural, surface and energetic properties of nanomaterials.

Besides PL spectrum, the PL quantum yield, measured as the intensity ratio of the emitted light over the absorbed light, provides important indications about the properties of the nanoparticles. For example, if the PL quantum yield is low, it usually means that the nanoparticles have a high density of trap states due to surface or internal defect of the materials. Such nanoparticles are not desired for applications that require high PL intensity.

PL can be measured from not only the pristine semiconductor or insulator but also from dopant in it. For example, if Mn^{2+} is doped into ZnS, the Mn^{2+} PL can be used to study the dopant as well as its interaction with the host semiconductor ZnS. In some sense, the dopant is acting like a

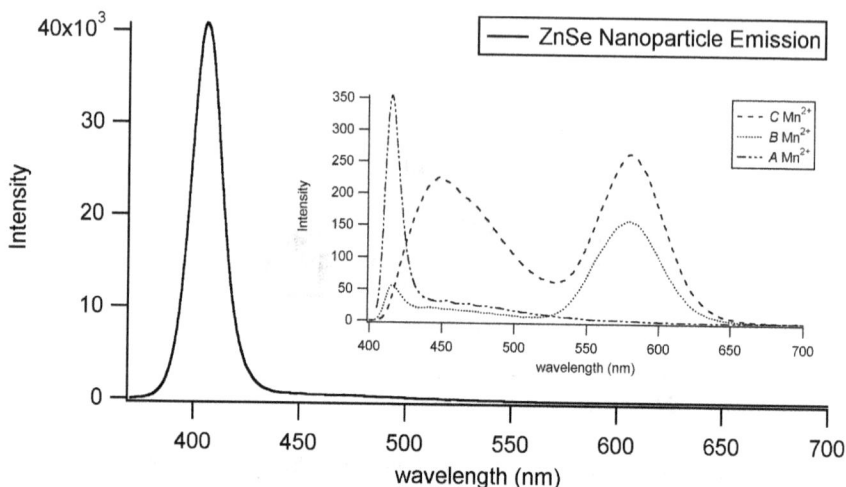

Fig. 2.6. Comparison of the luminescence spectra of undoped and Mn^{2+}-doped ZnSe nanoparticles samples A, B and C with increasing level of Mn^{2+} doping when going from A to B to C. The excitation wavelength was 357 nm for the undoped ZnSe and 390 nm for the doped nanoparticles (inset). Reproduced with permission from Ref. 2.

defect site, except that the dopant is often well-known and has well-defined properties that can be used as a probe for the nanomaterial. Figure 2.6 shows PL spectra of undoped and Mn^{2+}-doped ZnSe semiconductor nanoparticles, with the 580 nm band from Mn^{2+} ions while the bluer PL band is around 410 nm from the host ZnSe nanoparticles [2]. In this case, it was found, through combined structural and optical studies, that at low doping level (1% based on reactant ratio) the Mn^{2+} ions exists mainly on the surface of the particles and are not luminescent. At high doping levels (~6%), Mn^{2+} ions exist in two different sites, surface versus interior, that have different symmetry and optical properties, with the interior site luminescent while the surface site nonluminescent. The results suggest an important correlation between local structure of dopants and their luminescence properties. Doped materials are interesting for different applications including phosphors and they will be discussed further in Chapters 5 and 10.

Here we wish to offer a few words of advice for practical PL measurements. In PL measurements, one needs to be careful about possible appearance of Raman signal, as mentioned before and to be discussed further in the next Sec. 2.3. One also needs to be careful about high order Rayleigh scattering from the strong incident light. Mistaking Rayleigh or Raman scattering as PL signals can easily occur and some such mistakes have made into the literature. For example, if 400 nm light is used for excitation, multiples of 400 nm can appear in the measured PL spectrum at 800 nm or 1200 nm due to higher orders of the grating in a spectrometer, when the 400 nm Rayleigh scattered light is not completely blocked by optical filters. This can be understood from the grating equation:

$$d(\sin\theta_m + \sin\theta_i) = m\lambda \qquad (2.3)$$

where d is the grating constant or spacing per line, θ_i is the light incident angle, m is an integer (positive, negative or zero), λ is the wavelength of light, and θ_m is the angle at which the diffracted light has maxima. As can be seen from this equation, for a given wavelength of light, incident angle, and grating constant, there are multiple pairs of θ_m and m that can satisfy Eq. (2.3). If a strong peak at 800 nm is observed when 400 nm light is used for excitation, it is usually not real PL but likely due to the second order ($m = 2$) grating effect, since some of the 400 nm light can make into the spectrometer and be detected. This is also why the PL spectrum is usually measured by starting the scan from the red or lower energy compared to the incident light, to avoid the first order Rayleigh scattering due to the incident 400 nm light. However, this cannot eliminate the higher order scattering of the incident light if the scan covers that spectral range (e.g. 800 nm). Similarly, higher grating order Raman signal, besides first order, can also appear in a measurement. Therefore, care must be taken to identify and avoid such potential problems or artifacts in PL studies.

2.2.4. *Electroluminescence (EL)*

A technique related to PL is electroluminescence (EL), which underlies light emitting devices such as light emitting diodes (LEDs) [3]. The main difference is that in PL the excitation source is light while in EL the excitation is electrical or electronic [4–7]. In both cases, luminescence from the sample of interest is detected. In PL, the emission is a result of electron-hole recombination following the creation of the electron in the conduction and the hole in the valence band by a photon. In EL, the electron is injected into the conduction band and hole into the valence band, both electrically. Due to the different excitation mechanisms, PL and EL spectra for the same sample may be similar but are not expected to be identical in practice. For example, the contacts to the active nanomaterial in a real EL device, usually absence in PL measurement, could affect the relevant energy levels and the resulting EL spectra features.

Figure 2.7 shows a schematic of a single quantum dot LED device fabricated by using focused ion beam (FIB) doping and subsequent molecular beam epitaxy (MBE)-overgrowth (top) and EL spectra measured from this device (bottom) [8]. The EL spectra consist of sharp emission lines, which can be assigned to exciton recombination from a single quantum dot. Such devices are promising for developing electrically driven single-photon sources that are useful for many photonics applications including optical communication and computing.

2.3. Infrared (IR) and Raman vibrational spectroscopy

2.3.1. *IR spectroscopy*

Infrared and Raman are two common vibrational spectroscopy techniques useful for characterizing structural properties such as vibrational frequencies of molecules and phonons as well as crystal structures of solids. Since they often have different selection rules for transitions, they are complementary.

IR spectroscopy is based on the measurement of transmitted IR light through a sample. The absorbance measured as a function of frequency contains information about the vibration or phonon modes or frequencies of the sample. The key components for an IR spectrometer are similar to that of UV-visible spectrometer except that the light is in the IR and the detector and optical components such as gratings and mirrors all need to be appropriate for IR light. Various commercial spectrometers, including FTIR (Fourier transform IR), are usually available in most institutions.

The sample for IR spectroscopy measurement needs to be thin or dilute enough so Beer's law is valid or saturation can be avoided, similar to UV-visible spectroscopy. For molecules, the IR spectrum reflects their

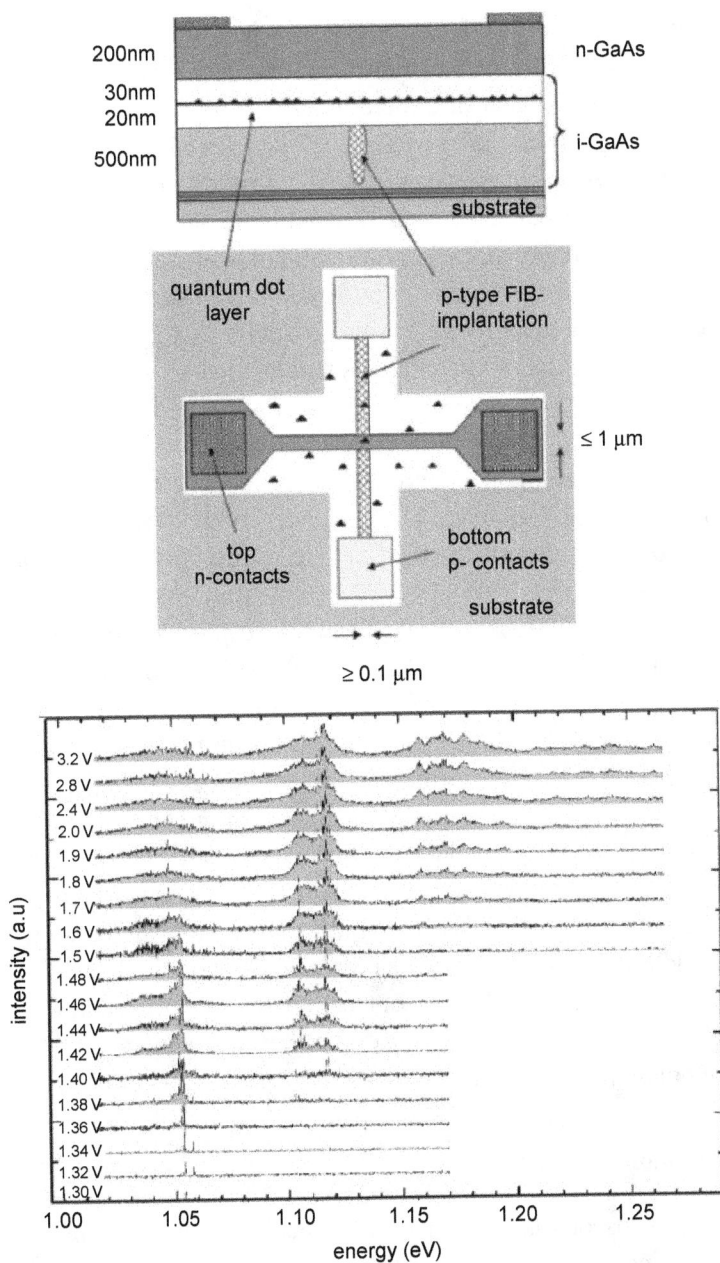

Fig. 2.7. (top) Schematic of a single-dot LED device, (bottom) EL-spectrum of single-dot LED, with increasing bias voltage the *s-*, *p-*, *d-*, *f*-exciton luminescence arise one after another, the shells are clearly separated and are saturated. Reproduced with permission from Ref. 8.

vibrational modes, determined by selection rules. For solid nanoparticles, the IR spectrum measures the phonon modes, also governed by appropriate selection rules that are determined by their crystal symmetry properties. In practice, most nanoparticles prepared chemically in solutions have molecules or ions on the particle surface, either intentionally or unintentionally introduced during synthesis. The IR spectrum therefore is composed of phonon modes of the solid nanoparticle and vibrational modes of the surface molecules or molecular ions. Interaction between the surface molecules or ions with the nanoparticles usually cause changes in the vibrational or phonon frequencies compared to isolated molecules and/or pure, naked nanoparticles. One can thus use the changes in frequencies measured in the IR spectrum to learn about the fundamental interaction between the nanoparticle and surface molecules [9–12]. This is important for both gaining a better understanding of their nature and strength of interaction as well as exploiting different applications of nanoparticles.

For example, FTIR spectroscopy has been applied to study the bonding of surfactant to FePt nanoparticles. Figure 2.8 shows the 800–1800 cm^{-1} region of the FTIR spectrum of FePt nanoparticles with no excess oleic acid and oleylamine surfactants [13]. This spectrum is due to oleic acid and oleylamine adsorbed on the surface of the FePt nanoparticles. The peak at 1709 cm^{-1} is due to the v(C=O) stretch mode and indicates that some fraction of the oleic acid is bonded to nanoparticles either in monodentate form or as an acid. The peak at 1512 cm^{-1} is due to the v_a(COO) mode and indicates the presence of bidentate carboxylate bonding to the nanoparticles. Similarly other peaks can also be assigned to the surfactant molecules associated with the FePt nanoparticles. This example shows that IR spectroscopy is indeed useful for probing interaction between surface bound molecules and nanoparticles.

2.3.2. *Raman spectroscopy*

Raman spectroscopy is somewhat more specialized than IR spectroscopy but is gaining in popularity. Raman spectroscopy is based on the phenomenon called Raman scattering, named after the Indian scientist Raman who first discovered it in 1928. In Raman scattering measurement, a single

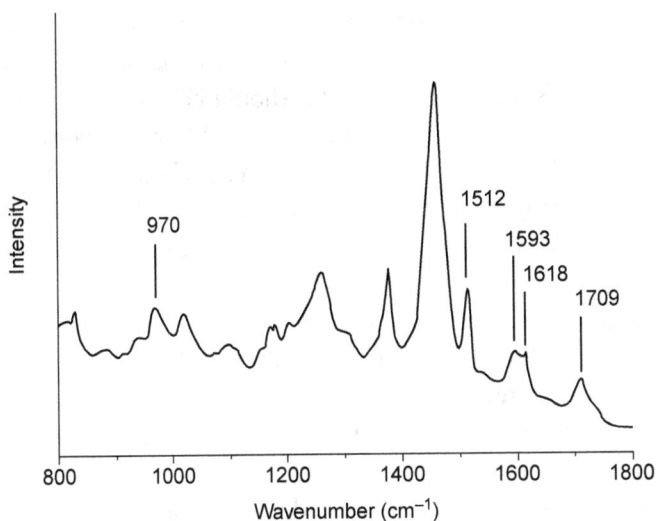

Fig. 2.8. FTIR spectra of surfactant (1:1 mixture, oleic acid and oleylamine) adsorbed on FePt nanoparticles during synthesis in 800–1800 cm^{-1} region. Reproduced with permission from Ref. 13.

frequency light, usually from a single mode laser source, shines on the sample and scattered light is measured off the angle with respect to the incident light to minimize Rayleigh scattering. The inelastically scattered light with lower (Stokes scattering) or higher (anti-Stokes scattering) frequencies can be measured with a photodetector. The energy difference between the scattered and incident light, so-called *Raman shift* (usually given in wave number cm^{-1} = $1/\lambda$ with wavelength λ expressed in cm), equals to the vibrational or phonon frequencies of the sample, as long as selection rules allow. The spectrum is usually presented in terms of the intensity of the Raman scattered light as a function of Raman shift.

Figure 2.5 illustrates the process of Stokes Raman scattering. There is substantial similarity between PL and Raman scattering, particularly resonance Raman scattering that involves electronic transitions between the ground and excited electronic states. Raman can be practically considered as PL involving extremely short-lived excited state, like the so-called "virtue state" for normal Raman [14]. If the excited state is very short lived (<100 fs), like in a dissociate state, resonant excitation would pro-duce primarily Raman scattering and no PL. On the other extreme, when the excited state is long lived (~ns), PL should be very strong and Raman would be comparatively very weak. A major difference between PL and Raman is that relaxation or energy loss in the excited state usually occurs for PL but not for Raman. As a result, Raman provides specific informa-tion about vibrational frequencies in the ground electronic state while PL is determined by the excited and ground electronic states.

It should be noted that even though the Raman shift typically appears in the region of a few hundred to a few thousand cm^{-1}, similar to that of IR measurement (a few to a few tens of microns), the signal detected is usually visible if the incident light is in the visible range. If UV light is used, the Raman signal would be in the UV or near UV range. Stokes Raman signal is most often detected and it always appears lower in energy or frequency compared to the incident light frequency. Since visible detectors are usually more sensitive than IR detectors, Raman detection is therefore considered more sensitive than IR detection. In addition, Raman is a zero-background measurement while IR is not, Raman should be more sensitive than IR in this regard as well. However, Raman scattering cross-section or quantum yield (QY) is often very low, on the order of 10^{-6} to 10^{-8}, a relatively high intensity light source, such as a laser, is usu-ally needed and the signal detection can be challenging sometimes or may require long data acquisition time.

Raman signal can be enhanced by using light frequency that is on res-onance with an electronic transition of the sample, namely *resonance Raman scattering*. While the signal can be enhanced by a factor of 100–1000, sample degradation often occurs due to the electronic absorp-tion of the incident light and subsequent photochemical reactions. The resonance enhancement is due to the fact that the Raman signal depends on the "detuning" or energy difference between the incident photon fre-quency and the energy difference between the electronic and ground electronic states. As this energy difference becomes smaller (more on res-onance), the Raman signal becomes larger [14].

Another method to enhance the Raman signal is surface enhanced Raman scattering (SERS) that occurs when a molecule is on or near the surface of a metal nanostructure (substrate) that absorbs the incident light and creates an effectively enhanced electromagnetic (EM) field near the surface. The enhancement is mainly due to enhanced absorption of photons, not so much in terms of scattering quantum yield. The enhancement is usually attributed to EM enhancement [15]. But "chemical effect" has been introduced to account for some of the enhancement observed. The chemical effect is less well defined but could be related to the surface effect. It is becoming clear that the property of the substrate surface is important for SERS, e.g. in influencing how strongly the molecules interact or bind to the substrate surface that in turn will influence the SERS signal. A more detailed discussion of SERS will be given in Chapter 7.

As mentioned earlier, Raman signal can often be observed in practical PL measurements, especially when the PL signal is weak. The Raman signal from the embedding medium, such as solvent, can be easily seen due to the very high concentration. Raman peaks are often narrower than PL peaks, especially for most nanoparticles. A simple way to verify if a sharp peak is Raman or PL is to change the excitation light wavelength a little bit, say 10–20 nm, and to observe if the peak position changes accordingly. If the peak shifts with a change of the excitation wavelength, it should be Raman scattering; if it does not, it should be PL.

Both IR and Raman have been used for characterizing vibration or phonon features of nanoparticles or molecules associated with nanoparticles such as surfactants or adsorbates. For example, UV Raman has been used to study phase transitions of TiO_2 nanoparticles. Figure 2.9 shows Raman spectra of TiO_2 nanoparticles with a mixture of anatase and rutile phases [16]. The ratio between the two phases varied with calcinating temperatures, with higher temperature favoring the rutile phase. This example demonstrates that Raman is useful for structural determinations.

2.4. Time-resolved optical spectroscopy

While most optical properties of nanomaterials have been characterized using frequency domain techniques, i.e. spectroscopy, time-domain studies based on pulsed lasers can provide useful and complementary information about the optical as well as dynamic properties. Lifetimes as short as tens

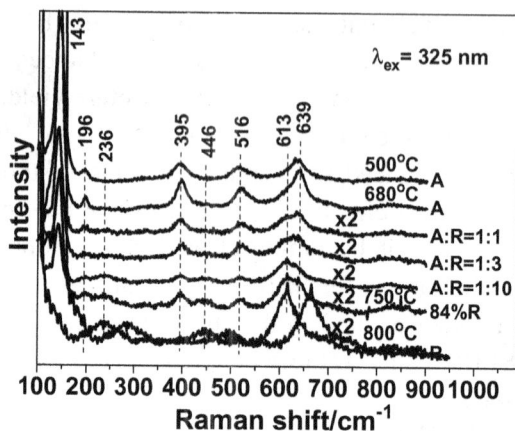

Fig. 2.9.   UV Raman spectra of TiO_2 calcinated at different temperatures and the mechanical mixture with 1:1, 1:3, and 1:10 ratios of pure anatase phase to pure rutile phase with the excitation line at 325 nm. The "A" and "R" denote the anatase and rutile phases, respectively. The symbol "84%R" represents that the weight fraction of rutile phase in the sample is 84% based on UV Raman spectrum. Reproduced with permission from Ref. 16.

of femtoseconds (fs) can be readily measured nowadays using modern ultrafast lasers. This is particularly important for condensed matter systems such as nanomaterials that involve ultrafast dynamic processes and significant inhomogeneous spectral broadening, for which conventional spectroscopy cannot be used to obtain dynamic information.

One common time-resolved laser technique is transient absorption (TA). TA measurement is based on two short laser pulses, one to excite or pump the sample to an excited state, and the second to interrogate or probe the excited state population by monitoring its change in absorbance. The second pulse is delayed in real time with respect to the first pulse using a translation stage [17]. Figure 2.6 shows a schematic of a typical experimental setup that includes an fs or ps laser, harmonic generation, white light generation, pump-probe configuration, and detection system. Typically, the fs pulses are generated by an fs laser system (often involving the use of amplified lasers for high peak power) with a specific center wavelength (e.g. 780 nm), repetition rate (e.g. 1 KHz), energy (e.g. 1 mJ/pulse), and pulse duration (e.g. 100 fs). The laser pulses can be used to generate

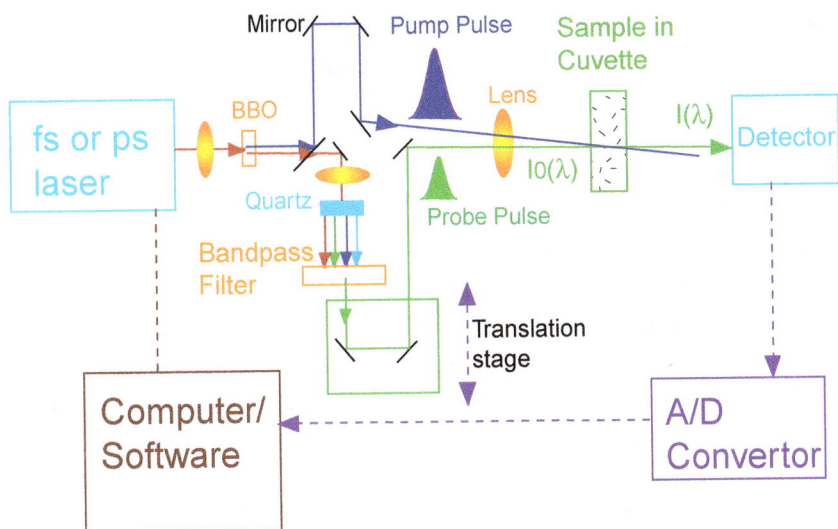

Fig. 2.10. Schematic of pump–probe setup with an fs or ps laser. The fundamental laser wavelength is doubled via second harmonic generation to produce a higher frequency laser pulse for excitation (pump). The remaining fundamental is used to produce a white light from which a specific frequency with certain bandwidth is selected using a bandpass filter as a probe pulse. The probe is delayed in time with respect to the pump pulse by a translation stage. The probe signal, often normalized to a reference beam without going through the sample (not shown), is detected, processed and plotted as a function of delay time between the pump and probe pulses.

tunable wavelengths using optical parametric oscillators (OPO) and amplifiers (OPA) for experiments that require or benefit from multiple colors for pump and/or probe. No OPO or OPA is shown in the example in Fig. 2.10. In this setup, the fs pulses are focused into a nonlinear optical crystal such as BBO (beta-BaB_2O_4) or KDP (potassium-dihydrogen-phosphate) to generate a second harmonic signal, e.g. 390 nm if the fundamental wavelength is 780 nm. The shorter wavelength of the second harmonic allows for more molecules or samples that do not absorb the long fundamental wavelength to be studied. The second harmonic and the fundamental can be separated by a dichotic mirror. The second harmonic is usually used as the excitation or pump pulse to excite the sample of interest. The remaining fundamental pulses or pulses selected from a particular wavelength from a while light generated by the fundamental pulses in a medium such as a quartz crystal, are used as a probe beam to interrogate the excited species created by the pump beam. The probe pulse is delayed in a controllable manner, usually by a translation stage, with respect to the pump pulse.

The excited state population change is manifested as a change of the absorbance of the probe pulse intensity as a function of the delay time between the probe and pump pulses. Typically, the second pulse is much weaker than the probe pulse in intensity, about 1/10, to avoid direct excitation of the sample in the ground state by the probe pulse. The two pulses must overlap spatially in the sample. It is important that the pump pulse never reaches the detector. To improve signal-to-noise (S/N) ratio, a reference probe beam is often used, which is split off the probe beam and detected by another detector without going through the sample. The probe signal is normalized by the reference signal for every probe laser pulse. An average is usually necessary for many pairs of pump and probe pulses for each time delay to get good S/N ratio. The experiment is repeated for different time delays, as illustrated in Fig. 2.11. Time zero is defined at the time when the pump and probe pulses overlap or arrive at the sample at the same time. The probe pulse is absorbed by the population in the excited state created by the pump pulse. The main advantage of TA is that the time resolution is very high, limited only by the cross-correlation of the pump and probe pulses. It can be used to study both luminescent and nonluminescent samples. The one limitation is that the probe pulse excites the population to another higher-lying excited state that is often not known or well-defined, making data interpretation somewhat challenging, and requiring certain assumptions to be made.

For example, Fig. 2.12 shows a typical transient absorption decay profile as a function of delay time between pump and probe pulses for a fixed probe wavelength for CdS nanoparticles [18]. The TA signal in the vertical axis is plotted as a function of time delay between the pump and probe pulses. At time = zero, the pump and probe pulses overlap in time. The time shown in the figure is the delay time between the pump and probe pulses with the probe pulse coming after the pump pulse. The information obtained is the delay profile of the excited state or transient species created by the pump pulse and interrogated by the probe pulse. Overall decay

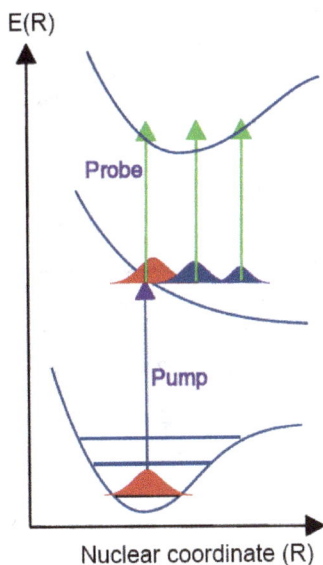

Fig. 2.11. Schematic of pump and probe pulses in a sample illustrated with electronic transitions in a diatomic molecule. The pump pulse excites the molecule from the ground to the first excited electronic state and the probe pulse monitors the first excited state population based on transition from the first excited state to a higher lying electronic state. Three different delay times are illustrated.

profile contains dynamic information of the excited metastable species, e.g. excited states of molecules or conduction band or trap states of semiconductor nanoparticles. Control experiments and data analysis are then carried out to extract useful dynamic information about the system. Physical models often need to be proposed or developed for proper data interpretation.

It is also possible to obtain a spectrum of the excited species by using a spectrally broad probe pulse at a fixed delay time. This usually is more involved, has poorer *S/N* ratio, but potentially contains more information. With recent advancement of ultrafast laser technique, e.g. OPO and OPAs, it is relatively easy to generate multiple frequencies in a broad spectral range, from UV to near IR. Thus, it is easier to obtain TA spectrum at different times, or to measure the transient absorption as a function of time and frequency and develop an image based on TA signal or intensity versus time and frequency. Figure 2.13 shows an example of time- and

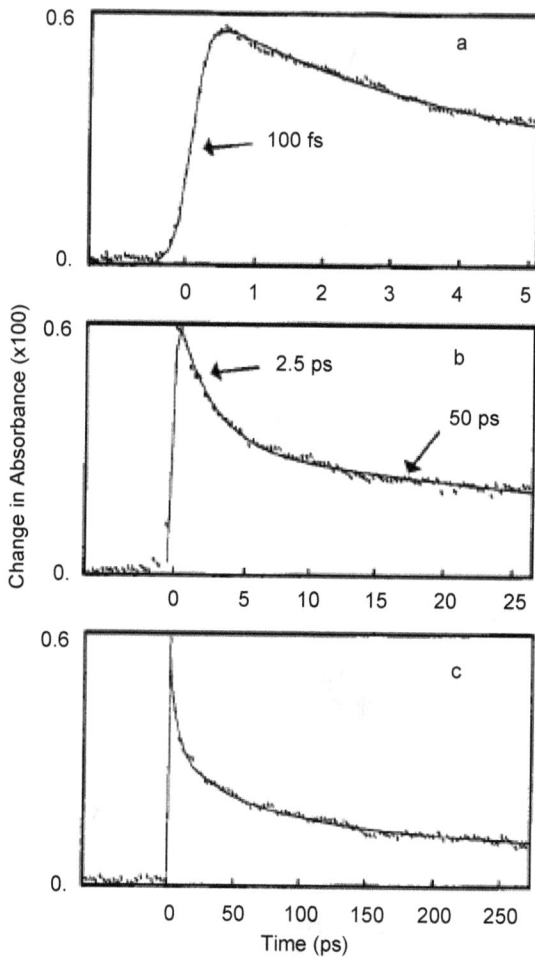

Fig. 2.12. Time evolution of the photoinduced electrons in CdS nanoparticles in aqueous solution (30-Å average particle diameter) probed at 780 nm following excitation at 390 nm with an amplified fs Ti-Sapphire laser system on three different time scales (a)–(c). The pH of the colloid is 8.0, and the pump fluence is 1.18 photons/Å². The dotted lines are the experimental data; the solid lines are fits using a sum of multiple exponentials. The fast rise time (<100 fs) is due to the cross overlap of the pump and probe pulses. Reproduced with permission from Ref. 18.

frequency-resolved transient absorption in Au nanorods upon excitation at 800 nm [19]. These types of images contain more information useful for understanding the dynamic process, and more studies using such techniques are expected in the future for molecules and nanomaterials.

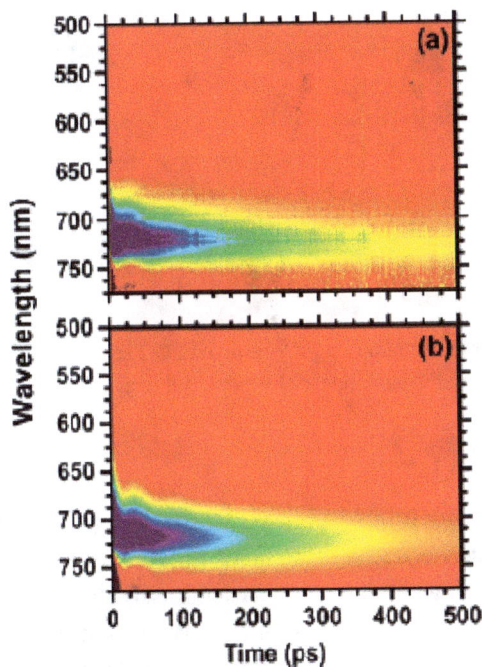

Fig. 2.13. (a) Experimental time-frequency-resolved transient absorbance in Au nanorods upon excitation at 800 nm. (b) Global fit according to a theoretical model. Reproduced with permission from Ref. 19.

Time-resolved fluorescence (TRF) is a time-resolved technique that is complementary to TA. TRF is based on the measurement of excited state photoluminescence. In contrast to TA that involves a higher-lying excited state, TRF has the advantage that only the ground state, which is often well-defined or known, and the excited state of interest are involved. TRF also tends to be more sensitive than TA since it is a zero-background measurement. The limitation is that the time resolution is often lower than that of TA, since it is usually limited by the detectors with time-resolution much longer than the laser pulses used. One solution to this problem is to use a scheme called *fluorescence up-conversion* (FUC) that allows significant improvement of time-resolution to that point where it is limited only by the laser pulses involved. However, the signal is usually very weak, and its practical application is thus limited. It is also only useful for luminescent samples.

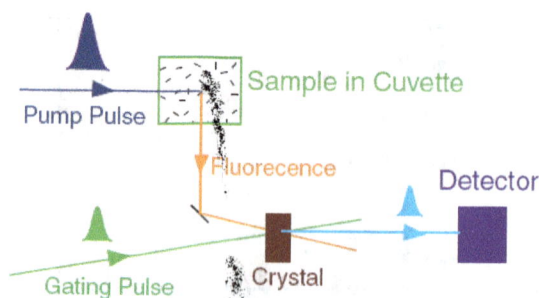

Fig. 2.14.   Schematic diagram of a time-resolved FUC setup. The gating pulse and fluorescence light are mixed in a nonlinear optical crystal (e.g. $LiIO_3$) appropriate for the right wavelength of light.

Figure 2.14 illustrates the key idea behind a typical time-resolved FUC setup. In this case, a short laser pulse is used to excite the sample, similar to that in TA. Fluorescence from the excited state is collected and mixed with a second short laser pulse, referred to as a "gating" pulse as in electronic gating, to generate an up-converted signal of interest. The frequency or energy of the up-converted signal is the sum of that of the fluorescence light and the second gating laser pulse, while the pulse width of the up-converted signal is determined by the gating laser pulse, since the fluorescence is not pulsed. By changing the time delay between the gating laser pulse with respect to the pump laser pulse and detecting the up-converted signal as a function of the delay time, one can obtain the excited state fluorescence decay profile with time resolution limited only by the cross-correlation between the pump and gating pulses. The time resolution of such a time-resolved FUC scheme is much higher than the actual photodetector resolution that is limited by the electronics of the detector rather than the laser.

For example, time-resolved FUC has been used to study charge carrier dynamics in CdSe nanoparticles [20]. Figure 2.15 shows a representative time decay profile of the observed FUC for CdSe QDs with different sizes.

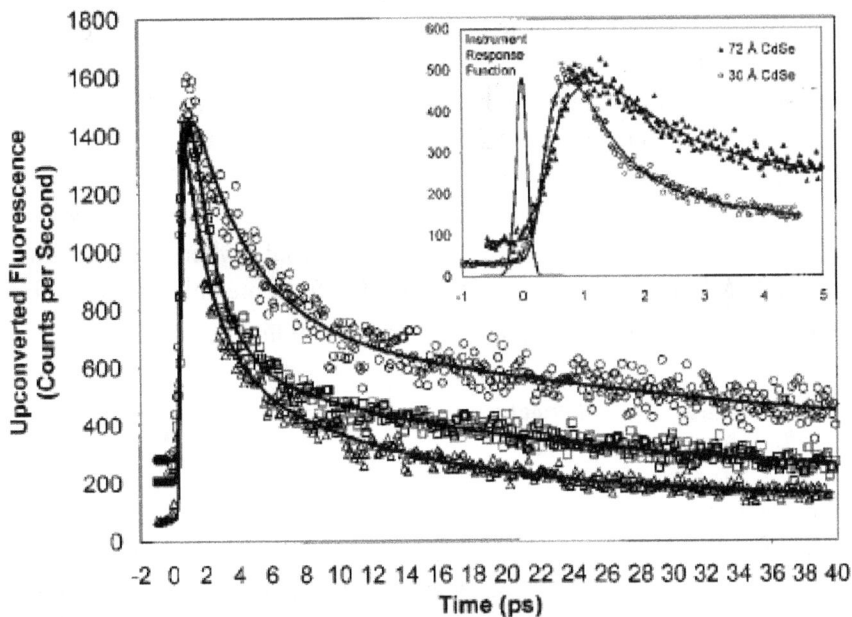

Fig. 2.15. Size dependence of the band edge emission for 30, 52 and 72 Å CdSe nanocrystals in toluene measured by FUC. Each plot is a sum of five individual scans and is normalized to the 30 Å CdSe trace. Inset: higher time resolution scan of 30 and 72 Å CdSe NC's indicating the difference in rise and decay time scales. Reproduced with permission from Ref. 20.

It is worth pointing out that the observed lifetime, τ_{ob}, is related to the radiative, τ_r, and nonradiative, τ_{nr}, lifetimes, by the following equation:

$$1/\tau_{ob} = 1/\tau_r + 1/\tau_{nr} \qquad (2.4)$$

The lifetimes are related to the PL quantum yield, Φ_{PL}, by:

$$\Phi_{PL} = \tau_{ob}/\tau_r. \qquad (2.5)$$

There is confusion sometimes between radiative lifetime (τ_r) and observed lifetime (τ_{ob}). These two are strictly speaking equal only in the limit that τ_{nr} is very long or the PL quantum yield is near unity or 100%, as in a perfect single crystal, according to Eqs. (2.4) and (2.5). Any lifetime measured experimentally based on time-resolved PL or TA measurements is just the observed lifetime τ_{ob}, which contains contributions from both τ_r and τ_{nr}.

Chapter 9 is devoted to a more detailed discussion of time-resolved studies of charge carrier dynamics in nanomaterials, where more specific examples will be discussed to highlight what fundamental dynamic information can be obtained from such time-resolved measurements.

2.5. Nonlinear optical spectroscopy: harmonic generation and up-conversion

Nonlinear optical spectroscopy refers to a class of experimental techniques that are based on nonlinear optical phenomenon such as second harmonic generation (SHG) or sum-frequency generation (SFG) [21–23]. These techniques complement conventional frequency domain spectroscopy and linear time-resolved techniques in probing different properties of materials [24–29].

SHG is a coherent two-photon process in which two photons with the same wavelength are converted into one photon with energy equal to that of the original two photons. For example, two 800 nm photons can be converted into a 400 nm photon in a nonlinear optical crystal such as BBO or KDP. Due to the nonlinear nature of this process, SHG is only efficient at high light intensity, which requires laser light. SHG is useful for generating different wavelengths of light for various applications. Similarly, higher order harmonics, such as third, fourth, or even higher, can be generated. The efficiency becomes weaker for higher order harmonics. Similar to SHG, SFG is a nonlinear process that involves converting two photons into one photon. Different from SHG, however, the two photons have different wavelengths or frequencies in SFG, as illustrated in Fig. 2.16. SFG is also useful for generating various wavelengths of light for applications.

Besides nonlinear light generation applications, SHG and SFG have become popular experimental techniques for probing surfaces and interfaces due to their surface-specific capability for materials with central

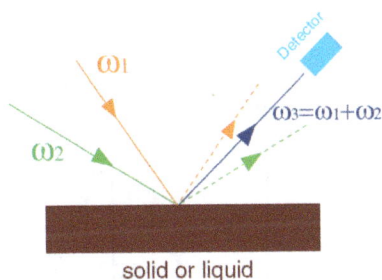

Fig. 2.16.   Illustration of the principle and setup for SFG that converts two photons with different frequencies into a photon of higher frequency that is equal to the sum of the frequencies of the two original photons.

symmetry [21]. Interfaces such as liquid–air, liquid–liquid, liquid–solid that have been challenging to study using conventional techniques have been studied extensively using nonlinear SHG and SFG techniques in the last decade [22–29].

The application of SHG and SFG in probing nanomaterials is somewhat limited. However, several studies have demonstrated their usefulness in gaining information about surface and dynamic properties of semiconductor, metal and metal oxide nanomaterials [30–34] or for wavelength conversion [35]. It was interesting to note that SHG can be produced with centro-symmetric particles due to asymmetric surface characteristics [24].

Another interesting nonlinear optical phenomenon is *luminescence up-conversion (LUC)*, in which the emitted light has a higher frequency or shorter wavelength than the incident light. Figure 2.17 shows schematically some of the possible mechanisms for luminescence up-conversion. In case A, a photon is emitted as a result of direct two-photon excitation that involves no relaxation or energy loss in the excited or ground state. The emitted photon energy is exactly equal to the sum of the two incident photons. This is essentially how SHG works. In case B, there is some relaxation or energy loss in the excited state, and as a result, the emitted photon energy is more than the single incident photon energy but less than the sum of the energy of two incident photons. Similarly, in cases B and C, the emitted photon has less than twice the incident photon energy but more than the single photon energy. In contrast to case B, however, both

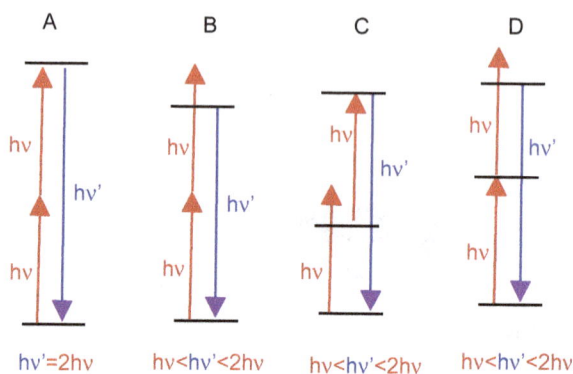

Fig. 2.17.   Illustration of different luminescence up-conversion (LUC) mechanisms.

cases C and D involve another excited state or intermediate state between the ground state and the emitting excited state. Energy relaxation occurs in the intermediate state for case C and in the emitting excited state for case D.

Luminescence up-conversion has been observed for doped and up-doped semiconductor as well as doped insulator nanomaterials, typically with transition metal or rare earth ions as dopants [36–42]. Most LUC can be explained using one of the above mechanisms. In some cases, more states are involved, especially when co-dopants are involved. LUC is useful for optical wavelength conversion as well as for probing the fundamental properties of nanomaterials. Because the efficiency for LUC is usually low, high intensity light, such as lasers, is often required for excitation. It should be cautioned that extreme care need to be taken to ensure that the observed LUC is not artifacts due to Raman or higher order scattered light of the incident light or residue light at the wavelength of detection from the light source.

2.6. Single nanoparticle and single molecule spectroscopy

Most spectroscopy studies of samples have been carried out on ensembles of a large number of particles or molecules. The properties measured are thus ensemble averages of the properties of individual particles. Due to heterogeneous distributions in size, shape, environment or embedding medium, and surface properties, the spectrum measured is thus inhomogeneously broadened. This results in loss of spectral information and complications in interpretation of the results [43]. For instance, it has been predicted by theory that nanocrystallites should have a spectrum of discrete, atomic-like energy states [44, 45]. However, transition line widths observed experimentally appear significantly broader than expected, even though the discrete nature of the excited states has been verified [46, 47]. This is true even when size-selective optical techniques are used to extract homogeneous line widths [43, 46–51].

One way to solve the above problem is to create particles with truly uniform size, surface, environment and shape. However, this is usually challenging to do [52]. Another, perhaps simpler approach to remove the heterogeneity is to conduct the measurement on one single particle. This approach is similar to that used in the field of single molecule spectroscopy [53, 54]. In this approach, the signal is detected only from one molecule or one nanoparticle at a time. A number of single nanoparticle studies have been reported on semiconductor nanoparticles, including CdS [55] and CdSe [56–61]. Most single nanoparticle spectroscopy studies have been based on detection of emission or scattering from single nanostructures while only a few studies based on absorption have been reported recently. Emission measurements are zero-background experiments so they are usually more sensitive than the nonzero background absorption measurements. Compared to ensemble averaged samples, single particle spectroscopy studies of nanoparticles such as CdSe revealed several new features, including fluorescence blinking, ultranarrow transition line width, a range of phonon couplings between individual particles, and spectral diffusion of the emission spectrum over a wide range of energies [61, 62].

For nonluminescent nanoparticles such as most metal nanostructures, single nanoparticle spectroscopy has also been demonstrated successfully based on the measurement of Rayleigh scattering [63]. One can anticipate more studies of nanostructures using single nanoparticle spectroscopy in the future.

2.7. Dynamic light scattering (DLS)

Dynamic light scattering is a relatively simple spectroscopic technique based on Rayleigh scattering to determine the size of nanoscale objects such as particles, aggregates and large molecules, e.g. polymers and proteins [64–68]. When light hits particles that are small compared to the wavelength of light, a major scattering mechanism observed is elastic Rayleigh scattering. DLS is based on measuring the speed at which particles move under Brownian motion by monitoring the intensity of light scattered by the sample at some fixed angle. Brownian motion causes constructive and destructive interference of the scattered light.

Technically, DLS uses the method of autocorrelation to uncover information contained in the light intensity fluctuations. Autocorrelation measures how well a signal matches a time-delayed version of itself, as a function of the amount of time delay. A quicker autocorrelation function decay indicates faster motion of the particles. The autocorrelation function can be related to the hydrodynamic radius (R_H) of the particles, which is the information of interest, by the Stokes Einstein equation [69]. Larger particles correspond to slower Brownian motion and slower delays of the autocorrelation function. DLS has been used frequently in the study of nanoparticles [68, 70, 71].

2.8. Summary

This chapter covers some of the most commonly used optical spectroscopy techniques including electronic absorption, luminescence, IR, Raman, as well as time-resolved techniques, such as transient absorption and time-resolved luminescence. More specialized techniques such as single molecular spectroscopy and nonlinear optical techniques have also been briefly discussed. Selection of these different techniques for applications depends on the information of interest. In later chapters, many specific examples of applications based on these various techniques will be discussed to demonstrate how to extract physical information from the measured spectrum.

References

1. T. Vossmeyer, L. Katsikas, M. Giersig, I.G. Popovic, K. Diesner, A. Chemseddine, A. Eychmuller and H. Weller, *J. Phys. Chem.* **98**, 7665 (1994).
2. T.J. Norman, D. Magana, T. Wilson, C. Burns, J.Z. Zhang, D. Cao and F. Bridges, *J. Phys. Chem. B* **107**, 6309 (2003).

3. S. Nakamura, K. Kitamura, H. Umeya, A. Jia, M. Kobayashi, A. Yoshikawa, M. Shimotomai and K. Takahashi, *Electron. Lett.* **34**, 2435 (1998).
4. M.Y. Gao, B. Richter, S. Kirstein and H. Mohwald, *J. Phys. Chem. B* **102**, 4096 (1998).
5. T. Brunhes, P. Boucaud, S. Sauvage, F. Aniel, J.M. Lourtioz, C. Hernandez, Y. Campidelli, O. Kermarrec, D. Bensahel, G. Faini and I. Sagnes, *Appl. Phys. Lett.* **77**, 1822 (2000).
6. I.E. Itskevich, S.T. Stoddart, S.I. Rybchenko, Tartakovskii, II, L. Eaves, P.C. Main, M. Henini and S. Parnell, *Phys Status Solidi A* **178**, 307 (2000).
7. X.L. Xu, A. Andreev, D.A. Williams and J.R.A. Cleaver, *Appl. Phys. Lett.* **89**, 091120 (2006).
8. R. Schmidt, M. Vitzethum, R. Fix, U. Scholz, S. Malzer, C. Metzner, P. Kailuweit, D. Reuter, A. Wieck, M.C. Hubner, S. Stufler, A. Zrenner and G.H. Dohler, *Physica E* **26**, 110 (2005).
9. Y.S. Kang, D.K. Lee and P. Stroeve, *Thin Solid Films* **329**, 541 (1998).
10. C.S. Yang, R.A. Bley, S.M. Kauzlarich, H.W.H. Lee and G.R. Delgado, *J. Am. Chem. Soc.* **121**, 5191 (1999).
11. A. Pradeep, P. Priyadharsini and G. Chandrasekaran, *J. Magn. Magn. Mater.* **320**, 2774 (2008).
12. T. Hikov, A. Rittermeier, M.B. Luedemann, C. Herrmann, M. Muhler and R.A. Fischer, *J. Mater. Chem.* **18**, 3325 (2008).
13. N. Shukla, C. Liu, P.M. Jones and D. Weller, *J. Magn. Magn. Mater.* **266**, 178 (2003).
14. J.L. McHale, *Molecular Spectroscopy*, 1st ed, Upper Saddle River, NJ: Prentice Hall. 463 (1998).
15. G.C. Schatz, M.A. Young and R.P. Van Duyne, in *Surface-Enhanced Raman Scattering: Physics and Applications*, 19 (2006).
16. W.G. Su, J. Zhang, Z.C. Feng, T. Chen, P.L. Ying and C. Li, *J. Phys. Chem. C* **112**, 7710 (2008).
17. G.R. Fleming, *Chemical Applications of Ultrafast Spectroscopy*, New York: Oxford University Press. 288 (1986).
18. J.Z. Zhang, R.H. Oneil and T.W. Roberti, *J. Phys. Chem.* **98**, 3859 (1994).
19. H.Y. Seferyan, R. Zadoyan, A.W. Wark, R.A. Corn and V.A. Apkarian, *J. Phys. Chem. C* **111**, 18525 (2007).
20. D.F. Underwood, T. Kippeny and S.J. Rosenthal, *J. Phys. Chem. B* **105**, 436 (2001).
21. Y.R. Shen, *The Principles of Nonlinear Optics*, New York: J. Wiley. 563 (1984).
22. P. Guyotsionnest, P. Dumas and Y.J. Chabal, *J. Electron Spectroscopy and Related Phenomena* **54**, 27 (1990).

23. A. Tadjeddine and P. Guyotsionnest, *Electrochimica Acta* **36**, 1849 (1991).
24. H. Wang, E.C.Y. Yan, E. Borguet and K.B. Eisenthal, *Chem. Phys. Lett.* **259**, 15 (1996).
25. S. Baldelli, A.S. Eppler, E. Anderson, Y.R. Shen and G.A. Somorjai, *J. Chem. Phys.* **113**, 5432 (2000).
26. G. Rupprechter, *Phys. Chem. Chem. Phys.* **3**, 4621 (2001).
27. R.D. Schaller, J.C. Johnson, K.R. Wilson, L.F. Lee, L.H. Haber and R.J. Saykally, *J. Phys. Chem. B* **106**, 5143 (2002).
28. J. Holman, S. Ye, D.J. Neivandt and P.B. Davies, *J. Am. Chem. Soc.* **126**, 14322 (2004).
29. M. Iwamoto, T. Manaka, E. Lim and R. Tamura, *Mole. Cryst. Liquid Cryst.* **467**, 285 (2007).
30. B. Lamprecht, A. Leitner and F.R. Aussenegg, *Appl. Phys. B-Lasers Opt.* **68**, 419 (1999).
31. M. Jacobsohn and U. Banin, *J. Phys. Chem. B* **104**, 1 (2000).
32. K.R. Li, M.I. Stockman and D.J. Bergman, *Phys. Rev. B* **72**, 153401 (2005).
33. A.M. Moran, J.H. Sung, E.M. Hicks, R.P. Van Duyne and K.G. Spears, *J. Phys. Chem. B* **109**, 4501 (2005).
34. C.F. Zhang, Z.W. Dong, G.J. You, R.Y. Zhu, S.X. Qian, H. Deng, H. Cheng and J.C. Wang, *Appl. Phys. Lett.* **89** (2006).
35. N. Thantu, R.S. Schley and B.L. Justus, *Opt. Commun.* **220**, 203 (2003).
36. H.X. Zhang, C.H. Kam, Y. Zhou, X.Q. Han, S. Buddhudu and Y.L. Lam, *Opt. Mater.* **15**, 47 (2000).
37. W. Chen, A.G. Joly and J.Z. Zhang, *Phys. Rev. B* **6404**, 1202 (2001).
38. W.X. Que, Y. Zhou, Y.L. Lam, Y.C. Chan, C.H. Kam, L.H. Gan and G.R. Deen, *J. Electron. Mater.* **30**, 6 (2001).
39. Y.P. Rakovich, J.F. Donegan, S.A. Filonovich, M.J.M. Gomes, D.V. Talapin, A.L. Rogach and A. Eychmuller, *Physica E* **17**, 99 (2003).
40. H.T. Sun, Z.C. Duan, C.L. Yu, G. Zhou, M.S. Liao, J.J. Zhang, L.L. Hu and Z.H. Jiang, *Solid State Commun.* **135**, 174 (2005).
41. H.T. Sun, C.L. Yu, G. Zhou, Z.C. Duan, M.S. Liao, J.J. Zhang, L.L. Hu and Z.H. Jiang, *Spectrochim Acta A* **62**, 1000 (2005).
42. C.G. Dou, Q.H. Yang, X.M. Hu and J. Xu, *Opt. Commun.* **281**, 692 (2008).
43. A.P. Alivisatos, A.L. Harris, N.J. Levinos, M.L. Steigerwald and L.E. Brus, *J. Chem. Phys.* **89**, 4001 (1988).
44. A.L. Efros and A.L. Efros, *Fizika i Tekhnika Poluprovodnikov* **16**, 1209 (1982).
45. L.E. Brus, *J. Chem. Phys.* **80**, 4403 (1984).
46. D.J. Norris and M.G. Bawendi, *Phys. Rev. B-Condensed Matter* **53**, 16338 (1996).

47. D.J. Norris and M.G. Bawendi, *J. Chem. Phys.* **103**, 5260 (1995).
48. D.M. Mittleman, R.W. Schoenlein, J.J. Shiang, V.L. Colvin, A.P. Alivisatos and C.V. Shank, *Phys. Rev. B-Condensed Matter* **49**, 14435 (1994).
49. U. Woggon, S. Gaponenko, W. Langbein, A. Uhrig and C. Klingshirn, *Phys. Rev. B-Condensed Matter* **47**, 3684 (1993).
50. H. Giessen, B. Fluegel, G. Mohs, N. Peyghambarian, J.R. Sprague, O.I. Micic and A.J. Nozik, *Appl. Phys. Lett.* **68**, 304 (1996).
51. V. Jungnickel and F. Henneberger, *J. Luminescence* **70**, 238 (1996).
52. J. Ouyang, M.B. Zaman, F.J. Yan, D. Johnson, G. Li, X. Wu, D. Leek, C.I. Ratcliffe, J.A. Tripmeester and K. Yu, *J. Phys. Chem. C*, in press, 1111 (2008).
53. W.E. Moerner, *Science* **265**, 46 (1994).
54. T. Basche, W.E. Moerner, M. Orrit and U.P. Wild, eds., VCH: Weinheim; Cambridge. 250 (1997).
55. J. Tittel, W. Gohde, F. Koberling, T. Basche, A. Kornowski, H. Weller and A. Eychmuller, *J. Phys. Chem. B* **101**, 3013 (1997).
56. S.A. Blanton, A. Dehestani, P.C. Lin and P. Guyot-Sionnest, *Chem. Phys. Lett.* **229**, 317 (1994).
57. S.A. Empedocles, D.J. Norris and M.G. Bawendi, *Phys. Rev. Lett.* **77**, 3873 (1996).
58. M. Nirmal, B.O. Dabbousi, M.G. Bawendi, J.J. Macklin, J.K. Trautman, T.D. Harris and L.E. Brus, *Nature* **383**, 802 (1996).
59. S.A. Blanton, M.A. Hines and P. Guyot-Sionnest, *Appl. Phys. Lett.* **69**, 3905 (1996).
60. S.A. Empedocles and M.G. Bawendi, *Science* **278**, 2114 (1997).
61. S.A. Empedocles, R. Neuhauser, K. Shimizu and M.G. Bawendi, *Adva. Mater.* **11**, 1243 (1999).
62. S. Empedocles and M. Bawendi, *Acc. Chem. Res.* **32**, 389 (1999).
63. M. Hu, H. Petrova, A.R. Sekkinen, J.Y. Chen, J.M. McLellan, Z.Y. Li, M. Marquez, X.D. Li, Y.N. Xia and G.V. Hartland, *J. Phys. Chem. B* **110**, 19923 (2006).
64. R. BarZiv, A. Meller, T. Tlusty, E. Moses, J. Stavans and S.A. Safran, *Phys. Rev. Lett.* **78**, 154 (1997).
65. A. Kamyshny, D. Danino, S. Magdassi and Y. Talmon, *Langmuir* **18**, 3390 (2002).
66. A. Shukla, M.A. Kiselev, A. Hoell and R.H.H. Neubert, *Pramana-J Phys.* **63**, 291 (2004).
67. S. Chakraborty, B. Sahoo, I. Teraoka and R.A. Gross, *Carbohydrate Polymers* **60**, 475 (2005).

68. H. Xie, K.L. Gill-Sharp and P. O'Neal, *Nanomed. Nanotechnol. Biol. Med.* **3**, 89 (2007).
69. W.P. Wuelfing, A.C. Templeton, J.F. Hicks and R.W. Murray, *Anal. Chem.* **71**, 4069 (1999).
70. N.G. Khlebtsov, V.A. Bogatyrev, B.N. Khlebtsov, L.A. Dykman and P. Englebienne, *Colloid J.* **65**, 622 (2003).
71. J.L. Dickson, S.S. Adkins, T. Cao, S.E. Webber and K.P. Johnston, *Indust. Engin. Chem. Res.* **45**, 5603 (2006).

Chapter 3

Other Experimental Techniques: Electron Microscopy and X-ray

Even though this book focuses on optical properties and related optical spectroscopy techniques, it seems necessary to briefly review some of the other relevant experimental techniques used for characterization of nanomaterials. This is partly because it is essential for nanomaterials to be thoroughly characterized as much as possible using a combination of experimental techniques due to the intrinsic complex nature of nano-materials. For example, two of the most basic characteristics are size and shape of nanostructures. Another important property is the surface of nanomaterials due to their extremely large surface-to-volume ratio. Furthermore, it is often important to know the crystal structures of the nanomaterials. To study these different properties requires many other experimental methods, besides optical spectroscopy. Examples include electron microscopy, electrochemistry, X-ray based methods such as X-ray diffraction (XRD), X-ray photoelectron spectroscopy (XPS), and extended X-ray absorption fine structure (XAFS). Of course, most of these techniques can be used for investigating bulk materials as well and are not exclusive for nanomaterials. Nonetheless, these tech-niques are necessary and powerful in disseminating information about nanomaterials.

3.1. Microscopy: AFM, STM, SEM and TEM

Structural determination is essential for nanomaterials research. Since the nanostructures are usually too small to be visualized with conventional optical microscopes, it is important to use appropriate tools to adequately characterize their structure and surface in detail at the molecular or atomic level. This is important not only for understanding their fundamental properties but also for exploring their functional and technical perform-ance in technological applications. There are several experi-mental techniques that can be used to characterize structural and surface properties of nanomaterials either directly or indirectly, e.g. XRD, STM (scanning tunneling microscopy), AFM (atomic force microscopy), SEM (scanning electron microscopy), TEM (transmission electron microscopy), XAS (X-ray absorption spectroscopy), EXAFS, XANES (X-ray absorption near edge structure), EDX (energy dispersive X-ray), XPS, IR (infrared), Raman, and DLS (dynamic light scattering) [1–5]. Some of these techniques are more surface sensitive than others. Some techniques are directly element-specific while others are not. The choice of techniques depends strongly on the information being sought about the material.

3.1.1. *Scanning probe microscopy (SPM): AFM and STM*

Scanning probe microscopy (SPM) represents a group of techniques, including scanning tunneling microscopy (STM), atomic force microscopy (AFM), and chemical force microscopy (CFM), that have been exten-sively applied to characterize nanostructures with atomic or subatomic spatial resolution [1, 4, 6–8]. A common characteristic of these techniques is that an atom sharp tip scans across the specimen surface and images are formed by either measuring the current flowing through the tip or the force acting on the tip. SPM can be operated in a number of environmen-tal conditions, in a variety of different liquids or gases, allowing direct imaging of inorganic surfaces and organic molecules. It allows viewing and manipulation of objects on the nanoscale and its invention is a major milestone in nanotechnology.

For nonconductive nanomaterials, atomic force microscopy (AFM) is a better choice [9, 10]. AFM is based on measuring the force between the tip and the solid surface. The interaction between two atoms is repulsive at short-range and attractive at long-range. The force acting on the tip reflects the distance from the tip atom(s) to the surface atom, thus images can be formed by detecting the force while the tip is scanned across the specimen. A more generalized application of AFM is scanning force microscopy, which can measure magnetic, electrostatic, frictional, or molecular interaction forces allowing for nanomechanical measurements.

A typical AFM setup involves a sharp tip mounted on a microscale cantilever, a laser, a position sensitive detector, a piezoelectric tube (PZT) scanner, and control electronics, as shown in Fig. 3.1 [8]. The cantilever is typically silicon or silicon nitride with a tip radius or curvature on the order of nanometers. When the tip is brought near a sample surface, forces between the tip and the sample lead to a deflection of the cantilever according to Hooke's law. Depending on the situation, forces that are

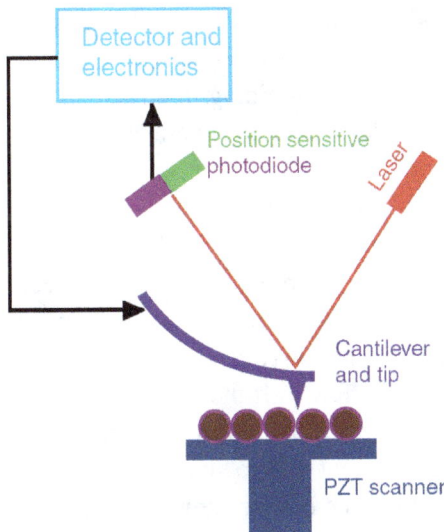

Fig. 3.1. Schematic illustration of the principle of operation underlying atomic force microscopy (AFM).

measured in AFM include mechanical contact force, electrostatic forces, chemical bonding, van der Waals forces, capillary forces and magnetic forces. Typically, the deflection is measured using a laser beam reflected from the top of the cantilever into an array of position sensitive photodiodes. To avoid possible collision between the tip and sample surface, a feedback mechanism is often employed to adjust the tip-to-sample distance to maintain a constant force between the tip and the sample. Traditionally, the sample is mounted on a PZT that can move the sample in the z direction for maintaining a constant force and in the x and y directions for scanning the sample. The resulting map of the area $s = f(x,y)$ represents the topography of the sample. The AFM can be operated in a number of modes, depending on the application. In general, possible imaging modes are divided into static (also called contact) modes and a variety of dynamic (or non-contact) modes.

Figure 3.2 shows a frequency modulation (FM)-AFM image obtained with a (111) oriented silicon tip [11]. This image was recorded at positive frequency shift, and thus the forces between the front atom and sample have been repulsive. An advantage of this mode is that the feedback can be set much faster, because the risk of feedback oscillations with a catastrophic tip crash is avoided in this mode. This example clearly shows that AFM is capable of directly providing unprecedentedly detailed information on the atomic scale, which is important for understanding chemical bonding and electronic structure of atoms and molecules.

Another powerful SPM technique is scanning tunneling microscopy (STM), discovered in the 1980s [12]. STM is based on measuring the tunneling current between a sharp metallic tip and specimen. The STM tip does not actually touch the surface of the sample measured. As shown in Fig. 3.3, a voltage is applied between the tip and the specimen, typically between a few mV and a few V. If the tip touches the surface of the specimen, the voltage will result in an electrical current. If the tip is far away from the surface, the current is zero since it is essentially an open circuit. STM operates in the regime of extremely small distances between the tip and the surface of only 0.5 to 1.0 nm, or only a few atomic diameters. At these distances, electrons can tunnel from the probe tip to the surface or vice versa, which is why it is named tunneling microscopy. Since tunneling is a weak process, the tunneling current is thus very low.

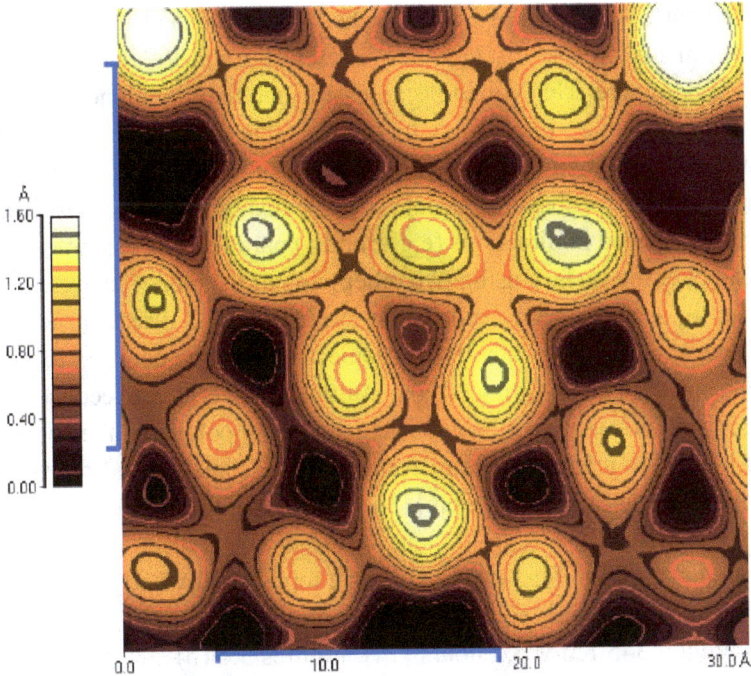

Fig. 3.2.   Topographic image of Si(111)–(7 × 7) observed by FM-AFM, imaged with a single crystal silicon tip roughly oriented in a (111) direction. Reproduced with permission from Ref. 11.

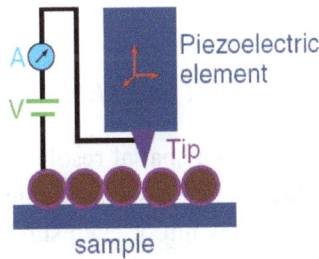

Fig. 3.3.   Schematic illustration of the basic operating principle of STM.

STMs typically operate at tunneling currents between a few picoAmperes (pA) and a few nanoAmperes (nA). The tunneling current depends strongly on the distance between the tip and the specimen. Therefore, tunneling current provides a highly sensitive measure of the distance between the tip and the surface. The STM tip is attached to a piezoelectric element used to control the precise distance between the tip and the surface with an electrical voltage. This voltage is adjusted such that the tunneling current is a constant in the z direction, which means that the distance between the tip and the specimen surface is kept at constant. This distance control is achieved using feedback electronics. While distance control is active, the tip can be moved in directions (e.g. x and y) parallel to the sample surface to scan over the surface using two other parts of the piezoelectric element. As the tip scans, the tip needs to be moved in the z direction perpendicular to x and y, through the piezoelectric voltage control, to ensure a constant current. In doing so, the control voltage in the feedback electronics contains information about the surface topology of the specimen. This information is usually transformed into 3D images and plotted on a computer for easy visualization [13]. STM is applicable mainly for conductive samples.

As an example, Fig. 3.4 shows some representative STM topographs of an undecanethiol self-assembled monolayer (SAM) upon UHV annealing at 345 K for 4.5 hours [14]. Figures 3.4 (b)–3.4 (d) are zoom-in images within the area marked in Fig. 3.4(a). The SAM consists of etch-pit free domains up to several hundreds of nanometers in size, with only occasional missing molecules or other point defects. This demonstrates the high spatial resolution capability of STM that is easily on the angstrom scale.

3.1.2. *Electron microscopy: SEM and TEM*

Optical microscopes have limited spatial resolution, usually on the order of a few hundred nm in the best case scenario due to the diffraction limit of light. Higher resolution, a few nm or even sub nm, is needed for many applications, especially in the study of nanomaterials. Scanning electron microscopy (SEM) is a powerful and popular technique for imaging the surfaces of almost any material with a resolution down to about 1 nm [2, 3]. The image resolution offered by SEM depends not only on the property of the electron probe, but also on the interaction of the electron probe with the specimen. The interaction of an incident electron beam with the specimen produces secondary electrons, with energies typically

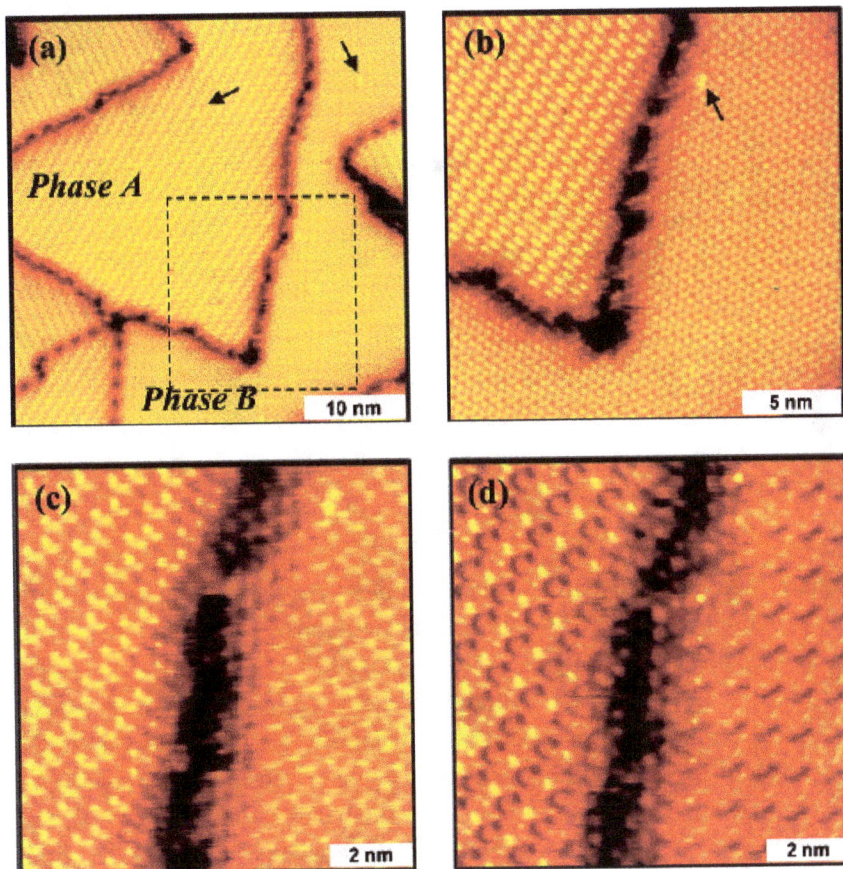

Fig. 3.4. STM topographic images of an undecanethiol self-assembled monolayer (SAM), after annealing at 345 K for 4.5 h under UHV. The images were recorded at (a) 0.8 V, 14 pA; (b) 0.8 V, 14 pA; (c) 0.3 V, 5 pA; (d) 0.3 V, 14 pA. Image (b) was taken inside the square area indicated in panel (a), and panels (c) and (d) are zoomed-in images of panel (b). Point defects in the SAM are indicated by arrows in panels (a) and (b). Two distinct phases, A and B, coexist on the surface, observed simultaneously under the same imaging conditions. Reproduced with permission from Ref. 14.

smaller than 50 eV, the emission efficiency of which sensitively depends on surface geometry, surface chemical characteristics and bulk chemical composition [15]. SEM can thus provide information about the surface topology, morphology and chemical composition. The high resolution

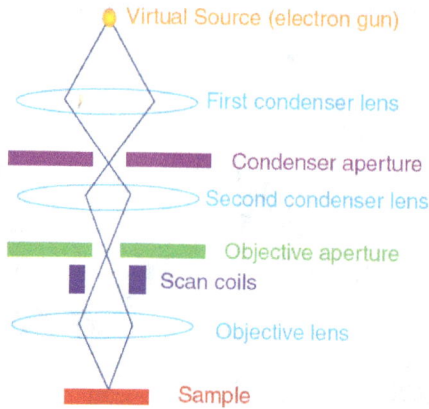

Fig. 3.5. Schematic diagram to illustrate the major components for image formation in a typical scanning electron microscope.

capability afforded by SEM makes it convenient for probing nanomaterials of which the structural features on the nanoscale are critical to their properties and functionalities.

Figure 3.5 shows a schematic of the imaging system in a typical scanning electron microscope starting from the electron source. A stream of monochromatic electrons generated by an electron gun is condensed by the first condenser lens to both form the beam and limit the amount of current, as well as, in conjunction with the condenser aperture, to eliminate the high-angle electrons from the beam. The second condenser lens focuses the electrons into a thin, tight, coherent beam and an objective aperture is used to further eliminate high-angle electrons from the beam. A set of coils is used to scan the beam in a grid fashion. The objective lens focuses the scanning beam onto the specimen desired, one point at a time. Interaction between the electron beam and the sample generates back scattered electrons (BSE), X-ray, secondary electrons (SE), and Auger electrons in a thick or bulk sample. These various electrons are detected and the signal detected contains information about the specimen under investigation. BSE is more sensitive to heavier elements than SE. The X-ray radiation can be detected in a technique called energy dispersive X-ray (EDX) spectroscopy that can be used to identify specific elements [16, 17].

As an example, Fig. 3.6 shows SEM images for a MgH_2–Ni nanocomposite material prepared by ball-milling [16]. Such material is of interest for potential hydrogen storage applications. In the secondary electron (SE) micrograph [Fig. 3.6(a)], various sized particles exist in the range of less than 1 μm to more than 5 μm, and the particle size is clearly not very homogeneous. In the back scattering electron (BSE) image [Fig. 3.6(b)], which is the same area as the SE image, small-size particles with <1 μm in diameter are bright and uniformly distributed on the large-size particles. The BSE signals for the heavier element or the composite containing heavier element can be detected as much brighter spots. Therefore, the bright part shows the distribution of Ni particles because Mg has much smaller atomic number. This indicates that the Ni particles homogeneously distribute in such a micrometer scale range on the surface of MgH_2. This example shows that SE and BSE images can differ in the information they can provide.

Transmission electron microscopy (TEM) is a high spatial resolution structural and chemical characterization tool [18]. A modern TEM has the capability to directly image atoms in crystalline specimens at resolutions close to 0.1 nm, smaller than interatomic distance. An electron beam can also be focused to a diameter smaller than ~0.3 nm, allowing quantitative chemical analysis from a single nanocrystal. This type of analysis is extremely important for characterizing materials at a length scale from atoms to hundreds of nanometers. TEM can be used to characterize nanomaterials to gain information about particle size, shape, crystallinity, and interparticle interaction [2, 19].

Figure 3.7 shows a schematic of important components of the image formation system in a typical TEM. Similar to SEM, a stream of monochromatic electrons produced by an electron gun is focused into a small, thin, coherent beam by two condenser lenses. The electron beam is restricted by the condenser aperture to remove high angle electrons before it reaches the specimen. It is important in this case that the specimen is thin enough to allow some electrons to transmit through the sample. Interaction between the electron beam and specimen generates elastically and inelastically scattered electrons, along with some unscattered electrons, in the forward direction after the sample has been detected. The detected signal contains information about the sample. The detection

Fig. 3.6. (a) Secondary electron (SE) and (b) back scattering electron (BSE) images of the MgH$_2$–Ni nanocomposite, in which both (a) and (b) micrographs are taken in the same field. Reproduced with permission from Ref. 16.

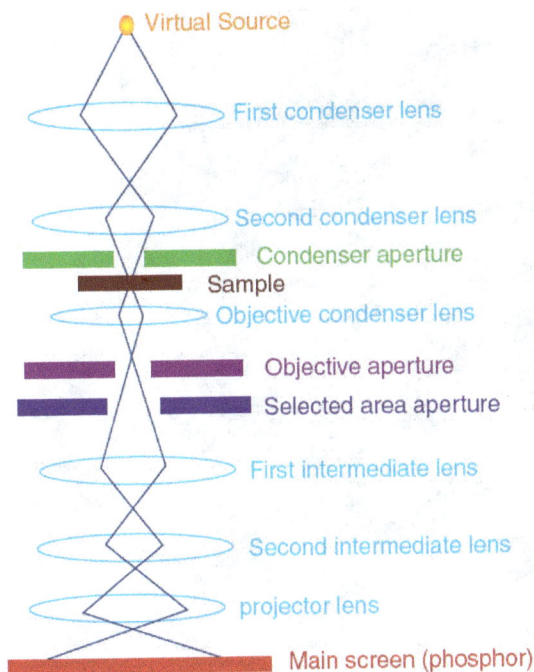

Fig. 3.7. Schematic of major components for image formation of a typical TEM.

involves several lenses to focus the electrons to be detected before they reach the detection phosphor screen. Also shown in the figure are optional objective and selected area metal aperture that can be used to restrict the beam, with the objective aperture enhancing contrast by blocking out high-angle diffracted electrons and the selected area aperture enabling the user to examine the periodic diffraction of electrons by ordered arrangements of atoms in the sample examined. Thus, one major difference between SEM and TEM is that TEM detects "transmitted electrons" and SEM detects "backscattered" and/or secondary electrons. While both techniques can provide topological, morphological and compositional information about the sample, TEM can provide crystallographic information as well. In addition, TEM allows for diffraction patterns to be detected that also contain useful crystallographic information about the sample.

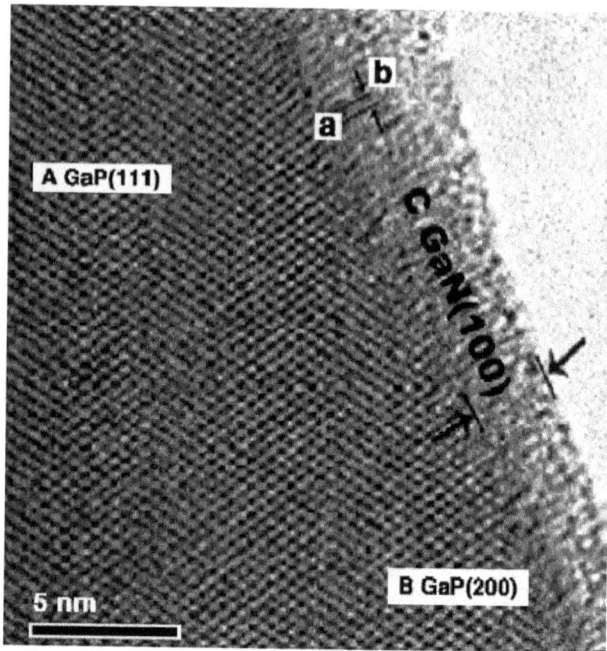

Fig. 3.8. HRTEM image of GaP/GaN core/shell nanocomposite structures clearly showing the boundary between the core and shell. Reproduced with permission from Ref. 20.

Figure 3.8 shows a magnified high resolution TEM (HRTEM) image of GaP/GaN core/shell nanostructures [20]. This image not only clearly shows the crystallographic planes of the core material (GaP) but also the boundary between the core and shell (GaN). Although the shell is mostly amorphous, the lattice planes of some of the areas (e.g. area C) of the shell near the core can also be clearly identified as (100) lattice planes of crystalline GaN. The location "a" measured from the HRTEM image is 0.34 nm, which is larger than that of the GaP(111) and attributed to crystalline grain distortion caused by lattice mismatch between GaP and GaN.

3.2. X-ray: XRD, XPS, and XAFS, SAXS

X-ray diffraction (XRD) is a powerful and routine technique for determining the crystal structure of crystalline materials [21–23]. By examining the diffraction pattern, one can identify the crystalline phase of the material.

Small angle scattering is useful for evaluating the average interparticle distance while wide-angle diffraction is useful for refining the atomic structure of nanoclusters [24]. The widths of the diffraction lines are closely related to the size, size distribution, defects and strain in nanocrystals. As the size of the nanocrystals decreases, the line width is broadened due to loss of long range order relative to the bulk. This XRD line width can be used to estimate the size of the particle via the Debye–Scherrer formula.

$$D = \frac{0.9\lambda}{\beta\cos\theta} \quad\quad (3.1)$$

where D is the nanocrystal diameter, λ is the wavelength of light, β is the full width half at maximum (FWHM) of the peak in radians, and θ is the Bragg angle.

Figure 3.9 shows XRD patterns of Fe^{3+} doped TiO_2 (xFe-TiO_2) as well as N and Fe^{3+} doped TiO_2 (N-xFe-TiO_2) with different Fe^{3+} doping concentration [25]. The major crystalline phase of TiO_2 was determined to be anatase, and the samples after doping with nitrogen show trace amount of brookite phase, which is attributed to chloride introduced in using ammonium chloride as nitrogen source. No iron oxide or Fe_xTiO_y peak could be observed in the XRD spectra, which suggests that all the iron ions were incorporated into the structures of titania and replaced titanium ion or located at interstitial site. The crystallite sizes of samples calculated using the Debye–Scherrer equation are about 10–12 nm.

X-ray based spectroscopies are useful in determining the chemical composition of materials. These techniques include X-ray absorption spectroscopy (XAS) such as extended X-ray absorption fine structure (EXAFS) and X-ray absorption near edge structure (XANES), X-ray fluorescence spectroscopy (XRF), energy dispersive X-ray spectroscopy (EDX), and X-ray photoelectron spectroscopy (XPS) [26, 27]. They are mostly based on detecting and analyzing radiation absorbed or emitted from a sample after excitation with X-rays, with the exception that electrons are analyzed in XPS. The spectroscopic features are characteristic of specific elements and thereby can be used for sample elemental analysis.

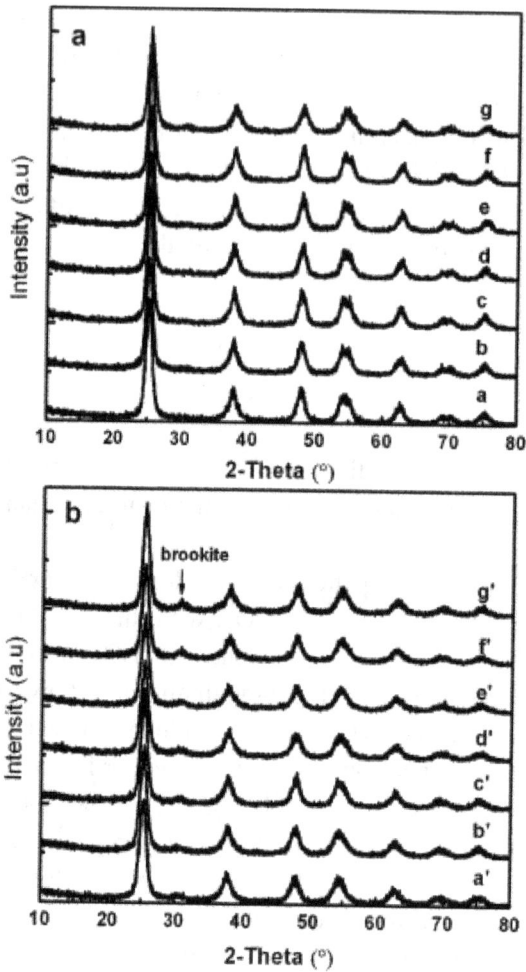

Fig. 3.9. X-ray diffraction patterns of xFe-TiO$_2$ (a) and N-xFe-TiO$_2$ (b) with different Fe^{3+}-doping concentration: (a, a′) 0; (b, b′) 0.05%; (c, c′) 0.1%; (d, d′) 0.5%; (e, e′) 1.0%; (f, f′) 1.5%; (g, g′) 2.0%. Reproduced with permission from Ref. 25.

XPS is based on the measurement of the kinetic energy of photoelectrons generated when the sample is illuminated with soft (1.5 kV) X-ray radiation in an ultrahigh vacuum (UHV) [28]. If one X-ray photon with energy hv is used to excite an atom in its initial state with energy E_i and to eject an electron with kinetic energy, *KE*, with the atom resulting in a final state with energy E_f, one would have the following equation based on total energy conservation:

$$hv + E_i = KE + E_f \qquad (3.2)$$

The difference between the photon energy and the electron kinetic energy is called the *binding energy* of the orbital from which the electron is ejected, which, based on Eq. (3.1), is equal to $E_f - E_i$. Since the photon energy is known from the X-ray radiation source used and the electron kinetic energy can be measured, the binding energy can be determined, which gives the energy difference between the final and initial states of the atom involved in the transition. This binding energy is characteristic for different orbitals of specific elements and is roughly equal to the Hartree–Fock energy of the electron orbital. Therefore, peaks in the photoelectron spectrum can be identified with specific atoms and surface composition can be analyzed. Because the photoelectrons are strongly attenuated by passage through the sample material itself, the information obtained comes from the sample surface, with a sampling depth on the order of 5 nm. Chemical bonding in molecules will cause binding energy shifts, which can thus be used to extract information of a chemical nature (such as atomic oxidation state) from the sample surface.

Figure 3.10 shows the XPS spectrum of Au/SiO_2 core/shell nanoparticles in the region corresponding to the binding energy range of 110–70 eV, which includes the Si 2p, Au 4f, and Cu 3p peaks recorded at 90° and 30° electron takeoff angles, respectively [29]. From the measured binding energies, the chemical elements can be easily identified as corresponding to $Au^{(0)}$ and Si^{4+}. To gain more information about the possible geometry-dependent composition, comparative studies were conducted on gold nanoparticles vapor-deposited onto a silicon substrate containing ca. 4-nm oxide layer. It was found that the XPS intensity ratio of peaks from elements of the core and the shell is independent of the electron takeoff angle, and bulk data are not very reliable for elucidating the composition, size, geometry of nanoparticles from XPS measurements.

XFS is similar to XPS in terms of the excitation mechanism but differs in its detection mechanism. While XPS detects photoelectrons, XFS detects

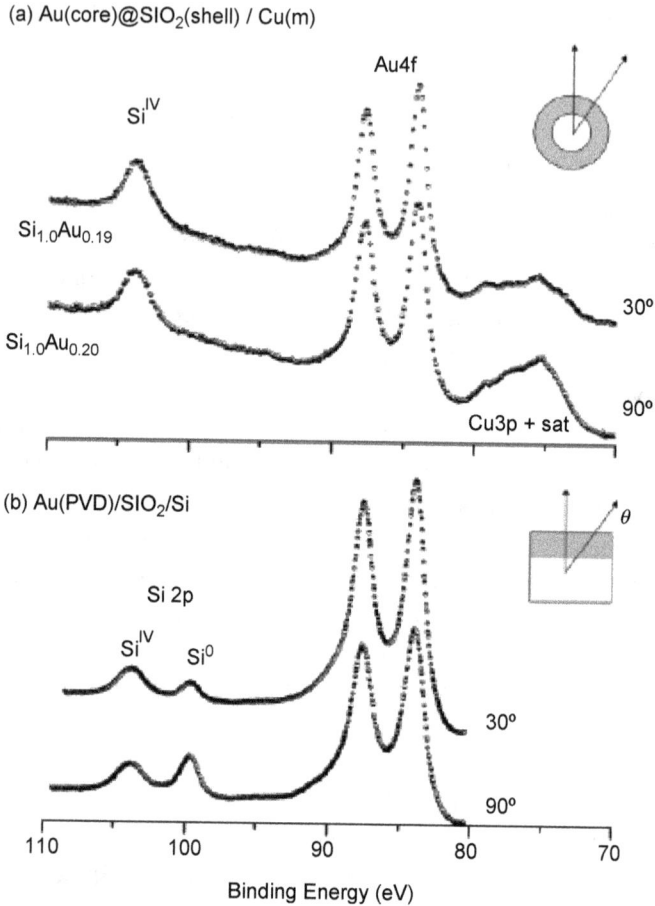

Fig. 3.10. The 110–70 eV region of the XPS spectrum recorded at 90° and 30° electron takeoff angles corresponding to (a) Au (core)/SiO_2 (shell) nanoparticles deposited on copper tape, (b) gold particles vapor-deposited (PVD) onto a silicon substrate containing ca. 4-nm oxide layer. Reproduced with permission from Ref. 29.

"secondary" or "fluorescent" X-rays from a material that has been excited by high-energy X-rays (or sometimes γ-rays) [30]. The principle behind XFS is relatively straightforward. When a material is exposed to high energy or short wavelength X-rays, ionization or electron ejection can take place if the X-ray photon energy is greater than its ionization energy. Due to the high energy of X-rays or γ-rays, tightly bound electrons in the inner, low energy orbitals of the atom in the material can be expelled. The resulting ionized atom is not unstable and electrons in outer, higher energy orbitals may fall or make a transition into the lower orbital to fill the hole left behind. In doing so, energy may be released in the form of a photon (usually with energy in the X-ray region still) with energy equal to the energy difference of the two orbitals involved. Because the orbitals are specific to individual atoms, the energy of the emitted photon that can be easily detected has energy characteristic of the atoms involved. The term "fluorescence" refers to the emitted X-ray photons, not visible light, even though visible light can also be generated and observed sometimes when a sample is subject to X-ray radiation.

Another powerful X-ray based spectroscopic technique is extended X-ray absorption fine structure (XAFS) [27]. EXFAS is based on measuring the fine structure near the absorption edge of a sample when subject to X-ray radiation. It is similar to UV-visible electronic absorption spectroscopy, in principle, except that the spectral range is in the X-ray region and EXAFS focuses on the fine structure specifically since it provides local structural information about specific atoms or ions. EXAFS relates to the details of how X-rays are absorbed by an atom at energies near and above the core-level binding energies of that atom. EXAFS measurements reflect the modulation of an atom's X-ray absorption probability due to the chemical and physical states of the atom. EXAFS spectra are especially sensitive to the formal oxidation state, coordination chemistry, and the local atomic structure of the selected element. One advantage of EXAFS is that it works for crystalline as well as noncrystalline or even highly disordered materials, including solutions. It is thus well suited for studying nanomaterials [31–34]. EXAFS measurements are relatively straightforward but require an intense and energy-tunable source of X-rays, which usually means the use of synchrotrons. While the experimental measurements can be simple, analysis of EXAFS data is somewhat involved and requires specific expertise and good knowledge of relevant physical principles. For convenience of data interpretation, the X-ray absorption spectrum is typically divided into two regimes: X-ray absorption near-edge spectroscopy (XANES) and extended X-ray absorption fine-structure spectroscopy (EXAFS). While XANES is more sensitive to formal oxidation state and coordination chemistry of the absorbing atom,

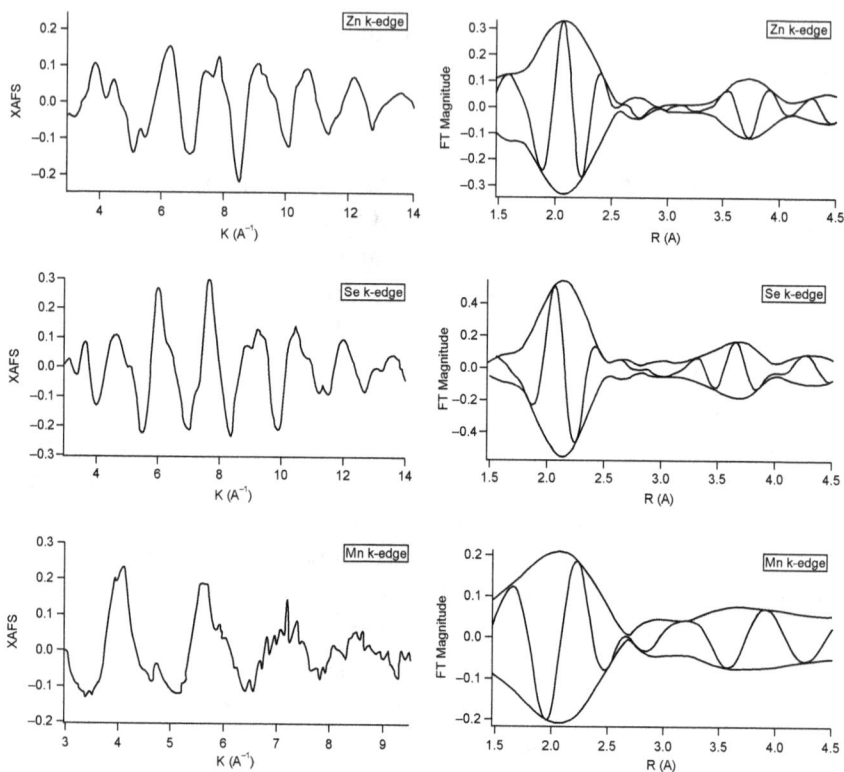

Fig. 3.11. XAFS (left) and FT-XAFS (right) traces for Mn^{2+}-doped ZnSe nanoparticles. The Fourier transform range is 3–14 Å^{-1} for the Zn K-edge and Se K-edge data and 3–9.5 Å^{-1} for the Mn K-edge. The XAFS data were collected at 20 K. Reproduced with permission from Ref. 35.

EXAFS is usually used to determine the distances, coordination number, and species of the neighbors of the absorbing atom.

Figure 3.11 shows some XAFS data for ZnSe:Mn nanoparticles [35]. By determining the local structure of the different ions in the nanoparticles, in conjunction with ESR data, and correlating to their optical properties, it was found that the Mn^{2+} ion is substitutional with tetrahedral symmetry inside the host ZnS nanoparticles and is luminescent around 580 nm, as expected, while Mn^{2+} ions on the nanoparticle surface have a different symmetry (octahedral) and are nonluminescent. This example shows the importance of simultaneous optical and structural studies in investigating nanomaterials.

A final example of X-ray related techniques useful for nanomaterials research is Small Angle X-ray Scattering (SAXS) [36, 37]. SAXS is an analytical technique often used for the structural characterization of solid and fluid materials in the nanometer range. In SAXS measurements, the sample is irradiated by a well-defined, monochromatic X-ray beam. Intensity distribution of the scattered beam at very small scattering angles is measured and it contains structural information of the scattering particles. SAXS can be used to study both monodisperse and polydisperse systems. Information about size, shape and internal structure of the particles can be determined for monodisperse systems and size distributions can be calculated for polydisperse systems. SAXS has been applied to investigate structural details of a variety of nanoparticles of inorganic, organic, as well as biological materials in the size range of 0.5 to 50 nm, usually with high intensity X-ray radiation.

3.3. Electrochemistry and photoelectrochemistry

Electrochemistry is another useful tool for characterizing electronic and electrochemical properties of nanomaterials [4, 38]. One major application is to determine the redox potentials or bandgaps with reference to a known potential or energy level, e.g. a known electrode such as Ag/AgCl or the standard hydrogen potential. This is more advantageous than optical measurement that usually only provide relative energy levels such as bandgap, but not on an absolute scale. Another key use is to determine charge transfer and transport properties in nanomaterials. Electrochemistry techniques have been used to study metal, semiconductor, and insulator or metal oxide nanostructures [39–45].

As example, electrochemistry techniques have been used to study the charging behavior of monodisperse gold nanoparticles, with 8 to 38 kilodaltons core mass or 1.1 to 1.9 nanometers in diameter, stabilized with short-chain alkanethiolate monolayers [46]. Figure 3.12 shows schematic STM double tunnel-junction model (a), electrochemical ensemble Coulomb staircase model (b), and differential pulse voltammograms for (c) butanethiolate (C4) and (d) hexanethiolate (C6) Au monolayer-protected metal clusters (MPCs) as a function of core size. The experimental results

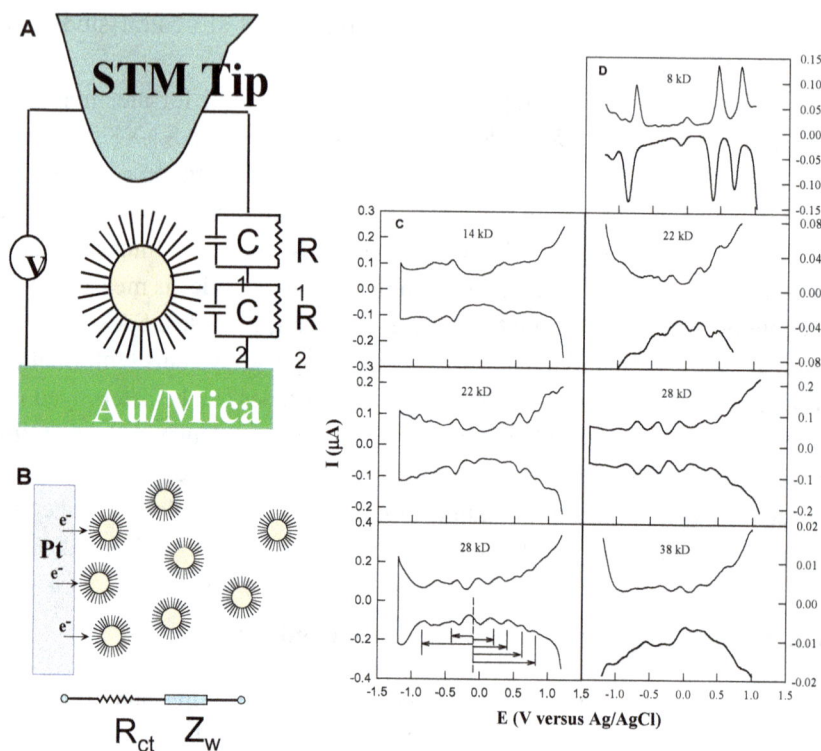

Fig. 3.12. Schematic STM double tunnel-junction model (a) and schematic electro-chemical ensemble Coulomb staircase model (b). R_{ct} is charge transfer resistance and Z_W is diffusional (Warburg) impedance for MPC transport through the solution. Differential pulse voltammograms for (c) butanethiolate (C4) and (d) hexanethiolate (C6) Au MPCs as a function of uniform core size, in 0.05 M Hex$_4$NClO$_4$/toluene/acetonitrile (2/1 v:v), at 9.5×10^{-3} cm^2 Pt electrode; DC potential scan 10 mV/s, pulse amplitude 50 mV. Concentrations are: (c) 14 kD, 0.086 mM; 22 kD, 0.032 mM; 28 kD, 0.10 mM; (d) 8 kD, 0.30 mM; 22 kD, 0.10 mM; 28 kD, 0.10 mM; 38 kD, 0.10 mM. Reproduced with permission from Ref. 46.

show a transition from metal-like double-layer capacitive charging to redox-like charging as the core size decreases. This example demonstrates the usefulness of electrochemistry methods in studying nanomaterials.

Related to electrochemistry and spectroscopy is a more specialized technique called photoelectrochemistry (PEC). It is based on electrochemical measurements when one electrode, usually the anode, is subject to light illumination [47–52]. PEC is useful for studying nanomaterials that are active in photoelectochemical reactions. One example is hydrogen genera-tion from water splitting that is attractive for solar energy conversion. The PEC technique involves measuring photocurrent in an electrochemical setup with light illumination on one or both electrondes (anode and cath-ode). A reference electrode is also used in a typical PEC setup. A more detailed discussion of PEC will be given in Chapter 10.

3.4. Nuclear magnetic resonance (NMR) and electron spin resonance (ESR)

3.4.1. *Nuclear magnetic resonance (NMR)*

Techniques based on magnetic resonance, such as NMR (nuclear mag-netic resonance) and ESR (electron spin resonance), are also useful tools for materials characterization. NMR is particularly powerful for structural determination and is usually used to characterize molecules on the surface of nanomaterials as well as molecule–nanoparticle interaction.

The principle behind NMR is nuclear spin transitions in an applied external magnetic field [53]. Many nuclei have spin and all nuclei are electrically charged. If an external magnetic field is applied, energy trans-fer or transition is possible between the different nuclear spin energy levels. The transition takes place at a wavelength corresponding to radio frequencies (typically in the 40–800 MHz range). The signal that matches the transition, at the resonance frequency, is measured and processed in order to yield an NMR spectrum for the nucleus concerned. The resonant frequency of the transition energy is dependent on the effective magnetic field at the nucleus. This field is affected by electron shielding which is, in turn, dependent on the chemical environment. As a result, information about the nucleus' chemical environment can be derived from its resonant frequency. The most common nucleus probe in NMR is the proton, ^1H. For convenience, a chemical shift, δ, is defined for proton NMR as:

$$\delta = (\nu - \nu_0)/\nu_0 \tag{3.3}$$

where ν is the detected resonance frequency for the proton of interest and $\nu_0 = 400.13000000$ MHz is the customarily adopted proton reference frequency of tetramethylsilane (TMS). For example, the NMR frequency

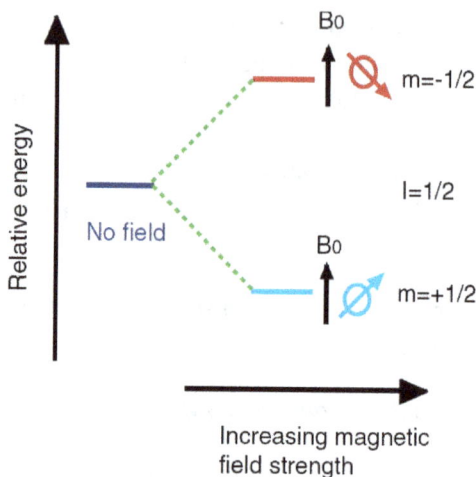

Fig. 3.13. Illustration of NMR transitions for a nuclear spin $I = 1/2$ system, e.g. a proton, under an applied external field.

of protons in benzene is 400.132869 MHz and its chemical shift is calculated to be 7.17 ppm based on Eq. (3.3). The chemical shift is usually expressed in ppm (part per million) for convenience.

The operating principle behind NMR is illustrated in Fig. 3.13, where transitions are shown between two nuclear spin states that split under an applied external magnetic field. The transition energy or the splitting between the two spins states is proportional to the applied magnetic field strength B_0. At higher frequency or field strength, the signal/noise ratio is better due to larger population difference between the ground and excited states.

NMR is a routine and powerful technique for structural determination and chemical analysis of organic and biological molecules including proteins and DNA. For nanomaterials research, NMR has often been used to identify or analyze molecules on the surface of nanomaterials or elements in the nanoparticle [54, 55]. For example, solid state ^{113}Cd, ^{77}Se, ^{13}C and ^{31}P NMR has been used to study a number of Cd chalcogenide nanoparticles synthesized in tri-n-octyl-phosphine (TOP) with different compositions and architectures. The pure CdSe and CdTe nanoparticles show a dramatic, size-sensitive broadening of the ^{113}Cd NMR line, which can be explained

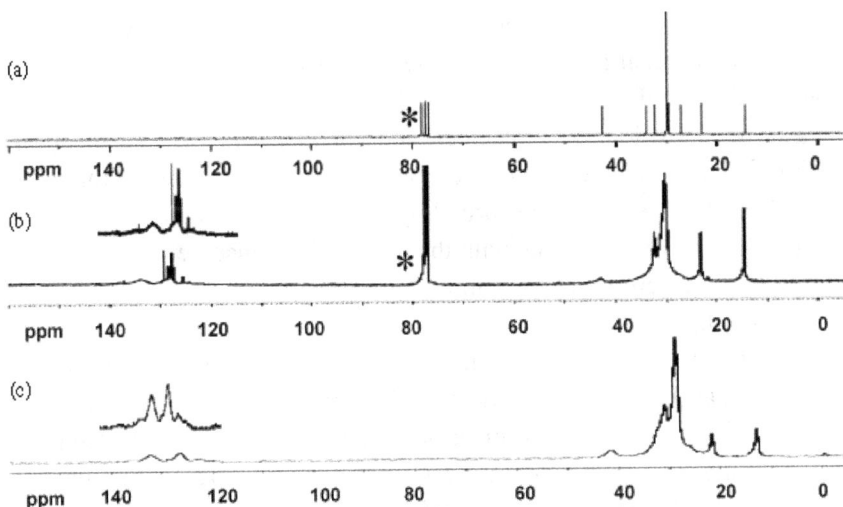

Fig. 3.14.   (a) Solution ^{13}C spectra of HDA in CDCl$_3$*. (b) Solution ^{13}C spectra of 2-nm CdSe-HDA in CDCl$_3$*. This spectrum shows HDA on the surface of the particle as well as bound and unbound thiophenol. (c) Solid-state ^{13}C MAS of 2-nm CdSe-HDA shows the presence of HDA and thiophenol. The R-carbon of HDA is more pronounced than in solution and is at the same chemical shift as free HDA. Spinning speed is 12 kHz. Reproduced with permission from Ref. 55.

in terms of a chemical shift distribution arising from multiple Cd environments [56]. As another example, Fig. 3.14 shows ^{13}C NMR data on hexadecylamine (HDA)-passivated 2-nm CdSe in solution and the solid state [55]. A detailed analysis of these data, in conjunction with ^{1}H-^{113}Cd and ^{1}H-^{77}Se cross-polarization magic angle spinning (MAS) experiments performed on the Cd and Se atoms, provides a picture about the surface properties of the HDA-capped CdSe nanoparticles. It has been found that both HDA and thiophenol bind selectively to specific sites of the CdSe nanocrystal surface.

3.4.2. *Electron spin resonance (ESR)*

Another, somewhat more specialized, technique based on magnetic resonance is ESR (electron spin resonance), also called EPR (electron paramagnetic resonance) [57]. ESR is similar to NMR except that the spin is based on the electron rather than nuclei and the frequency range for electron spin transitions is in the range of microwave frequency (GHz), as compared to NMR in the lower radio frequency (MHz). ESR is based on measuring transitions in molecules, ions or atoms possessing electrons with unpaired spins, i.e. electronic spin $S > 0$, under an external magnetic field. With ESR, energy is absorbed by the sample when the frequency of the radiation is on resonance with the energy difference between two spin states of the electrons in the sample and the appropriate selection rules are satisfied.

Most materials in a bulk form at normal conditions have net zero electronic spin and are thus ESR silent. ESR activity is observed in a number of situations where unpaired electrons are involved or the net electronic spin is not zero, e.g. transition-metal and rare-earth species that contain unpaired nd and/or mf electrons, organic free radicals, such as 1, 1'-Diphenyl-2-picryl-hydrazyl (DPPH) and 2, 2, 6, 6-tetramethyl-1-piperidinyloxyl (TEMPO), organic ion-radicals created during redox reactions, and triplet state organic molecules and biradicals.

The principle of transition in ESR is illustrated in Fig. 3.15 Two electron spin states are split into two nondegenerate states under an applied

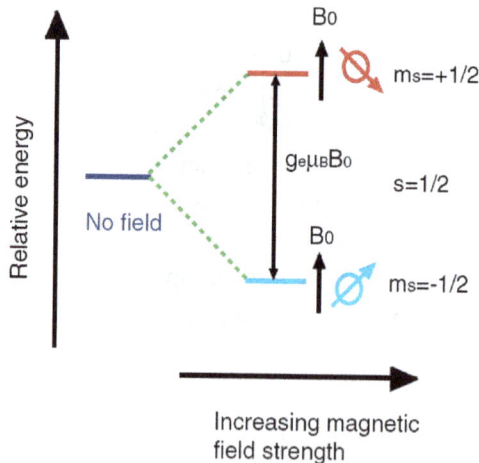

Fig. 3.15. Illustration of electron spin transitions under an external magnetic field B_0, where g_e is the so-called g-factor, a constant around 2 for the free electron, and μ_B is the Bohr magneton, which is equal to $eh/4\pi m_e = 9.2740 \times 10^{-24}$ J/T.

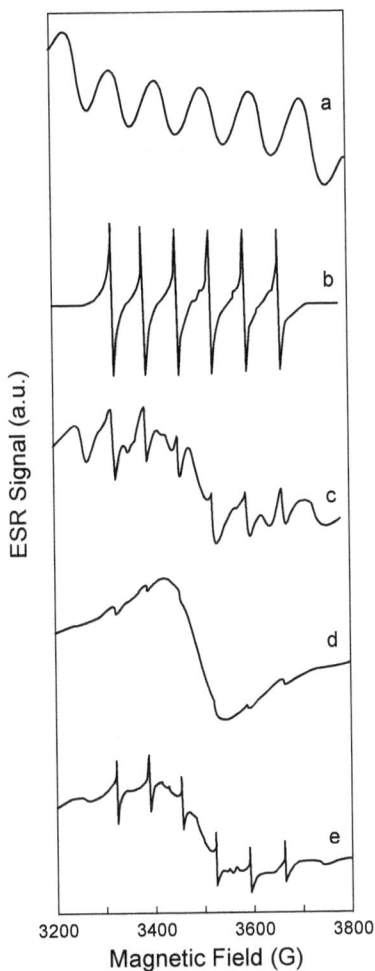

Fig. 3.16. ESR spectra of Mn^{2+}/USY (a), ZnS:Mn^{2+}/USY (b) and different sized ZnS:Mn nanoparticles: (c) 3.5 nm, (d) 4.5 nm and (e) 10 nm. USY is for ultrastable zeolite Y. Reproduced with permission from Ref. 59.

external magnetic field. Under resonance conditions, transitions can occur between these two states, resulting in absorption of electromagnetic radiation, usually in the microwave frequency range. Similar to NMR, the transition energy or the splitting between the two spins states is proportional to the applied magnetic field strength B_0. At higher frequency or field strength, the signal/noise ratio is better due to larger population difference between the ground and excited states. For similar reason, ESR is more sensitive than NMR due to the higher frequency or larger excited and ground state population difference in ESR than NMR.

In nanomaterials research, ESR is particularly useful for studying species with unpaired electrons such as dopants in nanomaterials [35, 58–61]. For example, Fig. 3.16 shows the ESR spectra of Mn^{2+} in different environment, ultrastable zeolite Y (USY) and ZnS nanoparticles in USY, and different sized ZnS nanoparticles [59]. It is clear that the ESR patterns observed for Mn^{2+} are very sensitive to their host environment.

Another interesting application of ESR is to study species with unpaired electrons on nanomaterials surface, such as radicals. For example, radicals on CdSe QD surfaces have been studied by ESR along with optical spectroscopy to understand electron transfer from QD to the radical [62, 63], as will be discussed in more detail in Chapter 4.

3.5. Summary

This chapter provides a brief survey of experimental techniques that are not based on optical spectroscopy but useful for characterizing different properties of nanomaterials, including X-ray (XRD, XAFS, XFS, XPS), microscopy (STM, AFM, TEM, and SEM), electrochemistry, NMR and ESR. They are important for determining topology, morphology, surface, chemical composition and crystal structures of nanomaterials. These techniques complement optical spectroscopy that provides optical, electronic as well as structural information. Given the complex nature of nanomaterials, it is common that a combination of these experimental techniques is used to gain a complete picture and understanding of their properties.

One example is the distinction between so-called "core/shell" structures versus aggregates in the case of gold nanostructures from the reaction of $HAuCl_4$ and Na_2S [64]. While the initial and several follow-up reports claimed Au_2S/Au core/shell structure mainly based on spectroscopic evidence [65, 66], more detailed subsequent studies using a combination of optical and structural characterization techniques, including EXAFS, SAXS and TEM, have shown that the reaction product is mostly aggregates of gold nanoparticles [67–69]. This example shows that high quality, direct structural studies are essential for unambiguously determining the structure of nanomaterials.

References

1. G.Y. Liu, S. Xu and Y.L. Qian, *Acc. Chem. Res.* **33**, 457 (2000).
2. Z.L. Wang, *J. Phys. Chem. B* **104**, 1153 (2000).
3. Z.L. Wang, P. Poncharal and W.A. de Heer, *Microscopy and Microanalysis* **6**, 224 (2000).
4. J.Z. Zhang, Z.L. Wang, J. Liu, S. Chen and G.-y. Liu, *Self-assembled Nanostructures*. Nanoscale Science and Technology, New York: Kluwer Academic/Plenum Publishers. **316** (2003).
5. G. Cao, *Nanostructures & Nanomaterials: Synthesis, Properties & Applications*, London: Imperial College Press. **433** (2004).
6. F.J. Giessibl, S. Hembacher, H. Bielefeldt and J. Mannhart, *Science* **289**, 422 (2000).
7. D. Bonnell, *Scanning Probe Microscopy and Spectroscopy: Theory, Techniques, and Applications*, New York: Wiley-VCH (2000).
8. E. Meyer, *Atomic Force Microscopy: Fundamentals to Most Advanced Applications* Vol. 1, New York: Springer-Verlag TELOS. **250** (2007).
9. G. Binnig, C.F. Quate and C. Gerber, *Phys. Rev. Lett.* **56**, 930 (1986).
10. H.J. Guntherodt, D. Anselmetti and E. Meyer, eds. NATO ASI series. Series E, Applied sciences. Vol. 286, Kluwer Academic. **644** (1995).
11. F.J. Giessibl, H. Bielefeldt, S. Hembacher and J. Mannhart, *Annalen Der Physik* **10**, 887 (2001).
12. C. Bai, *Scanning Tunneling Microscopy and Its Applications*, New York: Springer-Verlag TELOS. **366** (2007).
13. A.M. Baro, G. Binnig, H. Rohrer, C. Gerber, E. Stoll, A. Baratoff and F. Salvan, *Phys. Rev. Lett.* **52** 1304 (1984).
14. A. Riposan and G.Y. Liu, *J. Phys. Chem. B* **110**, 23926 (2006).
15. J.I. Goldstein, D.E. Newbury, P. Echlin, D.C. Joy, J. Romig, A.D., C.E. Lyman, C. Fiori and E. Lifshin, *Scanning Electron Microscopy and X-ray Microanalysis*. 2nd ed, New York: Plenum Press. **820** (1992).
16. N. Hanada, E. Hirotoshi, T. Chikawa, E. Akiba and H. Fujii, *J. Alloy Compd.* **450**, 395 (2008).
17. Z.L. Liu, J.C. Deng, J.J. Deng and F.F. Li, *Materials Science and Engineering B-Advanced Functional Solid-State Materials* **150**, 99 (2008).

18. Z.L. Wang, ed., Wiley-VCH: New York. **406** (2000).
19. Z.L. Wang, *Advan. Mater.* **10**, 13 (1998).
20. N.S. Yu, X.P. Hao, X.G. Xu and M.H. Jiang, *Mater Lett.* **61**, 523 (2007).
21. R.L. Whetten, J.T. Khoury, M.M. Alvarez, S. Murthy, I. Vezmar, Z.L. Wang, P.W. Stephens, C.L. Cleveland, W.D. Luedtke and U. Landman, *Advan. Mater.* **8**, 428 (1996).
22. C.B. Murray, C.R. Kagan and M.G. Bawendi, *Science* **270**, 1335 (1995).
23. C.B. Murray, C.R. Kagan and M.G. Bawendi, *Ann. Rev. Mater. Sci.* **30**, 545 (2000).
24. A.P. Alivisatos, *Science* **271**, 933 (1996).
25. Y. Cong, J.L. Zhang, F. Chen, M. Anpo and D.N. He, *J. Phys. Chem. C* **111**, 10618 (2007).
26. T.L. Barr, *Modern ESCA: The Principles and Practice of X-Ray Photoelectron Spectroscopy*, New York: CRC Press. **384** (1994).
27. D.C. Koningsberger and R. Prins, eds. A Series of Monographs on Analytical Chemistry and Its Applications, Wiley-Interscience: New York. **688** (1988).
28. T.L. Barr, *Modern ESCA: The Principles and Practice of X-Ray Photoelectron Spectroscopy*, Boca Raton, FL: CRC Press. **376** (2008).
29. I. Tunc, S. Suzer, M.A. Correa-Duarte and L.M. Liz-Marzan, *J. Phys. Chem. B* **109**, 7597 (2005).
30. J.A. Carlisle, S.R. Blankenship, R.N. Smith, A. Chaiken, R.P. Michel, T. van Buuren, L.J. Terminello, J.J. Jia, T.A. Callcott and D.L. Ederer, *J. Clust. Sci.* **10**, 591 (1999).
31. H. Hosokawa, H. Fujiwara, K. Murakoshi, Y. Wada, S. Yanagida and M. Satoh, *J. Phys. Chem.* **100**, 6649 (1996).
32. J. Rockenberger, L. Troger, A. Kornowski, T. Vossmeyer, A. Eychmuller, J. Feldhaus and H. Weller, *J. Phys. Chem. B* **101**, 2691 (1997).
33. R.E. Benfield, D. Grandjean, M. Kroll, R. Pugin, T. Sawitowski and G. Schmid, *J. Phys. Chem. B* **105**, 1961 (2001).
34. Y.V. Zubavichus, Y.L. Slovokhotov, M.K. Nazeeruddin, S.M. Zakeeruddin, M. Gratzel and V. Shklover, *Chem. Mater.* **14**, 3556 (2002).
35. T.J. Norman, D. Magana, T. Wilson, C. Burns, J.Z. Zhang, D. Cao and F. Bridges, *J. Phys. Chem. B* **107**, 6309 (2003).
36. O. Glatter and O. Kratky, eds., Acamemic Press: New York. **515** (1982).
37. S. Remita, P. Fontaine, E. Lacaze, Y. Borensztein, H. Sellame, R. Farha, C. Rochas and M. Goldmann, *Nuclear Instruments & Methods in Physics Research Section B-Beam Interactions with Materials and Atoms* **263**, 436 (2007).

38. G. Hodes, Wiley-VCH: Weinheim; New York. **326** (2001).

39. C. Lebreton and Z.Z. Wang, *Surface Sci.* **382**, 193 (1997).

40. C. Schonenberger, B.M.I. vanderZande, L.G.J. Fokkink, M. Henny, C. Schmid, M. Kruger, A. Bachtold, R. Huber, H. Birk and U. Staufer, *J. Phys. Chem. B* **101**, 5497 (1997).

41. W. Chen and S.G. Sun, *Spectroscopy and Spectral Analysis* **24**, 817 (2004).

42. J. Meier, J. Schiotz, P. Liu, J.K. Norskov and U. Stimming, *Chem. Phys. Lett.* **390**, 440 (2004).

43. X.H. Guan, Y.H. Qin, S.Y. Zhang, J. Guo and J.H. Li, *Chemical J. Chinese Universities-Chinese* **26**, 1825 (2005).

44. L. Zhang, Q. Zhang, X.B. Lu and J.H. Li, *Biosens. Bioelectron.* **23**, 102 (2007).

45. W. Sun, R.F. Gao and K. Jiao, *J. Phys. Chem. B* **111**, 4560 (2007).

46. S.W. Chen, R.S. Ingram, M.J. Hostetler, J.J. Pietron, R.W. Murray, T.G. Schaaff, J.T. Khoury, M.M. Alvarez and R.L. Whetten, *Science* **280**, 2098 (1998).

47. C. Nasr, S. Hotchandani, W.Y. Kim, R.H. Schmehl and P.V. Kamat, *J. Phys. Chem. B* **101**, 7480 (1997).

48. Y.Q. Wang, H.M. Cheng, Y.Z. Hao, J.M. Ma, B. Xu, W.H. Li and S.M. Cai, *J. Mater. Sci. Lett.* **18**, 127 (1999).

49. P.V. Kamat, S. Barazzouk, S. Hotchandani and K.G. Thomas, *Chem. Eur. J.* **6**, 3914 (2000).

50. L. Zhang, Y.Q. Wang, M.Z. Yang, E.Q. Gao and S.M. Cai, *Chemical J. Chinese Universities-Chinese* **21**, 1075 (2000).

51. I. Willner, F. Patolsky and J. Wasserman, *Angew. Chem. Int. Edit* **40**, 1861 (2001).

52. T. Lana-Villarreal and R. Gomez, *Electrochem. Commun.* **7**, 1218 (2005).

53. J.B. Lambert and E.P. Mazzola, *Nuclear Magnetic Resonance Spectroscopy: an Introduction to Principles, Applications, and Experimental Methods*, Upper saddle River, NJ: Pearson/Prentice Hall. **341** (2004).

54. O. Kohlmann, W.E. Steinmetz, X.A. Mao, W.P. Wuelfing, A.C. Templeton, R.W. Murray and C.S. Johnson, *J. Phys. Chem. B* **105**, 8801 (2001).

55. M.G. Berrettini, G. Braun, J.G. Hu and G.F. Strouse, *J. Am. Chem. Soc.* **126**, 7063 (2004).

56. C.I. Ratcliffe, K. Yu, J.A. Ripmeester, M.B. Zaman, C. Badarau and S. Singh, *Phys. Chem. Chem. Phys.* **8**, 3510 (2006).

57. M.R. Brustolon, *Principles and Applications of Electron Paramagnetic Resonance Spectroscopy*, New York: Blackwell Publishers. **400** (2008).

58. P.H. Borse, D. Srinivas, R.F. Shinde, S.K. Date, W. Vogel and S.K. Kulkarni, *Phys. Rev. B* (Condensed Matter), **60**, 8659 (1999).

59. W. Chen, R. Sammynaiken, Y.N. Huang, J.O. Malm, R. Wallenberg, J.O. Bovin, V. Zwiller and N.A. Kotov, *J. Appl. Phys.* **89**, 1120 (2001).
60. V.N. Andreev, S.E. Nikitin, V.A. Klimov, S.V. Kozyrev, D.V. Leshchev and K.F. Shtel'makh, *Phys. Solid State* **43**, 788 (2001).
61. P. Balaz, M. Valko, E. Boldizarova and J. Briancin, *Mater. Lett.* **57**, 188 (2002).
62. V. Maurel, M. Laferriere, P. Billone, R. Godin and J.C. Scaiano, *J. Phys. Chem. B* **110**, 16353 (2006).
63. E. Heafey, M. Laferriere and J.C. Scaiano, *Photochem. Photobiol. Sci.* **6**, 580 (2007).
64. J.Z. Zhang, A.M. Schwartzberg, T. Norman, C.D. Grant, J. Liu, F. Bridges and T. van Buuren, *Nano Lett.* **5**, 809 (2005).
65. H.S. Zhou, I. Honma, H. Komiyama and J.W. Haus, *Phys. Rev. B-Condensed Matter* **50**, 12052 (1994).
66. R.D. Averitt, D. Sarkar and N.J. Halas, *Phys. Rev. Lett.* **78**, 4217 (1997).
67. T.J. Norman, C.D. Grant, A.M. Schwartzberg and J.Z. Zhang, *Opt. Mater.* **27**, 1197 (2005).
68. A.M. Schwartzberg, A. Wolcott, T. Willey, T. Van Buuren and J.Z. Zhang, *SPIE Proc.* **5513**, 213 (2004).
69. A.M. Schwartzberg, C.D. Grant, T. van Buuren and J.Z. Zhang, *J. Phys. Chem. C* **111**, 8892 (2007).

Chapter 4

Synthesis and Fabrication of Nanomaterials

Synthesis and fabrication of nanomaterials are not the primary focus of this book. However, in order to make the book self-contained and make it easier to appreciate the optical properties and the use of spectroscopy for their study, we will briefly cover the most common synthetic and fabrication techniques for nanomaterials and nanostructures. Roughly, the synthesis and fabrication methods can be divided into solution-based (usually chemical) and gas- or vapor-based (chemical or physical). Commonly used methods include mechanical milling (powder mixing), precipitated hydrolysis, hydrothermal, melting and rapid quenching, and thermal decomposition. Instead of providing a comprehensive review or compilation of all the synthetic methods, we use specific well-documented examples to illustrate the diversity of methods and types of nanomaterials that can be produced.

4.1. Solution chemical methods

4.1.1. *General principle for solution-based colloidal nanoparticle synthesis*

Solution-based colloidal chemistry method is arguably the most popular and convenient technique for synthesizing nanocrystals (NCs) or nanoparticles (NPs) of metals, semiconductors, and insulators. Two critical steps

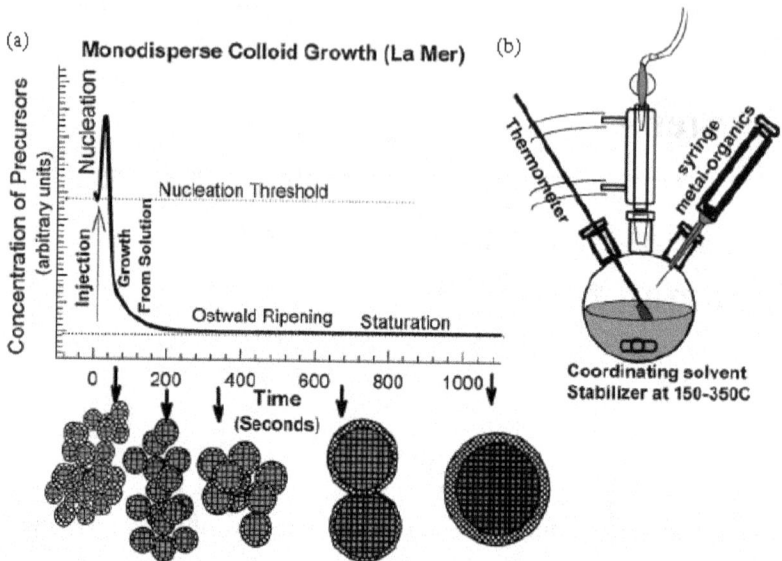

Fig. 4.1. (a) Cartoon depicting the stages of nucleation and growth for the preparation of monodisperse NCs in the framework of the La Mer model. As NCs grow with time, a size series of NCs may be isolated by periodically removing aliquots from the reaction vessel. (b) Representation of the simple synthetic apparatus employed in the preparation of monodisperse NC samples. Reproduced with permission from (Fig. 1) of Ref. 1.

involved in the synthesis are initial nucleation and subsequent growth [1]. Earlier studies by La Mer and Dinegar showed that the production of monodisperse colloids requires initial temporally discrete nucleation followed by slower controlled growth on the existing nuclei [Fig. 4.1(a)] [2]. Rapid addition of reagents to the reactor raises the precursor concentration above the nucleation threshold and quick nucleation starts and partially relieves the supersaturation. As long as the rate of feedstock consumption by the growing colloidal nanoparticles is faster than the rate of precursor addition to solution, no new nuclei would form. Since the growth of all the nanoparticles is similar, their initial size distribution is mainly determined by the time over which the nuclei are formed and begin to grow. If the growth during the nucleation period is small compared with subsequent growth, the nanoparticles can become more uniform over time, which is a phenomenon referred to as focusing of the size distribution [3].

Many systems exhibit a second growth phase called Ostwald ripening [4, 5]. In this process, small nanoparticles with high surface energy dissolve and larger nanoparticles grow. Therefore, over time the number of nanoparticles decreases while the average nanoparticle size increases. Ostwald ripening can be exploited in the preparation of nanoparticles with different sizes [6]. For instance, portions of the reaction mixture can be removed at increments in time, as depicted in Fig. 4.1(a). Raw material can often be extracted with initial distributions of 10% < σ < 15% in diameter, which are then narrowed to ≤5% through size-selective processing. This can be done during the reaction or reflux process using a syringe in a simple setup like the one shown in Fig. 4.1(b).

4.1.2. *Metal nanomaterials*

Chemical synthesis of metal nanostructures usually involves reduction of metal salts or metal ions by appropriate reducing agents in liquids. Simple reduction reactions usually produce nearly spherical nanoparticles. For more complex shaped structures, such as nanorods, aggregates, hollow nanospheres, nanowires, nanocages, nanoprisms, and nanoplates, the synthesis is typically more involved. Electrochemistry techniques have also been used for producing metal nanostructures such as nanorods.

The most popular and, perhaps, simplest method for producing gold nanoparticles is the one originally developed by Turkevich [7] and later improved by Frens and others [8–10]. The synthesis involves simply boiling a solution of gold chloride in the presence of a weak reducing agent, sodium citrate. A similar approach was introduced by Lee and Meisel for making silver nanoparticles [11]. While the Turkevich method normally produces relatively small, monodisperse spherical particles, the Lee–Meisel method generates large, inhomogeneous silver nanoparticles. Uniform or monodisperse size and shape of nanoparticles are important for certain applications and fundamental studies, and are thus usually desired. In order to stabilize the nanoparticles, a capping agent is often used. In the Turkevich and Lee–Meisel syntheses, sodium citrate acts as both a capping material and as the reducing agent.

As expected, metal nanoparticle synthesis involves two stages, nucleation (or small "seed" formation) and growth [12]. In the growth stage, the capping material is very important in determining the size and shape of the resulting nanoparticles. If the material is too weakly associated with the metal, there will be little or no protection and the formed nanocrystals will continue to grow into very large crystals or begin to aggregate. However, if the capping material is too strongly bound to the surface, it can limit or stop growth. It is thus possible to control the particle shape and size by selecting the appropriate capping material. In some cases, the capping materials can induce interparticle aggregation [13–15]. In the case of citrate, it is only weakly bound to the surface, mostly electrostatically. However, there is enough association and ionic repulsion to prevent aggregation. Aggregation is usually undesirable but can be useful and is thereby desired in some cases like for SERS or for achieving near IR absorption [15].

For some systems, the capping material can also limit the size of the particle, as in micelle or reverse micelle directed reactions [16, 17]. In the case of citrate or similar reducing agents, the concentration is an important factor [10]. The concentration of the initial "seed" clusters will be entirely determined by the amount of reducing agent present, which in turn determines the total number of resulting particles. Therefore, with little citrate there will be fewer seeds formed, leading to fewer, large particles as compared to the small sized, but higher concentration particles formed with higher citrate concentration.

In general, as particle size increases, the size dispersion becomes larger, which is undesirable for some applications. Because nucleation and particle growth can overlap in time, seed clusters continue to form while the initial particles are already growing. As a solution to this problem, the technique of seed-mediated growth has been developed [18–21]. Small seed particles are produced by the Turkevich method, and then they are diluted and mixed with additional gold ions. By adding a weak reducing agent (e.g. hydrazine or ascorbic acid), the gold will only grow on the surface of the particles and few or no additional seed particles will form. By separating the seed formation stage and the particle growth stage in time, it is possible to produce relatively large particles (>100 nm) with high homogeneity [22]. This method has also been found useful as a starting point for structural or shape control.

Because optical properties of metal nanostructures strongly depend on their shape, it is often highly desired to control shape to achieve different optical properties for both fundamental studies and applications. Depending on the surfactant (capping material) used, it is possible to produce shapes other than spheres during seed-mediated growth. The most notable example is the generation of nanorods via seed-mediated growth in the presence of cetyltrimethylammonium bromide (CTAB). There are several theories as to the nature of nanorod formation, which normally center on the preferential adsorption of the surfactant to certain crystal facets on seed particles. In the case of nanorods, CTAB binds to the surface radially, but not axially [23, 24]. This blocks crystal growth on surfaces with CTAB and the particle can only grow in the axial direction. The ratio of nanorods over nanoparticles can be significantly enhanced by simply adding silver ions to the growth medium [25–28]. Figure 4.2 shows representative TEM images and UV-vis spectra of gold nanorods

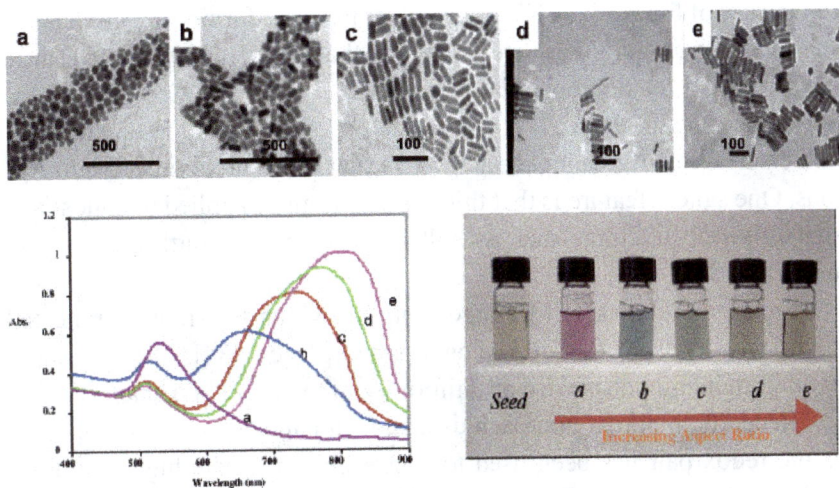

Fig. 4.2. Transmission electron micrographs (top), optical spectra (left), and photographs of (right) aqueous solutions of gold nanorods of various aspect ratios. Seed sample: aspect ratio 1; sample (a) aspect ratio 1.35 ± 0.32; sample (b) aspect ratio 1.95 ± 0.34; sample (c) aspect ratio 3.06 ± 0.28; sample (d) aspect ratio 3.50 ± 0.29; sample (e) aspect ratio 4.42 ± 0.23. Scale bars: 500 nm for (a) and (b) 100 nm for (c)–(e). Reproduced with permission from Ref. 25.

with different aspect ratios produced using CTAB as a surfactant [25]. It has also been found possible to control particle shape and size merely by altering CTAB and ionic concentrations in solution.

In addition to producing particles with nonspherical shapes, it is also possible to alter the internal structure of a particle to manipulate their optical properties. Dielectric/metal core/shell and hollow metal structures are good examples. Synthesis of dielectric/metal core/shells structures often start with a dielectric, e.g. SiO_2, nanoparticles and the metal "nanoshell" is grown onto the surface of the dielectric nanoparticles. It should be made clear that "nanoshells" made this way are usually not single crystals as one ideally would like to have, but rather aggregates in most cases and polycrystals in some rare cases. Hollow polycrystal nanospheres can be made using a technique called galvanic replacement. As the metal ions interact with the surface of nanoparticles of another metal, the metal ions can be reduced into the metal while the original metal nanoparticles are oxidized into metal ions due to their difference in redox potentials. In the case of the work of Xia *et al.*, silver particles formed are used as the "template" and gold ions are added to the solution [29–31]. Because the standard reduction potential of the $AuCl_4^-$/Au pair (0.99 V vs. SHE) is higher than that of the Ag^+/Ag pair (0.80 V vs. SHE) gold ions are reduced at the surface of Ag nanoparticles, and silver metal are oxidized into Ag^+ ions. One salient feature is that this approach can be applied to almost any shape of nanostructures such as hollow rods, spheres, rattles, cubes and wires [32, 33].

Figure 4.3 shows TEM images and XRD data for slightly truncated silver nanocubes synthesized using a polyol process [34]. The nanocubes exhibit high monodispersity or uniformity in their size and shape.

Recently, a similar approach using cobalt nanoparticles and gold ions as the redox pair has been used to successfully produce highly uniform hollow gold nanospheres (denoted as HGN or HAuNS) with tunable optical properties [15, 35]. It was found possible to tune the plasmon absorption in the entire visible and near IR region of the spectrum by simply controlling the shell thickness and sphere diameter. The UV-vis spectra and colors of some HGN samples with various combinations of shell thickness and sphere diameter are shown in Fig. 1.2 in Chapter 1. Again, it should be noted that these HGN are not single crystals. While there are large single

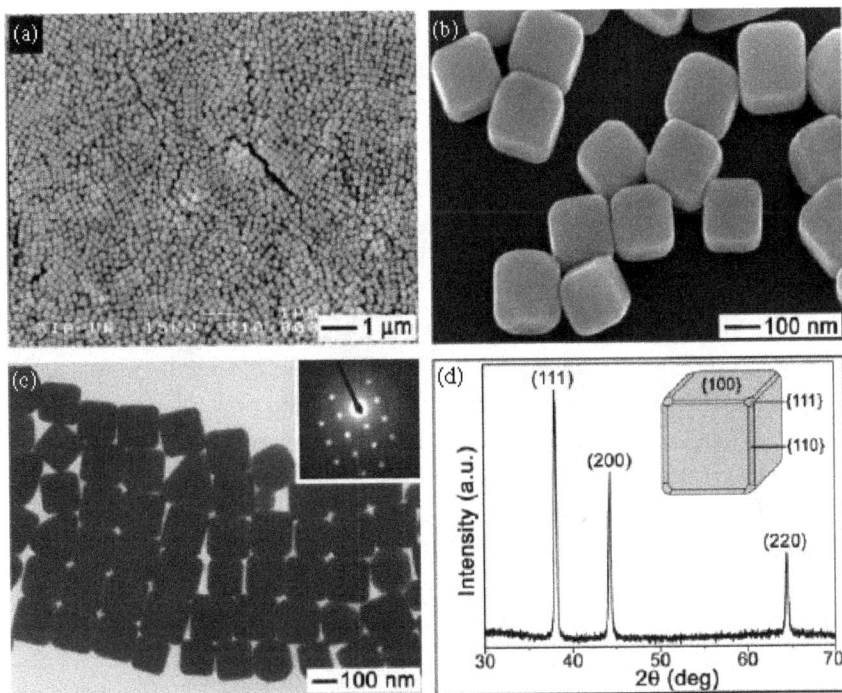

Fig. 4.3. (a) Low- and (b) high-magnification SEM images of slightly truncated silver nanocubes synthesized with a polyol process. The image shown in (b) was taken at a tilting angle of 20°. (c) A TEM image of the same batch of silver nanocubes. The inset shows the diffraction pattern recorded by aligning the electron beam perpendicular to one of the square faces of an individual cube. (d) An XRD pattern of the same batch of sample, confirming the formation of pure fcc silver (a.u. arbitrary units). Reproduced with permission from Ref. 34.

crystal domains, there is significant twinning when the entire shell is examined closely. The shell is apparently polycrystalline, as is likely true for most "shell" structures reported in the literature to date. It is extremely challenging, if possible at all, to produce a spherical shell that is entirely single crystalline. Structural characterization at the atomic scale is essential for nanomaterials. Lack of detailed structural study can often lead to controversies regarding proper identification of the true structure [15].

Electrochemistry techniques have also been used to produce metal nanoparticles and nanowires [36–42]. The mechanism behind such synthesis differs from that of normal chemical reduction by a reducing agent in that direct reduction of metal ions by electrons is primarily responsible for the metal nanostructure formation. Current density and metal ion concentration are factors that affect the nanostructures formed. Other methods for producing metal nanostructures include reduction of metal ions using various sources of radiation including light [43–47], microwave [48, 49], radio frequency (RF) sputtering [50], and γ-ray [51], or even ultrasound wave [52]. Generally, the structures of the nanomaterials produced by these methods tend to have broad size and shape distributions.

For certain applications, surface modification of metal nanoparticles is desired. One approach is silinization, i.e. coating of a silica layer on the metal surface using a seeded polymerization technique [53–55]. A more detailed explanation about surface modification of nanoparticles in general will be given in the next section when we discuss semiconductor nanomaterials.

4.1.3. *Semiconductor nanomaterials*

For semiconductors such as Si and Ge that involve only one element, their synthesis usually involve reduction of a salt of the corresponding element with appropriate reducing agent or decomposition of precursor compounds. This is similar to the synthesis of metal nanomaterials. For example, Si nanoparticle can be synthesized from porous silicon by electrochemical etching of bulk Si wafer using hydrogen fluoride [56], solution reduction [57], or reduction of $SiCl_4$ and $RSiCl_3$ (R=H, octyl) by sodium metal in a nonpolar organic solvent at high temperatures (385°C) and high pressures (>100 atmospheres). Similarly, Ge nanoparticles have been synthesized via ultrasound mediated reduction of mixtures of Ge precursors [58], metathesis reaction between NaGe and $GeCl_4$ in solution [59], or thermal decomposition of germane precursors such as trichlorogermane and tetraethylgermane in hot organic solvents [60].

In a typical synthesis of binary semiconductors such as II-VI and III-V semiconductors, reactants or molecular precursors containing the desired cation and anion components are mixed in an appropriate solvent to produce the nanostructured product of interest. The result of the synthesis depends strongly on a number of factors such as concentration, temperature,

mixing rate, and pH if in aqueous solution [6, 61–67]. Surfactant or capping molecules are often used to stabilize the nanoparticles and can even
direct particle growth along a particular crystal plane to produce nanorods
or other structures [68]. Truly bare or naked nanoparticles are not ther-
modynamically stable because of high surface energy or dangling
chemical bonds on the surface. Impurities either from starting materials or
introduced from some other source during synthesis can have profound
effects on the formation of the desired nanostructure and its properties. In
light of this, extreme care should be taken to ensure that high purity reactants are used and that the synthetic process is as clean as possible.

Earlier synthesis of semiconductor nanomaterials have been primarily conducted in aqueous solution, which lend themselves conveniently
for biomedical applications that usually involve aqueous media. For
example, 1-thioglycerol-capped CdS clusters have been prepared by
adding different amounts of 1-thioglycerol and H_2S to a solution of
$Cd(ClO_4)_2 \cdot 6H_2O$ in water and followed by different heat treatment [69].
The nanoparticles are usually stabilized by ions in the solution or surfactants added on purpose. Nanoparticles synthesized in aqueous solutions
tend to have a high density of defects or trap states and low overall PL
yield. The PL is also sensitive to pH and the reason is likely related to
surface oxide and/or hydroxide that are insulators or large bandgap semiconductors serving as a passivating layer. This is consistent with the
typical observation that the QDs are more stable and more luminescence
at higher pH than at lower pH, since lower pH likely removes the oxide
and/or hydroxide layer [70].

In recent years, synthesis in organic solvents has become popular
partly because higher quality nanoparticles, in terms of size monodispersity, surface properties, and PL yield, can be readily produced. The
organic solvents are usually mixtures of long chain molecules such as
alkylphosphines, R_3P, alkylphosphine oxides, R_3PO (R = butyl or octyl),
and alkylamines [6]. For example, highly monodisperse and luminescent
II–VI semiconductors such as ME (M for metal such as Cd and Zn and
E for S, Se, and Te) can be synthesized by injecting liquid metal-organic
precursors containing M and E into hot (around 150–350°C) coordinating
organic solvents. The M and E atoms are released from the precursor
molecules by thermal decomposition and then react to form ME nuclei

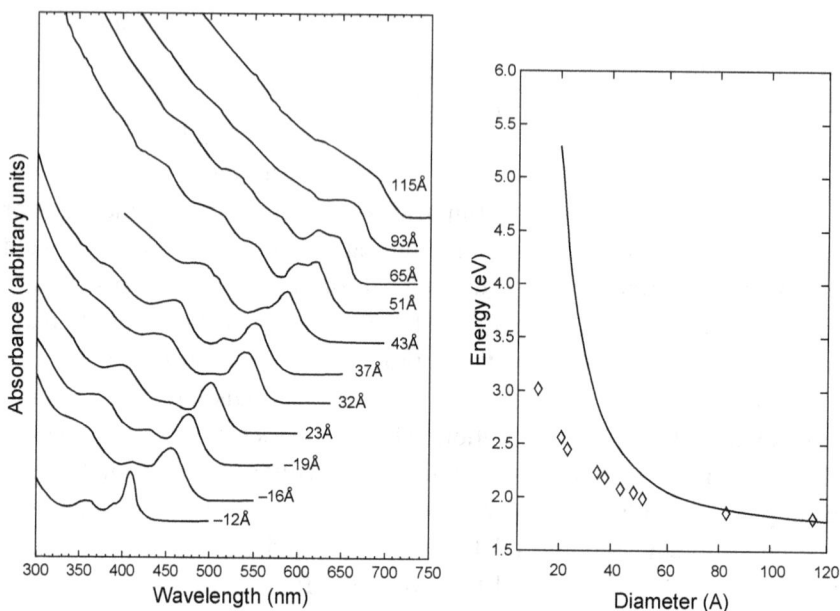

Fig. 4.4. Room temperature optical absorption spectra of CdSe nanocrystallites dispersed in hexane and ranging in size from ~12 to 115 Å (left). HOMO-LUMO transition energy of CdSe crystallites as a function of size (diamonds) compared with the prediction of the effective mass approximation (solid line) (right). Reproduced with permission from Ref. 6.

that subsequently grow into nanocrystals or nanoparticles. The growth is controlled by the coordinating solvent or surfactant molecules and can be easily monitored by the change in UV-vis absorption spectra, as shown in Fig. 4.4. Examples of precursors for group II elements include metal alkyls (dimethylcadmium, diethylcadmium, diethylzinc, dibenzylmercury), while, for group VI elements, organophosphine chalcogenides (R_3PE) or bistrimethylsilylchalcogenides TMS_2E (TMS, trimethylsilyl) are often used [1]. Some of the precursor molecules used are toxic and alternative methods have been explored in recent years involving less toxic molecules as precursors, as discussed in more detail in the last section of this chapter.

While QDs synthesized in organic solvents tend to have higher quality in terms of lower density of surface trap states and thus higher PL yield, their hydrophobic surfaces are not compatible with applications that require QDs with hydrophilic surfaces, e.g. biological applications involving aqueous environment. In order to render QDs with hydrophobic surface hydrophilic, a number of strategies have been developed in recent years. These include ligand exchange, silinization, and surface coating using amphiphilic polymers or surfactants such as phospholipids. Figure 4.5 provides a summary of these different methods [71]. The ligand exchange approach makes use of molecules that contain two functional groups, one reactive towards or binds to the QD surface, e.g. –SH, and another hydrophilic to ensure water solubility, e.g. -COOH. Examples of such ligand molecules include mercaptoacetic acids and mercaptobenzoic acid [72–76]. While the ligand exchange approach is relatively simple to carry out, instability of the ligand-surface binding is an issue, especially over an extended period of time [77].

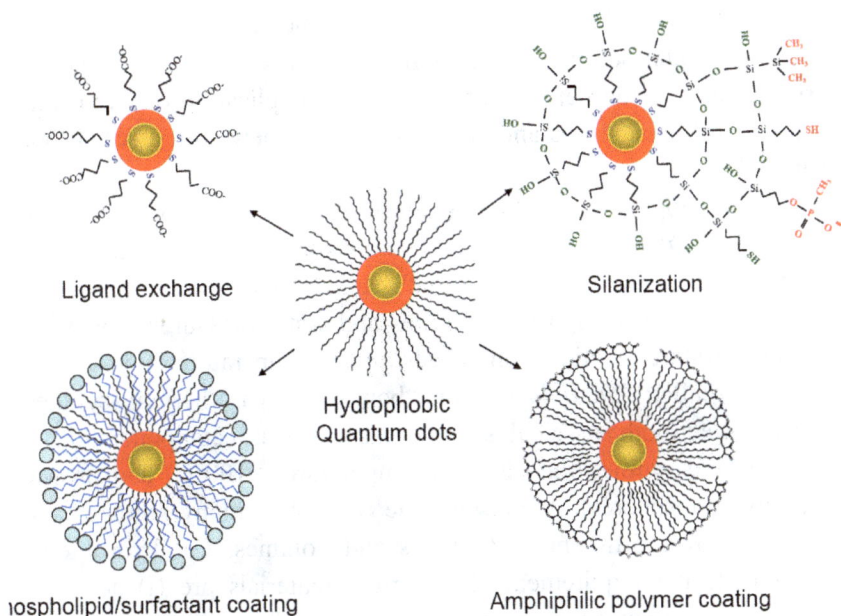

Fig. 4.5. Different strategies for hydrophilic modification of quantum dots: ligand exchange, surface silinization, amphiphilic polymer or surfactant coating. Reproduced with permission from Ref. 71.

Silinization provides a more stable system but tend to be more laborious and harder to control in synthesis [78, 79]. The silinization process involves first replacing the original hydrophobic surfactant with bifunctional ligands, e.g. mercaptopropyltrimethoxysilane (MTS), that allow for binding to the QD surface and formation of silica. In the second step, hydrophilic trimethoxysilane molecules are added to provide water solubility to the silica shell [80]. The hydrophilic groups stabilize the nanoparticles in aqueous solution.

As an alternative approach, QD surface can be modified by coating with polymer or surfactant molecules [81–84]. The hydrophobic tails of the polymer or surfactant intercalate the hydrophobic molecules on the nanoparticle surface while the hydrophilic groups of the polymer or surfactant protrude outwards to facilitate the water solubility of particles. In some cases more complex polymers such as triblock copolymers have been designed to provide better wrapping of the original hydrophobic nanoparticles [85]. Micelles based on molecules such as phospholipids have also been used to encapsulate hydrophobic nanoparticles and make the system water soluble or hydrophilic [86]. The choice of method for surface modification depends on the specific application of interest and each method has its pros and cons, as nicely reviewed in a recent book chapter [71].

Another popular method for synthesizing semiconductor NCs or QDs is *hydrothermal synthesis* based on reactions under elevated temperature and pressure. The term "hydrothermal" came from earth science, where it implied a regime of high temperatures and water pressures. For typical high temperature reactions, one needs a high temperature, high-pressure apparatus called *"autoclaves"* or "bombs". Hydrothermal synthesis involves water both as a catalyst and occasionally as a component of solid phases in the synthesis at elevated temperature ($>100°C$) and pressure ($>$ a few atmospheres). At present, one can obtain many kinds of autoclaves to cover different $P–T$ ranges and volumes. For hydrothermal experiments the requirements for starting materials are (i) accurately known composition, (ii) as homogeneous as possible, (iii) as pure as possible and (iv) as fine as possible [87]. For example, CdTe NCs have been generated using the hydrothermal method based on the reaction between NaHTe and $CdCl_2$ [88]. Figure 4.6 shows the UV-vis spectra of various

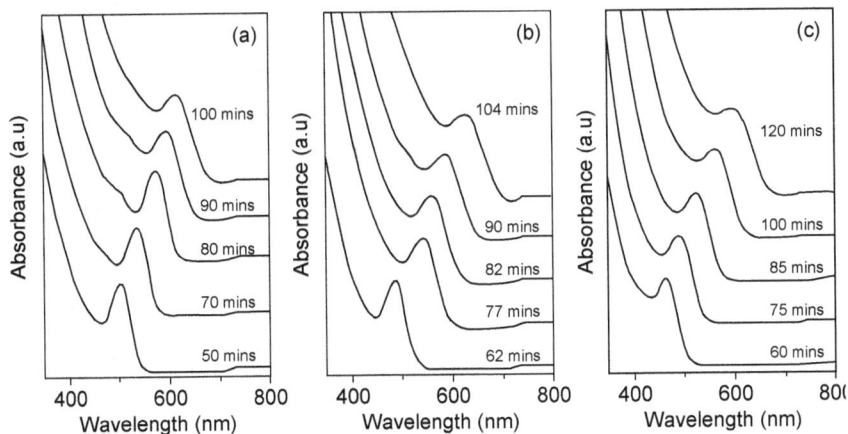

Fig. 4.6. Temporal evolution of UV-vis absorption spectra of CdTe NCs grown at 180°C in aqueous solution stabilized with thiolglycolic acid (TGA) (a), 3-mercaptopropionic acid (b), and 1-thioglycerol (c). Reproduced with permission from Ref. 88.

CdTe NCs synthesized in aqueous solution with different capping molecules and over reaction time. The NCs are highly luminescent (>10% PL quantum yield), especially considering that they are in aqueous solution and can be used directly for biomedical applications.

A special class of semiconductor nanoparticles are doped semiconductors [89]. Doping refers to the process of intentionally introducing a small amount of a foreign or different element into a host material. For example, B or N can be introduced into Si as a way to alter its chemical composition and, more importantly, to influence its properties, e.g. optical and electronic. Another major use of doping is in phosphors based on binary semiconductors such as ZnS with dopants such as Cu, Mn, and Ag ions [89–101]. For doped nanomaterials, the method is only slightly altered from that used for undoped semiconductor synthesis. The main modification is that the dopant element is introduced during synthesis. For doping to be successful, the dopant needs to be compatible to the host material. For example, if the dopant is a cation, its physical size and valence should ideally be similar to that of the cation of the host semiconductor. The same is true for anion dopants. Doping is usually determined by techniques such as PL, XPS, EXAFS, and ESR (if active due to unpaired electrons), as discussed in Chapter 3.

As an example, Mn^{2+}-doped Si nanoparticles capped with hydrogen have been synthesized based on the reaction between Mn^{2+}-doped Zintl salts ($NaSi_{1-x}Mn_x$, $x = 0.05, 0.1, 0.15$) as precursors and ammonium bromide [102]. The hydride-capped nanoparticles can be modified further via a hydrosilylation process to form chemically robust Si-C bonds on the surface and to protect the silicon particles from oxidation. The doping with Mn^{2+} was confirmed using ESR. Figure 4.7 shows TEM, XRD, and PLE and PL spectra of Mn^{2+}-doped Si nanoparticles capped with octyne [102].

4.1.4. *Metal oxides*

The general methodology for synthesizing insulator and metal oxide (MO) nanomaterials is similar to that used for semiconductor nanomaterials. Solution methods, such as hydrolysis, are most commonly used for making MO nanoparticles or other shaped nanomaterials, e.g. TiO_2, ZnO, WO_3, and SnO_2 [103–107]. Hydrothermal method, as discussed in

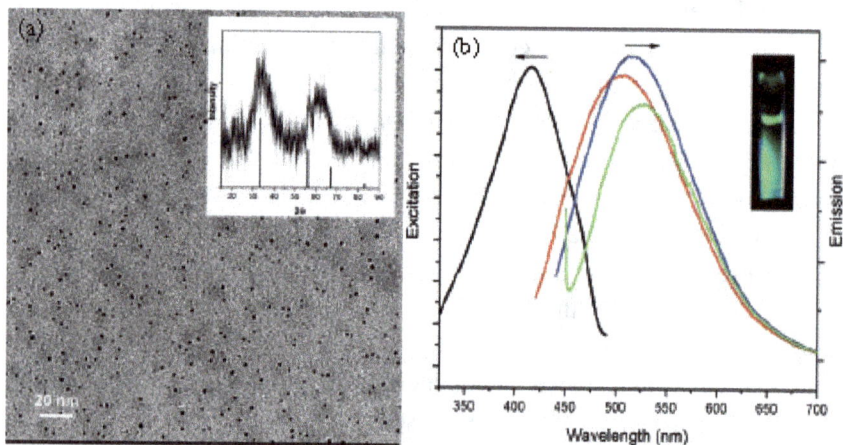

Fig. 4.7. (a) TEM image of 5% Mn^{2+}-doped Si nanoparticles, with XRD pattern shown in the inset, where the lines indicate the X-ray diffraction peak position for diamond structure silicon. (b) Photoluminescence excitation (PLE) spectra (black) and room-temperature PL spectra of Si nanoparticles in chloroform at different excitation wavelengths: 400 nm (red), 420 nm (blue), and 440 nm (green). The inset shows fluorescence from a cuvette of Mn^{2+}-doped Si nanoparticles when excited with a handheld UV lamp. Reproduced with permission from Ref. 102.

the last section, is also often used for making MO nanomaterials via gas phase methods [108–113]. Gas phase methods are particularly useful for making nonspherical particles such as one-dimensional (1D) nanostructures [114–116].

Similar to doped semiconductors discussed in the last section, many MO semiconductors and insulators have been successfully doped with different elements such as Gd^{3+} [104], N [117–121], and C [122]. As a specific example, Cu-doped ZnO NCs have been synthesized using copper acetate and zinc acetate as the Cu and Zn sources, respectively, through hydrolysis using sodium hydroxide and with isopropanol as a solvent [123]. Polyvinyl pyrollidone (PVP) was used as a capping agent in the case of capped nanocrystals. Atomic absorption spectroscopy was used to estimate the percentage of copper actually doped in the NC. Figure 4.8 shows fluorescence emission and excitation spectra for various concentrations of Cu-doped, capped, and free-standing ZnO NCs. Undoped ZnO passivated with PVP shows a spectral feature at ~360 nm, attributed to bandedge emission, with no trap state emission [Fig. 4.8(a)]. However, the free-standing ZnO NCs show strong PL between 500 and 600 nm [Fig. 4.8(b)], attributed to surface trap states or oxygen vacancies in the core of the NCs. Cu-doped ZnO NCs additionally show two clear peaks at ~410 nm and ~430 nm [Figs. 4.8(a) and 4.8(b)], which are apparently associated with transitions involving the Cu d energy levels [inset of Fig. 4.8(b)]. These peaks increase in intensity at the cost of the ZnO bandedge emission with increase in copper doping level. PLE spectrum shows that these two blue peaks originate from excitation near the absorption bandedge of ZnO.

4.1.5. *Complex nanostructures*

Advancement in synthetic methodology has made it possible to synthesize and fabricate complex and composite nanostructures. To make the discussion simpler, we will focus on the complex structure of single-component materials in this subsection and defer discussion of complex structures of multicomponent materials into composite materials in the next subsection.

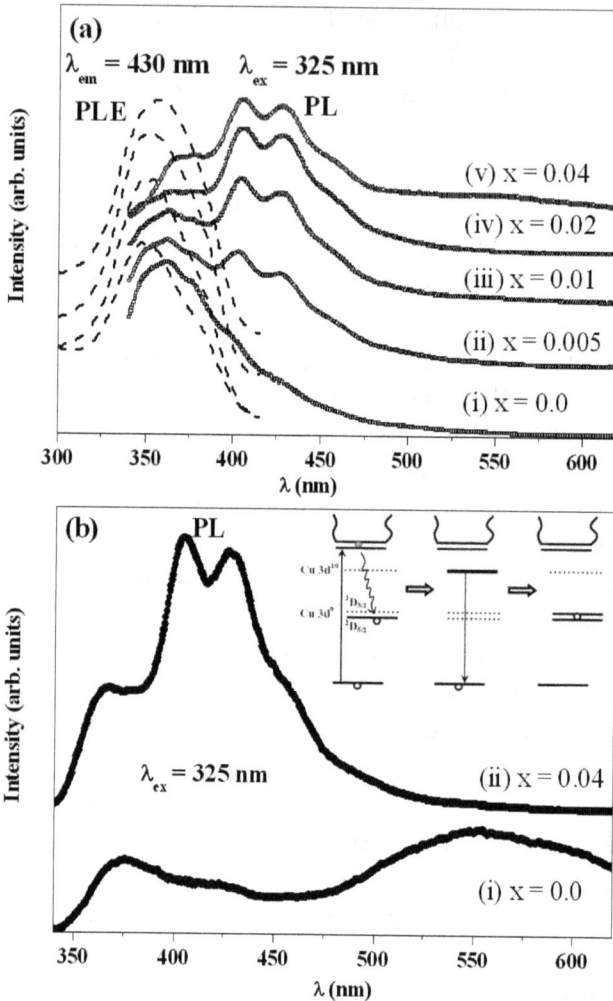

Fig. 4.8. (a) Fluorescence excitation (λ_{em} = 430 nm) and emission spectra (λ_{ex} = 325 nm) of $Zn_{1-x}Cu_xO$ NCs. (b) Fluorescence emission spectra of free-standing $Zn_{1-x}Cu_xO$ NCs (without PVP capping). The energy level diagram of the process involved is shown in the inset. Reproduced with permission from Ref. 123.

Complex nanostructures refer to nanostructures that involve architectures beyond isolated simple systems such as spherical nanoparticles, nanotubes, nanorods, nanowires and usually consist of multiple components with defined internal and external structures. Examples include aligned or ordered arrays of 0 D, 1D or 2D materials or mixtures of different dimensional nanostructures. Of course, simple and complex are relative and there is no strict scientific distinction between simple and complex structures. The reality is a broad range of nanostructures from very simple to very complex and the degree of complexity evolves as new synthetic and fabrication techniques develop.

For example, ZnO is one of the most structurally rich materials to have been produced in almost countless number of different nanostructures including nanorods, nanobelts, nanoflowers, nanowires, nanodisks, nanotowers, and so on. Figure 4.9 shows some examples of ZnO nanostructures [124]. The rich nanostructures produced originate from its unique crystal structures.

Similar to semiconductors, MO can be doped as a means to alter their optical and electronic as well as magnetic properties. For example, Co-doped ZnO and TiO_2 NCs has been extensively studied and the Co-doping has significant effect on the optical, structural and magnetic properties, including enhanced visible absorption [125, 126]. Co^{2+}-doped TiO_2 NCs have been synthesized using an inverse micelle procedure [126]. In this synthesis, an inverse micelle solution was

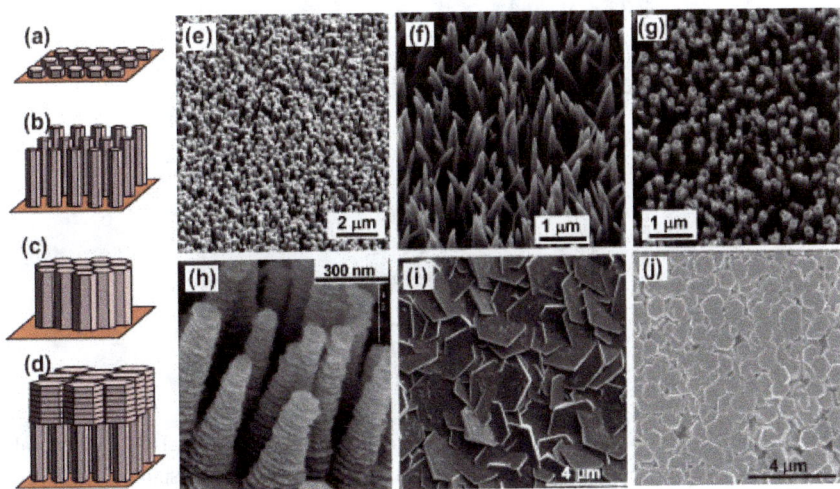

Fig. 4.9. Electron micrographs of examples of complex ZnO nanostructures. Reproduced with permission from Ref. 124.

prepared from cyclohexane, dioctyl sulfosuccinate, sodium salt (AOT), water and Co^{2+} salt $Co(NO_3)_2 \cdot 6H_2O$. An anaerobic solution containing 1-hexanol and titanium isopropoxide was slowly transferred to the Co_{2+}/inverse micelle solution under N_2. The mixture was stirred and then aged in air, producing an amorphous product containing Co^{2+} that could be isolated from the micelle solution. To induce crystallization, solutions of the amorphous product were sealed under air and heated at 150°C for 24 h in a Teflon-lined stainless steel autoclave. The resulting transparent tan solution contained nanocrystalline Co^{2+}-doped TiO_2, which was subsequently isolated and purified. One of the isolation and purification procedures used introduced trioctylphosphine oxide (TOPO) as part of the procedure that results in capping of the nanocrystals with TOPO on the surface [126]. Figure 4.10 shows some representative UV-visible and MCD spectra of the prepared Co^{2+}-doped TiO_2 nanocrystals.

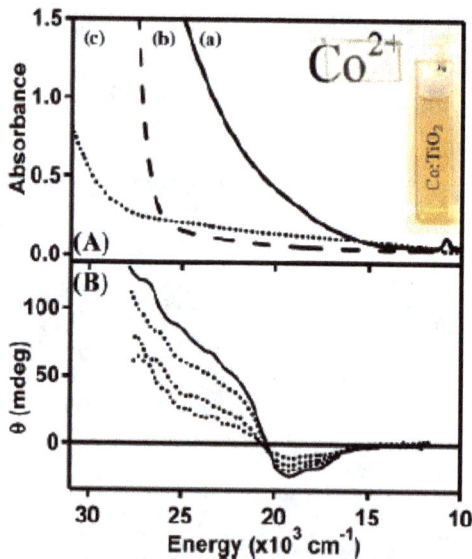

Fig. 4.10. (A) 300 K absorption spectra: (a) TOPO-capped 3% Co^{2+}:TiO_2 nanocrystals (NCs) (labeled as NC1) in toluene, (b) undoped TiO_2 NCs in toluene, (c) spin-coated film of 3% Co^{2+}:TiO_2 NCs (film 1). Inset: Photos of NC1 suspension and film 1, with lettering behind each to illustrate transparency. (B) 5.5 T variable-temperature (5 (solid), 10, 20, 40 K) MCD spectra of 1.3% Co^{2+}:TiO_2 NCs. Reproduced with permission from Ref. 126.

It is clear that Co^{2+}-doping substantially enhanced visible absorption [on comparing Figs. 4.10 (Aa) and (Ab)].

4.1.6. *Composite and hetero-junction nanomaterials*

Composite materials refer to materials that contain more than one type of material and the different materials are not homogeneously mixed or dispersed at the atomic scale. If materials are dispersed homogeneously at the atomic scale, they are considered as doped materials or alloys. While there are many important composite nanomaterials involving organic and/or biological molecules, we will focus on nanocomposites composed of only inorganic components.

One class of unique nanostructures that can be considered as composite materials are so-called core/shell structures that contain two different types of materials, one as the core and another as the shell. Various combinations of core/shell structures can result from using metal, semiconductor or insulator for the core or shell, or just two different metals, semiconductors or insulators for the core and shell. We will discuss a few examples based on semiconductor/semiconductor, metal/metal, semiconductor/metal, metal/semiconductor, metal/insulator, semiconductor/insulator, insulator/metal, insulator/semiconductor core/shell structures. They are of interest for different fundamental studies as well as applications.

One common core/shell structure is based on semiconductor/semiconductor that have different bandgaps so one can manipulate their electronic band structures and optical properties. For example, CdSe/ZnS has been studied as a system to enhance PL stability and yield of CdSe as well as to lower the toxicity since ZnS is less toxic than CdSe [127–129]. The shell semiconductor is usually chosen to have a larger bandgap and lower toxicity compared to the core semiconductor. Methods for producing such core/shell structures have been well developed. For example, CdSe nanocrystals have been overcoated with ZnS [127–133], ZnSe [134], and CdS [64, 135], which resulted in dramatic improvements in luminescence efficiency, exemplified by the work on CdSe/ZnS [128]. In some cases, the shell is a smaller bandgap semiconductor than the core. For instance, CdS NCs have been overcoated with lower bandgap materials such as HgS [136, 137], and, similarly, CdTe cores have been coated with HgTe shells [138]. More complex structures, such as core/shell/shell, can also be produced. For example, the HgS shell in CdS/HgS systems has, in turn, been buried by the deposition of new CdS shell to produce a shell of low-bandgap HgS nested in a wider bandgap NC structure [139, 140]. The detailed crystal structures of the shells are usually not as well identified as the core due to their thinness. In many cases, the shells are polycrystalline.

There are many other types of nanocomposite or hetero-junction nanomaterials, including CdS/TiO_2 [141], CdS/Au [142], TiO_2/Pt [143], TiO_2/Au [144], $TiO_2/CdSe$ [145], $CdSe/SiO_2$ [146], $CdTe/SiO_2$ [147], and Au/SiO_2 [148–150]. A detailed discussion of some examples of nanocomposites and their properties is given in Chapter 8. In terms of synthesis, the same synthetic methods discussed above for producing single component nanostructures can be used for making nanocomposite materials. In many cases, the individual components are synthesized separately first and mixed to produce composites subsequently. In other cases, the different reactants are introduced simultaneously or sequentially during the reaction [141]. Other complex structures include nanotubes, notably carbon nanotubes, with nanoparticle inside, outside or at the tips [45, 151–156]. The constituent components can again be synthesized simultaneously or, more commonly, sequentially. Changes of the properties of such composite or hetero-junction materials compared to the isolated components depend strongly on the nature of the individual components, as will be further discussed in Chapter 8.

4.2. Gas or vapor-based methods of synthesis: CVD, MOCVD and MBE

Complementing solution chemical methods, several other methods, often gas or vapor-based, have been developed to synthesize and fabricate nanostructures of metal, semiconductor and insulators. In general, gas or vapor-based methods are more involved but produce nanomaterials with higher quality in terms of purity as well as structural control. Examples of such techniques include chemical vapor deposition CVD, metal-organic CVD (MOCVD) [157–165], and molecular beam epitaxy (MBE) [166, 167]. We will discuss these techniques briefly and give a few specific examples of nanostructures synthesized or fabricated by these techniques.

In CVD, a precursor, often diluted in a carrier gas or gasses, is delivered into a reaction chamber at approximately ambient temperatures. As it passes over or comes into contact with a heated substrate, it reacts or decomposes to form a solid phase that is deposited onto the substrate. The substrate temperature is critical and can influence what reactions take place. The crystal structure of the substrate surface, along with other experimental parameters, determines what nanostructures can be generated. In MOCVD, atoms to be incorporated in a crystal of interest are combined with complex organic gas molecules and passed over a hot semiconductor wafer. The heat decomposes the molecules and deposits the desired atoms onto the substrate's surface. By controlling the composition of the gas, one can vary the properties of the crystal on the atomic scale. The crystal structure of the fabricated materials is dictated by the crystal structure of the substrate.

In CVD and MOCVD synthesis, the growth of nanostructures, especially 1D structures such as nanowires, is often based on the so-called VLS (vapor-liquid-solid) mechanism that was first discovered in the mid-1960s [168–171]. The mechanism consists of small metal particle catalysts deposited on a substrate. The substrate is then heated and vapor of the material of choice is introduced, e.g. Si, ZnO and GaN. The vapor diffuses into the metal until a saturated solution is generated and the material of choice precipitates to form nanowires. There are several recent examples using VLS to grow many different types of nanowires or other 1D nanostructures [172–176]. One variation to heating is laser ablation of a substrate containing a metal/semiconductor mixture to create a semiconductor/metal molten alloy [177–179]. The resulting nanowires undergo VLS growth. Nanowires made by the laser assisted catalytic growth have lengths up to several μm [180–183]. ZnS/SiO_2 core/shell nanowires have also been grown based on the VLS process [184]. Figure 4.11 shows some representative SEM images of the ZnS/SiO_2 core/shell nanowires as a function of growth time. The nanowires are initially randomly aligned and over time, after about 45 minutes, become aligned in the direction of the flow of the gas.

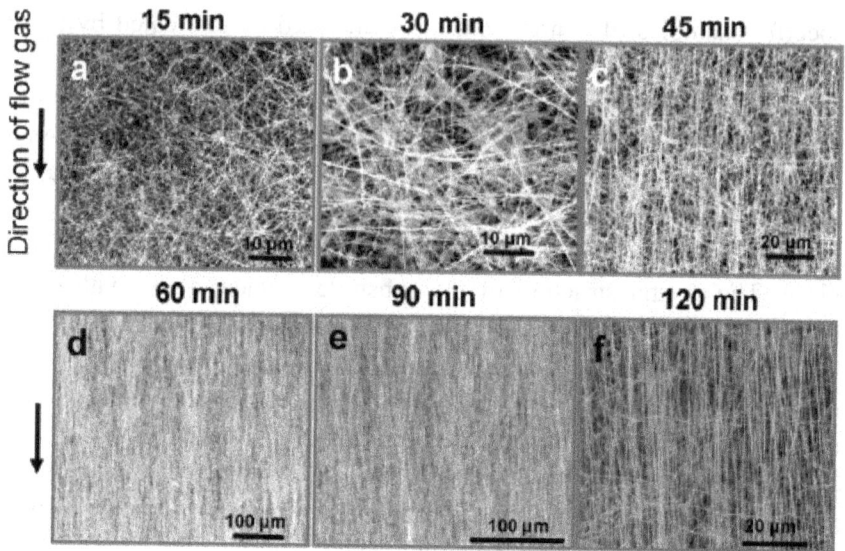

Fig. 4.11. SEM images of ultralong ZnS/SiO$_2$ core/shell nanowires as a function of growth time as indicated in the figure. Reproduced with permission from Ref. 184.

Molecular beam epitaxy (MBE) is a technique for epitaxial growth, via the interaction of one or several molecular or atomic beams, that occurs on a surface of a heated crystalline substrate. The solid source materials are placed in evaporation cells to provide an angular distribution of atoms or molecules in a beam. The substrate is heated to the necessary temperature and, when needed, continuously rotated to improve the growth homogeneity. MBE has often been used to grow single crystalline or polycrystalline films and superlattice quantum well structures under ultra high vacuum (UHV) environment [166, 167].

All the above synthetic techniques are generally considered bottom-up approaches where atoms and molecules are brought together to produce larger nanostructures. An opposite approach is top-down where large bulk scale structures are fabricated into smaller nanostructures. Lithographic techniques such as E-beam or photo-lithography are examples that allow the creation of nanostructures on the micron and nanometer scales, easily down to tens of nanometers [185]. Such techniques become critical conveniently for mass production of high quality and high purity structures in the microelectronics and computer industry. It is currently a challenge to create structures on a few nm scale using typical lithographic methods. There is an urgent need for developing new technologies that can meet the demand, especially for the microelectronics or nanoelectronics and computer industry. The combination of top-down and bottom-up approaches may hold the key to solving this problem in the future.

The following provides a few specific examples of various nano-structures produced by gas- or vapor-based techniques.

4.2.1. Metals

Gas phase techniques for metal nanostructure fabrication often involve deposition of a metal directly from the same metal source that is evaporated and then deposited on appropriate substrates. For example, silver nanorods have been fabricated using GLAD (glancing angle deposition) and demonstrated to have excellent SERS activities [186, 187]. Nanostructures produced in this way are usually cleaner and have better shape and size control. To date, the use of CVD, MOCVD or MBE for single component metal nanostructure fabrication has been limited. Pt-TiO_2 nanocomposites have been prepared using MOCVD [188].

4.2.2. Semiconductors

Examples of semiconductor nanostructures synthesized using CVD and MOCVD include GaAs/AlGaAs nanowires [189], InP QDs [190], InGaN/GaN [157], GaInP nanowires [161], GaN nanorods [191], and Si [163, 192]. Similarly, MBE techniques have been successfully applied to generate GaAs nanowires [193, 194], GaN thin films [195–197], IV–VI semiconductor nanostructures [198], $In_xGa_{1-x}N$ [199], ZnSe nanowires [200], InAs quantum dot arrays [201], Ge [202], and GaAs:Zn [203]. Figure 4.12 shows AFM images of InAs quantum dot arrays grown using templates fabricated by a nano-jet probe method [201].

4.2.3. Metal oxides

Similar to semiconductors, many MO have been synthesized using CVD, MOCVD and MBE techniques, including TiO_2-Pt [204], TiO_2-SiO_2

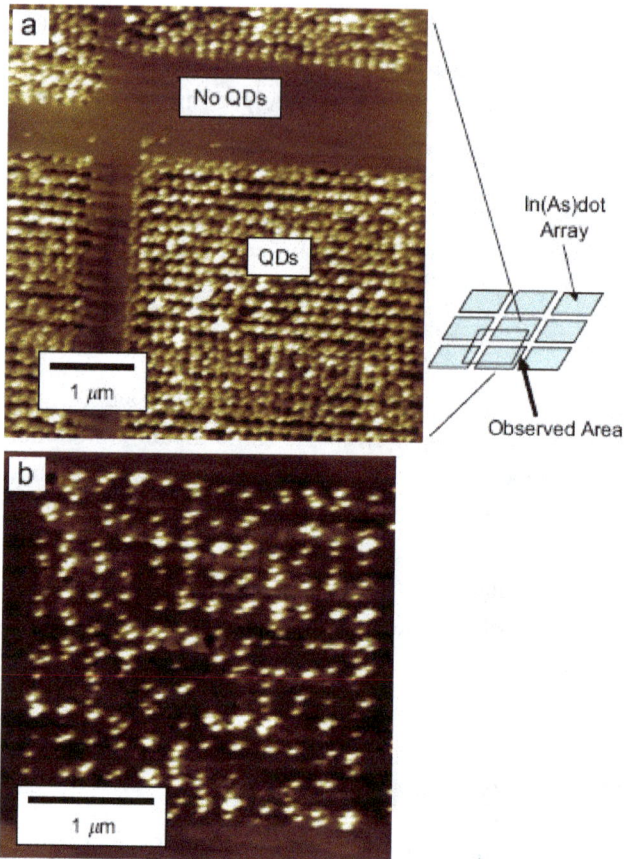

Fig. 4.12. (a) AFM image of stacked InAs dot arrays. The image was taken on an area encompassing four-square regions, as shown in the illustration. (b) AFM image of low-density stacked InAs dot arrays. Reproduced with permission from Ref. 201.

[158], ZnO [160, 164, 205], TiO_2 [159], and $MgAl_2O_4$ [206]. For example, nanostructured composite TiO_2-Pt thin films at different Pt/Ti ratios were obtained by CVD using titanium tetraisopropoxide (TTIP) and platinum bisacetylacetonate ($Pt(acac)_2$) as precursors in the presence of oxygen [204]. The single growth process was kinetically controlled for TiO_2 and diffusion limited for Pt under the same experimental conditions. The atomic Pt/Ti ratio in the films was varied in the range of 1:1 to 1:20 by changing the evaporation rates of the precursors. The obtained

composite films are biphasic and nanostructured with a regular and a smooth surface. As mentioned before, compared to solution phase synthesis, nanostructures fabricated using CVD, MOCVD or MBE tend to have less chemical impurities and are easier to generate in more sophisticated structures.

4.2.4. *Complex and composite structures*

The gas or vapor-based techniques lend themselves conveniently for fabricating complex and composite structures. Examples include thin film diamond [207–210], iridium-carbon nano composite films [211], carbon nanotubes [212, 213], carbon nanosheets [165], carbon nanofilms [214, 215], nano-graphenes [216], and Cr_3C_2 nanoparticles [162]. Some nanostructures involve complex or composite nanostructures that exhibit properties and functionalities very different from single or simple component nanostructures. Their synthesis tends to be somewhat more complex. Examples include ceramic nanomaterials that usually involve more than two components of materials [217–220].

4.3. Nanolithography techniques

As mentioned above, instead of creating or generating nanostructures from individual atoms and molecules in bottom-up approaches, lithographic methods, based on top-down approaches, have been successfully used to fabricate various nanostructures. The classic example are nanostructures fabricated in electronic chips. These approaches are very well established technologically and well understood scientifically. Many specialized books and reviews exist and such approaches will not be elaborated in this book [221–227]. With demand for smaller and smaller materials structures in device applications and with the advancement of technology, lithographic approaches are taking fabrication to finer and finer features, down to about 45 nm in current chip production and even smaller in research labs. As can be seen, the relevant features that can be created by the top-down or bottom-up approaches are merging on the lengths scales of a few to a few tens of nanometers.

4.4. Bioconjugation

One area of special interest and promising applications of nanomaterials is in biomedical detection, imaging and therapy. These applications require nanomaterials such as semiconductor QDs with proper surface properties and conjugation to biological molecules such as proteins and DNA. As a part of the synthetic effort, the surface needs to be adequately prepared and ultimately conjugated. There are a number of protocols for bioconjugation, based on thiol chemistry, electrostatic interaction, adsorption, and covalent linkage. Specific examples include the use of SMCC (Succinimidyl-4-(N-maleimidomethyl) cyclohexane-1-carboxylate), oxidized Fc-carbohydrate groups, amine-carboxylic acid coupling, and electrostatic interaction for conjugating proteins to QDs [131, 228–232]. These approaches vary in their ease of conduction, coupling strength, and stability of the conjugation product. The choice depends on the chemical details of the biological molecules and QDs as well as the media, e.g. pH of aqueous solution.

For example, in the SMCC protocol, reaction of sulfo-SMCC (4-(N-maleimidomethyl) cyclohexane-1-carboxylic 3-sulfo-n-hydroxysuccinimide ester) with antibodies creates amide bonds between the crosslinker and primary amine groups on the target proteins. The thiol reactive maleimide groups are then ready in the solution for bonding to the QD system, e.g. CdTe coated with silica, as shown schematically in Fig. 4.13 [147]. The maleimide ring retains binding activity of 50% after 64 hours due to the cyclohexane ring in its structure, and can be readily reacted with thiol groups on the surface of the CdTe nanoparticles [233]. PEG groups with their polymer chains reduce nonspecific binding, and stabilize the particles in solution through increased hydrogen bonding.

Successful bioconjugation can be confirmed by several methods, including using streptavidin/biotin technology, spectroscopic techniques and simple agarose gel electrophoresis. In agarose gel electrophoresis, the conjugated and unconjugated proteins often have different mobilities through the gel, which can thus be used to determine if conjugation is successful [147]. New protocols of conjugation are still being developed, especially in terms of better control to reduce nonspecific interactions [234].

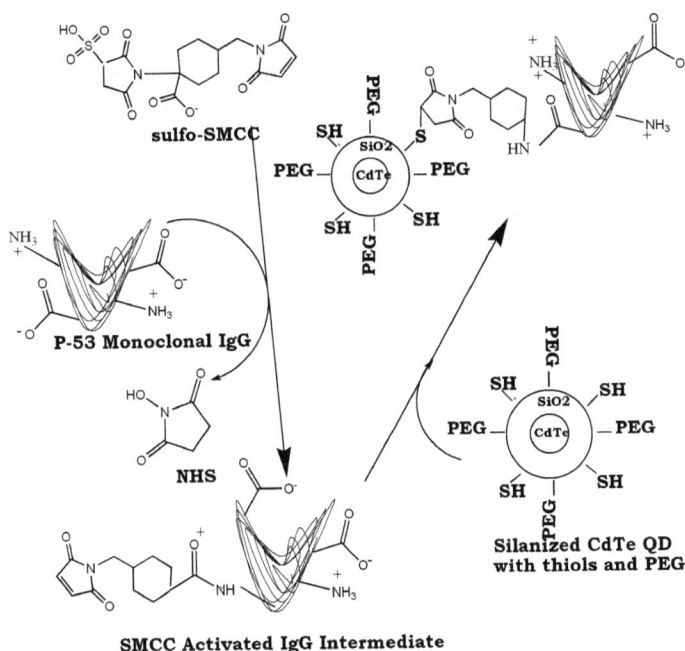

Fig. 4.13.   Schematic illustration of IgG protein conjugation to SiO_2 coated CdTe QDs using SMCC. Reproduced with permission from Ref. 147.

4.5. Toxicity and green chemistry approaches for synthesis

With the increasing awareness of environmental and health issues, there have been considerable concerns about the toxicity of nanomaterials and growing interest in developing methods of synthesis that are more environmentally friendly and can generate nanomaterials that are less toxic. Towards these goals, efforts have been made to use reactants that are less toxic and less expensive, to develop procedures that use less energy or other resources such as solvent, and generate products that can be easily handled and discarded when no longer needed.

For example, instead of using expensive and toxic organo-metallic precursors, less toxic compounds, such as CdO, have been explored for synthesizing high quality II–VI semiconductor QDs [235]. For II–VI semiconductor QDs, it would be ideal not to use Cd but to use less toxic ions such as Zn. However, Zn-based QDs usually have weak or no visible to near IR absorption and emission that are often desired for biomedical and optical applications. One way to get around this dilemma is to use core/ shell structures to encapsulate the more toxic component with desired optical properties inside as core and have the less toxic but optically inactive component as shell. For example, some toxic components could be encapsulated into a nontoxic or less toxic material for applications, such as CdSe/ZnS core/shell structure or $CdSe/SiO_2$ core/shell structures, as long as the properties of the core of interest are not altered noticeably by the shell [79].

Another potential approach is to use dopants to introduce optical activity into nontoxic but optically inactive host materials, e.g. Mn or Cu doped ZnS NCs. These doped nanomaterials, as will be discussed in Chapter 5, have traditionally been useful for phosphor and solid state lighting applications. They are also potentially useful for biomedical applications [89].

4.6. Summary

Many new synthetic and fabrication techniques have been developed over the last two decades for producing a large variety of different nanostructures. New nanostructures are reported on a daily basis. There are a number of review articles and books devoted to nanomaterial synthesis [1, 89, 236–239], and interested readers are referred to these resources. This chapter provides an overview of some of the commonly used synthetic methods. A number of specific examples of nanomaterials are presented to illustrate how the synthetic methods are applied and what they can generate. As one may expect, new synthetic methods and new nanostructures will continue to be developed. To produce nanostructures in a controlled manner and in large quantities is often challenging and will likely be an active area of research and development in the future.

References

1. C.B. Murray, C.R. Kagan and M.G. Bawendi, *Ann. Rev. Mater. Sci.* **30**, 545 (2000).

2. V.K. La Mer and R.H. Dinegar, *J. Am. Chem. Soc.* **72**, 4847 (1950).
3. H. Reiss, *J. Chem. Phys.* 19, 482 (1951).
4. Y. De Smet, L. Deriemaeker and R. Finsy, *Langmuir* **13**, 6884 (1997).
5. H. Gratz, *Scripta Materialia* **37**, 9 (1997).
6. C.B. Murray, D.J. Norris and M.G. Bawendi, *J. Am. Chem. Soc.* **115**, 8706 (1993).
7. J. Turkevich, P.C. Stevenson and J. Hiller, *Discussions of the Faraday Society* **11**, 55 (1951).
8. G. Frens, *Nature Phys. Sci.* **241**, 20 (1973).
9. J. Kimling, M. Maier, B. Okenve, V. Kotaidis, H. Ballot and A. Plech, *J. Phys. Chem. B* **110**, 15700 (2006).
10. X.H. Ji, X.N. Song, J. Li, Y.B. Bai, W.S. Yang and X.G. Peng, *J. Am. Chem. Soc.* **129**, 13939 (2007).
11. P.C. Lee and D. Meisel, *J. Phys. Chem.* **86**, 3391 (1982).
12. A. Henglein, *Israel J. Chem.* **33**, 77 (1993).
13. T.J. Norman, C.D. Grant, D. Magana, J.Z. Zhang, J. Liu, D.L. Cao, F. Bridges and A. Van Buuren, *J. Phys. Chem. B* **106**, 7005 (2002).
14. A.M. Schwartzberg, C.D. Grant, A. Wolcott, C.E. Talley, T.R. Huser, R. Bogomolni and J.Z. Zhang, *J. Phys. Chem. B* **108**, 19191 (2004).
15. A.M. Schwartzberg, C.D. Grant, T. van Buuren and J.Z. Zhang, *J. Phys. Chem. C* **111**, 8892 (2007).
16. J. Park, J. Joo, S.G. Kwon, Y. Jang and T. Hyeon, *Angew Chem. Int. Edit* **46**, 4630 (2007).
17. A.B. Smetana, J.S. Wang, J. Boeckl, G.J. Brown and C.M. Wai, *Langmuir* **23**, 10429 (2007).
18. N.R. Jana, L. Gearheart and C.J. Murphy, *J. Phys. Chem. B* **105**, 4065 (2001).
19. N.R. Jana, L. Gearheart and C.J. Murphy, *Chem. Commun.* **617** (2001).
20. K.R. Brown, D.G. Walter and M.J. Natan, *Chem. Mater.* **12**, 306 (2000).
21. K.R. Brown, L.A. Lyon, A.P. Fox, B.D. Reiss and M.J. Natan, *Chem. Mater.* **12**, 314 (2000).
22. J. Rodriguez-Fernandez, J. Perez-Juste, F.J.G. de Abajo and L.M. Liz-Marzan, *Langmuir* **22**, 7007 (2006).
23. C.J. Johnson, E. Dujardin, S.A. Davis, C.J. Murphy and S. Mann, *J. Mater. Chem.* **12**, 1765 (2002).
24. A. Gole and C.J. Murphy, *Chem. Mater.* **16**, 3633 (2004).
25. C.J. Murphy, T.K. San, A.M. Gole, C.J. Orendorff, J.X. Gao, L. Gou, S.E. Hunyadi and T. Li, *J. Phys. Chem. B* **109**, 13857 (2005).
26. L.F. Gou and C.J. Murphy, *Chem. Mater.* **17**, 3668 (2005).

27. C.J. Murphy, T.K. Sau, A. Gole and C.J. Orendorff, *MRS Bulletin* **30**, 349 (2005).
28. B. Nikoobakht and M.A. El-Sayed, *Chem. Mater.* **15**, 1957 (2003).
29. Y.G. Sun, B. Mayers and Y.N. Xia, *Advan. Mater.* **15**, 641 (2003).
30. Y.G. Sun, B. Wiley, Z.Y. Li and Y.N. Xia, *J. Am. Chem. Soc.* **126**, 9399 (2004).
31. C.H. Wang, D.C. Sun and X.H. Xia, *Nanotechnology* **17**, 651 (2006).
32. Y.G. Sun, B.T. Mayers and Y.N. Xia, *Nano Lett.* **2**, 481 (2002).
33. Y.G. Sun, B. Mayers and Y.N. Xia, *Nano Lett.* **3**, 675 (2003).
34. Y. Sun and Y. Xia, *Science* **298**, 2176 (2002).
35. A.M. Schwartzberg, T.Y. Olson, C.E. Talley and J.Z. Zhang, *J. Phys. Chem. B* **110**, 19935 (2006).
36. J.V. Zoval, R.M. Stiger, P.R. Biernacki and R.M. Penner, *J. Phys. Chem.* **100**, 837 (1996).
37. X.H. Liao, J.J. Zhu, X.N. Zhao and H.Y. Chen, *Chemical J. Chinese Universities-Chinese* **21**, 1837 (2000).
38. X.Q. Lin and S.Q. Wang, *Indian J. Chem. Section a-Inorganic Bio-Inorganic Physical Theoretical & Analytical Chemistry* **44**, 1016 (2005).
39. M. Starowicz, B. Stypula and J. Banas, *Electrochem. Commun.* **8**, 227 (2006).
40. M.M. Wadkar, V.R. Chaudhari and S.K. Haram, *J. Phys. Chem. B* **110**, 20889 (2006).
41. L. Shang, H.J. Chen and S.J. Dong, *J. Phys. Chem. C* **111**, 10780 (2007).
42. S.C. Lin, S.Y. Chen, Y.T. Chen and S.Y. Cheng, *J. Alloy Compd.* **449**, 232 (2008).
43. M. Maillard, P.R. Huang and L. Brus, *Nano Lett.* **3**, 1611 (2003).
44. M. Muniz-Miranda, *J. Raman Spectroscopy* **35**, 839 (2004).
45. D. Guin, S.V. Manorama, J.N.L. Latha and S. Singh, *J. Phys. Chem. C* **111**, 13393 (2007).
46. S. Tan, M. Erol, A. Attygalle, H. Du and S. Sukhishvili, *Langmuir* **23**, 9836 (2007).
47. B.M. Sergeev and G.B. Sergeev, *Colloid J.* **69**, 639 (2007).
48. Y. Hayashi, T. Fujita, T. Tokunaga, K. Kaneko, T. Butler, N. Rupesinghe, J.D. Carey, S.R.P. Silva and G.A.J. Amaratunga, *Diamond and Related Mater.* **16**, 1200 (2007).
49. J.Y. Chen, Z.Y. Li, D.N. Chao, W.J. Zhang and C. Wang, *Mater. Lett.* **62**, 692 (2008).
50. Y.Q. Xiong, H. Wu, Y. Guo, Y. Sun, D.Q. Yang and D. Da, *Thin Solid Films* **375**, 300 (2000).

51. K.M. Nie, J.L. Hu, W.M. Pang and Q.R. Zhu, *Mater. Lett.* **61**, 3567 (2007).
52. V.G. Pol, H. Grisaru and A. Gedanken, *Langmuir* **21**, 3635 (2005).
53. E. Mine, A. Yamada, Y. Kobayashi, M. Konno and L.M. Liz-Marzan, *J. Colloid Interf. Sci.* **264**, 385 (2003).
54. Y.S. Park, L.M. Liz-Marzan, A. Kasuya, Y. Kobayashi, D. Nagao, M. Konno, S. Mamykin, A. Dmytruk, M. Takeda and N. Ohuchi, *J. Nanosci. Nanotechnol.* **6**, 3503 (2006).
55. I. Pastoriza-Santos, J. Perez-Juste and L.M. Liz-Marzan, *Chem. Mater.* **18**, 2465 (2006).
56. R.A. Bley, S.M. Kauzlarich, J.E. Davis and H.W.H. Lee, *Chem. Mater.* **8**, 1881 (1996).
57. R.K. Baldwin, K.A. Pettigrew, E. Ratai, M.P. Augustine and S.M. Kauzlarich, *Chem. Commun.* **1822** (2002).
58. J.R. Heath, J.J. Shiang and A.P. Alivisatos, *J. Chem. Phys.* **101**, 1607 (1994).
59. B.R. Taylor, S.M. Kauzlarich, G.R. Delgado and H.W.H. Lee, *Chem. Mater.* **11**, 2493 (1999).
60. N. Zaitseva, Z.R. Dai, C.D. Grant, J. Harper and C. Saw, *Chem. Mater.* **19**, 5174 (2007).
61. C.B. Murray, M. Nirmal, D.J. Norris and M.G. Bawendi, *Z. Phys. D. Atom Mol. Cl.* **26**, S231 (1993).
62. A.P. Alivisatos, *Science* **271**, 933 (1996).
63. A.A. Guzelian, J.E.B. Katari, A.V. Kadavanich, U. Banin, K. Hamad, E. Juban, A.P. Alivisatos, R.H. Wolters, C.C. Arnold and J.R. Heath, *J. Phys. Chem.* **100**, 7212 (1996).
64. X.G. Peng, M.C. Schlamp, A.V. Kadavanich and A.P. Alivisatos, *J. Am. Chem. Soc.* **119**, 7019 (1997).
65. X.G. Peng, *Chem-Eur J.* **8**, 335 (2002).
66. J. Joo, H.B. Na, T. Yu, J.H. Yu, Y.W. Kim, F. Wu, J.Z. Zhang and T. Hyeon, *J. Am. Chem. Soc.* **125**, 11100 (2003).
67. A.Y. Nazzal, X.Y. Wang, L.H. Qu, W. Yu, Y.J. Wang, X.G. Peng and M. Xiao, *J. Phys. Chem. B* **108**, 5507 (2004).
68. L. Manna, E.C. Scher and A.P. Alivisatos, *J. Am. Chem. Soc.* **122**, 12700 (2000).
69. T. Vossmeyer, L. Katsikas, M. Giersig, I.G. Popovic, K. Diesner, A. Chemseddine, A. Eychmuller and H. Weller, *J. Phys. Chem.* **98**, 7665 (1994).
70. F. Wu, J.Z. Zhang, R. Kho and R.K. Mehra, *Chem. Phys. Lett.* **330**, 237 (2000).

71. C.-A. Lin, J.K. Li, R.A. Sperling, L. Manna, W.J. Parak and W.H. Chang, in *Annual Review of Nano Research*, G.Z. Cao and C.J. Brinker, Editors. World Scientific: Singapore, **467** (2007).

72. W.C.W. Chan and S.M. Nie, *Science* **281**, 2016 (1998).

73. C.C. Chen, C.P. Yet, H.N. Wang and C.Y. Chao, *Langmuir* **15**, 6845 (1999).

74. C.Y. Zhang, H. Ma, S.M. Nie, Y. Ding, L. Jin and D.Y. Chen, *Analyst* **125**, 1029 (2000).

75. J.O. Winter, T.Y. Liu, B.A. Korgel and C.E. Schmidt, *Advan. Mater.* **13**, 1673 (2001).

76. J.A. Kloepfer, R.E. Mielke, M.S. Wong, K.H. Nealson, G. Stucky and J.L. Nadeau, *Appl. Environ. Microb.* **69**, 4205 (2003).

77. J. Aldana, Y.A. Wang and X.G. Peng, *J. Am. Chem. Soc.* **123**, 8844 (2001).

78. M. Bruchez, M. Moronne, P. Gin, S. Weiss and A.P. Alivisatos, *Science* **281**, 2013 (1998).

79. D. Gerion, F. Pinaud, S.C. Williams, W.J. Parak, D. Zanchet, S. Weiss and A.P. Alivisatos, *J. Phys. Chem. B* **105**, 8861 (2001).

80. W.J. Parak, D. Gerion, D. Zanchet, A.S. Woerz, T. Pellegrino, C. Micheel, S.C. Williams, M. Seitz, R.E. Bruehl, Z. Bryant, C. Bustamante, C.R. Bertozzi and A.P. Alivisatos, *Chem. Mater.* **14**, 2113 (2002).

81. X.Y. Wu, H.J. Liu, J.Q. Liu, K.N. Haley, J.A. Treadway, J.P. Larson, N.F. Ge, F. Peale and M.P. Bruchez, *Nat. Biotechnol.* **21**, 41 (2003).

82. T. Pellegrino, L. Manna, S. Kudera, T. Liedl, D. Koktysh, A.L. Rogach, S. Keller, J. Radler, G. Natile and W.J. Parak, *Nano Lett.* **4**, 703 (2004).

83. M.A. Petruska, A.P. Bartko and V.I. Klimov, *J. Am. Chem. Soc.* **126**, 714 (2004).

84. B. Ballou, B.C. Lagerholm, L.A. Ernst, M.P. Bruchez and A.S. Waggoner, *Bioconjugate Chem.* **15**, 79 (2004).

85. X.H. Gao, Y.Y. Cui, R.M. Levenson, L.W.K. Chung and S.M. Nie, *Nat. Biotechnol.* **22**, 969 (2004).

86. B. Dubertret, P. Skourides, D.J. Norris, V. Noireaux, A.H. Brivanlou and A. Libchaber, *Science* **298**, 1759 (2002).

87. S. Somiya and R. Roy, *Bull. Mater. Sci.* **23**, 453 (2000).

88. H. Zhang, L.P. Wang, H.M. Xiong, L.H. Hu, B. Yang and W. Li, *Advan. Mater.* **15**, 1712 (2003).

89. W. Chen, J.Z. Zhang and A. Joly, *J. Nanosci. Nanotechnol.* **4**, 919 (2004).

90. M.C. Deng, T.S. Chin and F.R. Chen, *J. Appl. Phys.* **75**, 5888 (1994).

91. M.A. Chamarro, V. Voliotis, R. Grousson, P. Lavallard, T. Gacoin, G. Counio, J.P. Boilot and R. Cases, *J. Cryst. Growth* **159**, 853 (1996).

92. S. Kishimoto, T. Hasegawa, H. Kinto, O. Matsumoto and S. Iida, *J. Cryst. Growth* **214**, 556 (2000).
93. B.A. Smith, J.Z. Zhang, A. Joly and J. Liu, *Phys. Rev. B* **62**, 2021 (2000).
94. H.S. Yang, P.H. Holloway and B.B. Ratna, *J. Appl. Phys.* **93**, 586 (2003).
95. W.G. Lu, P.X. Gao, W. Bin Jian, Z.L. Wang and J.Y. Fang, *J. Am. Chem. Soc.* 126, 14816 (2004).
96. R. Viswanatha, S. Sapra, S. Sen Gupta, B. Satpati, P.V. Satyam, B.N. Dev and D.D. Sarma, *J. Phys. Chem. B* **108**, 6303 (2004).
97. M. Ghosh, R. Seshadri and C.N.R. Rao, *J. Nanosci. Nanotechnol.* **4**, 136 (2004).
98. N.S. Norberg, G.L. Parks, G.M. Salley and D.R. Gamelin, *J. Am. Chem. Soc.* **128**, 13195 (2006).
99. N. Pradhan, D.M. Battaglia, Y.C. Liu and X.G. Peng, *Nano Lett.* 7, 312 (2007).
100. N. Janssen, K.M. Whitaker, D.R. Gamelin and R. Bratschitsch, *Nano Lett.* **8**, 1991 (2008).
101. R. Beaulac, P.I. Archer and D.R. Gamelin, *J. Solid State Chem.* **181**, 1582 (2008).
102. X. Zhang, M. Brynda, R.D. Britt, E.C. Carroll, D.S. Larsen, A.Y. Louie and S.M. Kauzlarich, *J. Am. Chem. Soc.* **129**, 10668 (2007).
103. L. Spanhel, A. Henglein and H. Weller, *Ber. Bunsen-Ges. Phys. Chem.* **91**, 1359 (1987).
104. W.Y. Zhou, Y. Zhou and S.Q. Tang, *Mater. Lett.* **59**, 3115 (2005).
105. J.J. Qiu, W.D. Yu, X.D. Gao and X.M. Li, *Nanotechnology* **17**, 4695 (2006).
106. G. Oskam and F.D.P. Poot, *J. Sol-Gel Sci. Techn.* **37**, 157 (2006).
107. A. Wolcott, T.R. Kuykendall, W. Chen, S.W. Chen and J.Z. Zhang, *J. Phys. Chem. B* **110**, 25288 (2006).
108. H.M. Cheng, J.M. Ma, Z.G. Zhao and L.M. Qi, *Chemical J. Chinese Universities-Chinese* **17**, 833 (1996).
109. S.W. Lu, B.I. Lee, Z.L. Wang and W.D. Samuels, *J. Cryst. Growth* **219**, 269 (2000).
110. S. Jeon and P.V. Braun, *Chem. Mater.* **15**, 1256 (2003).
111. X. Wang and Y.D. Li, *Mater. Chem. Phys.* **82**, 419 (2003).
112. K. Lee, N.H. Leea, S.H. Shin, H.G. Lee and S.J. Kim, *Mat. Sci. Eng. B-Solid* **129**, 109 (2006).
113. Y. Ding, Y. Wan, Y.L. Min, W. Zhang and S.H. Yu, *Inorg. Chem.* **47**, 7813 (2008).
114. H.F. Wang, Y.Q. Ma, G.S. Yi and D.P. Chen, *Mater. Chem. Phys.* **82**, 414 (2003).

115. J. Yang, C.K. Lin, Z.L. Wang and J. Lin, *Inorg. Chem.* **45**, 8973 (2006).

116. J.X. Wang, X.W. Sun, Y. Yang, H. Huang, Y.C. Lee, O.K. Tan and L. Vayssieres, *Nanotechnology* **17**, 4995 (2006).

117. J.L. Gole, J.D. Stout, C. Burda, Y.B. Lou and X.B. Chen, *J. Phys. Chem. B* **108**, 1230 (2004).

118. H. Tokudome and M. Miyauchi, *Chem. Lett.* **33**, 1108 (2004).

119. M. Sathish, B. Viswanathan, R.P. Viswanath and C.S. Gopinath, *Chem. Mater.* **17**, 6349 (2005).

120. C. Burda and J. Gole, *J. Phys. Chem. B* **110**, 7081 (2006).

121. H.Y. Chen, A. Nambu, W. Wen, J. Graciani, Z. Zhong, J.C. Hanson, E. Fujita and J.A. Rodriguez, *J. Phys. Chem. C* **111**, 1366 (2007).

122. J.H. Park, S. Kim and A.J. Bard, *Nano Lett.* **6**, 24 (2006).

123. R. Viswanatha, S. Chakraborty, S. Basu and D.D. Sarma, *J. Phys. Chem. B* **110**, 22310 (2006).

124. T.L. Sounart, J. Liu, J.A. Voigt, J.W.P. Hsu, E.D. Spoerke, Z. Tian and Y.B. Jiang, *Adv. Funct. Mater.* **16**, 335 (2006).

125. W.K. Liu, G.M. Salley and D.R. Gamelin, *J. Phys. Chem. B* **109**, 14486 (2005).

126. J.D. Bryan, S.M. Heald, S.A. Chambers and D.R. Gamelin, *J. Am. Chem. Soc.* **126**, 11640 (2004).

127. A.R. Kortan, R. Hull, R.L. Opila, M.G. Bawendi, M.L. Steigerwald, P.J. Carroll and L.E. Brus, *J. Am. Chem. Soc.* **112**, 1327 (1990).

128. M.A. Hines and P. Guyot-Sionnest, *J. Phys. Chem.* **100**, 468 (1996).

129. B.O. Dabbousi, J. RodriguezViejo, F.V. Mikulec, J.R. Heine, H. Mattoussi, R. Ober, K.F. Jensen and M.G. Bawendi, *J. Phys. Chem. B* **101**, 9463 (1997).

130. S.L. Cumberland, K.M. Hanif, A. Javier, G.A. Khitrov, G.F. Strouse, S.M. Woessner and C.S. Yun, *Chem. Mater.* **14**, 1576 (2002).

131. A.R. Clapp, E.R. Goldman and H. Mattoussi, *Nature Protocols* **1**, 1258 (2006).

132. M. Dybiec, G. Chornokur, S. Ostapenko, A. Wolcott, J.Z. Zhang, A. Zajac, C. Phelan, T. Sellers and D. Gerion, *Appl. Phys. Lett.* **90**, 263112 (2007).

133. R.E. Anderson and W.C.W. Chan, *ACS Nano* **2**, 1341 (2008).

134. M. Danek, K.F. Jensen, C.B. Murray and M.G. Bawendi, *Chem. Mater.* **8**, 173 (1996).

135. J. Tang, H. Birkedal, E.W. McFarland and G.D. Stucky, *Chem. Commun.*, 2278 (2003).

136. A. Mews, A. Eychmuller, M. Giersig, D. Schooss and H. Weller, *J. Phys. Chem.* **98**, 934 (1994).

137. F. Koberling, A. Mews and T. Basche, *Phys. Rev. B* (*Condensed Matter*) **60**, 1921 (1999).

138. S.V. Kershaw, M. Burt, M. Harrison, A. Rogach, H. Weller and A. Eychmuller, *Appl. Phys. Lett.* **75**, 1694 (1999).

139. A. Mews, A.V. Kadavanich, U. Banin and A.P. Alivisatos, *Phys. Rev. B-Condensed Matter.* **53**, 13242 (1996).

140. M. Braun, C. Burda and M.A. El-Sayed, *J. Phys. Chem. A* **105**, 5548 (2001).

141. H. Fujii, K. Inata, M. Ohtaki, K. Eguchi and H. Arai, *J. Mater. Sci.* **36**, 527 (2001).

142. S.L. Cumberland, M.G. Berrettini, A. Javier and G.F. Strouse, *Chem. Mater.* **15**, 1047 (2003).

143. Y.M. Wang, X.J. Li and S.J. Zheng, *J. Inorg. Mater.* **22**, 729 (2007).

144. M.G. Manera, J. Spadavecchia, D. Buso, C.D. Fernandez, G. Mattei, A. Martucci, P. Mulvaney, J. Perez-Juste, R. Rella, L. Vasanelli and P. Mazzoldi, *Sensor Actuat. B-Chem.* **132**, 107 (2008).

145. L.J. Diguna, Q. Shen, A. Sato, K. Katayama, T. Sawada and T. Toyoda, *Materials Science & Engineering C-Biomimetic and Supramolecular Systems* **27**, 1514 (2007).

146. J.P. Ge, S. Xu, J. Zhuang, X. Wang, Q. Peng and Y.D. Li, *Inorg. Chem.* **45**, 4922 (2006).

147. A. Wolcott, D. Gerion, M. Visconte, J. Sun, A. Schwartzberg, S.W. Chen and J.Z. Zhang, *J. Phys. Chem. B* **110**, 5779 (2006).

148. L.M. Liz-Marzan and P. Mulvaney, *New J. Chem.* **22**, 1285 (1998).

149. B.P. Zhang, H. Masumoto, Y. Someno and T. Goto, *Mater. Trans.* **44**, 215 (2003).

150. H.W. Lee, S. Cho, S. Lee, F. Rotermund, J. Lee and H. Lim, *J Korean Phys. Soc.* **51**, 390 (2007).

151. X.X. Zhang, G.H. Wen, S.M. Huang, L.M. Dai, R.P. Gao and Z.L. Wang, *J. Magn. Magn. Mater.* **231**, L9 (2001).

152. M.A. Correa-Duarte, N. Sobal, L.M. Liz-Marzan and M. Giersig, *Advan. Mater.* **16**, 2179 (2004).

153. S.J. Han, T.K. Yu, J. Park, B. Koo, J. Joo, T. Hyeon, S. Hong and J. Im, *J. Phys. Chem. B* **108**, 8091 (2004).

154. M.A. Correa-Duarte, M. Grzelczak, V. Salgueirino-Maceira, M. Giersig, L.M. Liz-Marzan, M. Farle, K. Sierazdki and R. Diaz, *J. Phys. Chem. B* **109**, 19060 (2005).

155. J.P. Li, Q.L. Yu and T.Z. Peng, *Anal. Sci.* **21**, 377 (2005).

156. S.G. Chen, M. Paulose, C. Ruan, G.K. Mor, O.K. Varghese, D. Kouzoudis and C.A. Grimes, *J. Photoch Photobio A* **177**, 177 (2006).
157. J. Wang, M. Nozaki, M. Lachab, R.S.Q. Fareed, Y. Ishikawa, T. Wang, Y. Naoi and S. Sakai, *J. Cryst. Growth* **200**, 85 (1999).
158. S.Y. Lu and S.W. Chen, *J. Am. Ceram. Soc.* **83**, 709 (2000).
159. J.W. Wang, Y.P. Sun, Z.H. Liang, H.P. Xu, C.M. Fan and X.M. Chen, *Rare Metal Mat. Eng.* **33**, 478 (2004).
160. Y.J. Li, P.B. Shi, R. Duan, B.R. Zhang, Y.P. Qiao, G.G. Qin and L. Huang, *J. Infrared and Millimeter Waves* **23**, 176 (2004).
161. M. Sacilotti, J. Decobert, H. Sik, G. Post, C. Dumas, P. Viste and G. Patriarche, *J. Cryst. Growth* **272**, 198 (2004).
162. H.T. Lin, J.L. Huang, S.C. Wang and C.F. Lin, *J. Alloy Compd.* **417**, 214 (2006).
163. D.Y. He, X.Q. Wang, Q. Chen, J.S. Li, M. Yin, A.V. Karabutov and A.G. Kazanskii, *Chinese Sci. Bull.* **51**, 510 (2006).
164. Z.Z. Ye, J.Y. Huang, W.Z. Xu, J. Zhou and Z.L. Wang, *Solid State Commun.* **141**, 464 (2007).
165. J.Y. Wang and T. Ito, *Diamond and Related Materials* **16**, 589 (2007).
166. J. Huerta, M. Lopez and O. Zelaya-Angel, *J. Vacuum Sci. Technol. B* **18**, 1716 (2000).
167. J. Huerta-Ruelas, M. Lopez-Lopez and O. Zelaya-Angel, *Jpn. J Appl. Phys. Part 1-Regular Papers Short Notes & Review Papers* **39**, 1701 (2000).
168. R.L. Barns and W.C. Ellis, *J. Appl. Phys.* **36**, 2296 (1965).
169. R.S. Wagner and W.C. Ellis, *Trans. Metallurgical Soc. Aim.* **233**, 1053 (1965).
170. J.P. Sitarik and W.C. Ellis, *J. Appl. Phys.* **37**, 2399 (1966).
171. R.S. Wagner and C.J. Doherty, *J. Electrochem. Soc.* **115**, 93 (1968).
172. P.X. Gao and Z.L. Wang, *J. Phys. Chem. B* **108**, 7534 (2004).
173. Y. Ding, P.X. Gao and Z.L. Wang, *J. Am. Chem. Soc.* **126**, 2066 (2004).
174. F.M. Kolb, H. Hofmeister, R. Scholz, M. Zacharias, U. Gosele, D.D. Ma and S.T. Lee, *J. Electrochem. Soc.* **151**, G472 (2004).
175. M.C. Yang, J. Shieh, T.S. Ko, H.L. Chen and T.C. Chu, *Jpn. J. Appl. Phys.* **1**, 44, 5791 (2005).
176. J.L. Taraci, T. Clement, J.W. Dailey, J. Drucker and S.T. Picraux, *Nuclear Instruments & Methods in Physics Research Section B-Beam Interactions with Materials and Atoms* **242**, 205 (2006).
177. A.M. Morales and C.M. Lieber, *Science* **279**, 208 (1998).
178. Z.W. Pan, Z.R. Dai, C. Ma and Z.L. Wang, *J. Am. Chem. Soc.* **124**, 1817 (2002).

179. J.R. Morber, Y. Ding, M.S. Haluska, Y. Li, P. Liu, Z.L. Wang and R.L. Snyder, *J. Phys. Chem. B* **110**, 21672 (2006).

180. X.F. Duan and C.M. Lieber, *J. Am. Chem. Soc.* **122**, 188 (2000).

181. Y.Q. Chen, K. Zhang, B. Miao, B. Wang and J.G. Hou, *Chem. Phys. Lett.* **358**, 396 (2002).

182. J.C. Johnson, H.J. Choi, K.P. Knutsen, R.D. Schaller, P.D. Yang and R.J. Saykally, *Nat. Mater.* **1**, 106 (2002).

183. C.Y. Meng, B.L. Shih and S.C. Lee, *J. Nanopart. Res.* **9**, 657 (2007).

184. D. Moore, J.R. Morber, R.L. Snyder and Z.L. Wang, *J. Phys. Chem. C* **112**, 2895 (2008).

185. M. Gentili, C. Giovannella and S. Selci, eds. NATO Science Series E, Kluwer Academic Publishers: New York. **228** (1994).

186. S. Shanmukh, L. Jones, J. Driskell, Y.P. Zhao, R. Dluhy and R.A. Tripp, *Nano Lett.* **6**, 2630 (2006).

187. J.D. Driskell, S. Shanmukh, Y. Liu, S.B. Chaney, X.J. Tang, Y.P. Zhao and R.A. Dluhy, *J. Phys. Chem. C* **112**, 895 (2008).

188. S. Daniele, C. Bragato, G.A. Battiston and R. Gerbasi, *Electrochimica Acta* **46**, 2961 (2001).

189. J. Christen, E. Kapon, E. Colas, D.M. Hwang, L.M. Schiavone, M. Grundmann and D. Bimberg, *Surface Sci.* **267**, 257 (1992).

190. M.C. Hanna, Z.H. Lu, A.F. Cahill, M.J. Heben and A.J. Nozik, *J. Cryst. Growth* **174**, 605 (1997).

191. Z. Yu, Z.M. Yang, S. Wang, Y. Jin, G. Liu, M. Gong and X.S. Sun, *Chemical Vapor Deposition* **11**, 433 (2005).

192. M. Kambara, Y. Hamai, H. Yagi and T. Yoshida, *Surface & Coatings Technology* **201**, 5529 (2007).

193. D.I. Lubyshev, J.C. Rossi, G.M. Gusev and P. Basmaji, *J. Cryst. Growth* **132**, 533 (1993).

194. Y. Nomura, Y. Morishita, S. Goto and Y. Katayama, *J. Electron Mater.* **23**, 97 (1994).

195. M. Yoshizawa, A. Kikuchi, M. Mori, N. Fujita and K. Kishino, in *Compound Semiconductors 1996.* **187** (1997).

196. K. Kusakabe, A. Kikuchi and K. Kishino, *J. Cryst. Growth* **237**, 988 (2002).

197. W. Tong, M. Harris, B.K. Wagner, J.W. Yu, H.C. Lin and Z.C. Feng, *Surface & Coatings Technol.* **200**, 3230 (2006).

198. G. Springhotz and G. Bauer, *Physica Status Solidi B-Basic Solid State Physics* **244**, 2752 (2007).

199. A. Pretorius, T. Yamaguchi, C. Kubel, R. Kroger, D. Hommel and A. Rosenauer, *J. Cryst. Growth* **310**, 748 (2008).

200. S.K. Chan, Y. Cai, I.K. Sou and N. Wang, *J. Cryst. Growth* **278**, 146 (2005).
201. S. Ohkouchi, Y. Sugimoto, N. Ozaki, H. Ishikawa and K. Asakawa, *Physica E* **40**, 1794 (2008).
202. G.E. Cirlin, V.A. Egorov, B.V. Volovik, A.F. Tsatsul'nikov, V.M. Ustinov, N.N. Ledentsov, N.D. Zakharov, P. Werner and U. Gosele, *Nanotechnology* **12**, 417 (2001).
203. J. Hirose, I. Suemune, A. Ueta, H. Machida and N. Shimoyama, *J. Cryst. Growth* **214**, 524 (2000).
204. G.A. Battiston, R. Gerbasi, M. Porchia and A. Gasparotto, *Chemical Vapor Deposition* **5**, 13 (1999).
205. R. Groenen, J. Loffler, P.M. Sommeling, J.L. Linden, E.A.G. Hamers, R.E.I. Schropp and M.C.M. van de Sanden, *Thin Solid Films* **392**, 226 (2001).
206. R. Winter, M. Quinten, A. Dierstein, R. Hempelmann, A. Altherr and M. Veith, *J. Appl. Crystallogr.* **33**, 507 (2000).
207. M. Deguchi, M. Kitabatake, H. Kurokawa, T. Shiratori and M. Kitagawa, *Diamond Films and Technology* **7**, 273 (1997).
208. J.K. Kruger, J.P. Embs, S. Lukas, U. Hartmann, C.J. Brierley, C.M. Beck, R. Jimenez and P. Alnot, *Diamond and Related Materials* **9**, 123 (2000).
209. Y.J. Chen, T.F. Young, S.L. Lee, H.J. Huang, T.S. Hsi, J.G. Chu and D.J. Jang, *Vacuum* **80**, 818 (2006).
210. S.J. Askari, F. Akhtar, G.C. Chen, Q. He, F.Y. Wang, X.M. Meng and F.X. Lu, *Mater. Lett.* **61**, 2139 (2007).
211. T. Goto, T. Ono and T. Hirai, *Scripta Materialia* **44**, 1187 (2001).
212. S. Bhaviripudi, E. Mile, S.A. Steiner, A.T. Zare, M.S. Dresselhaus, A.M. Belcher and J. Kong, *J. Am. Chem. Soc.* **129**, 1516 (2007).
213. W. Wunderlich, *Diamond and Related Materials* **16**, 369 (2007).
214. A.N. Obraztsov, G.M. Mikheev, Y.P. Svirko, R.G. Zonov, A.P. Volkov, D.A. Lyashenko and K. Paivasaari, *Diamond and Related Materials* **15**, 842 (2006).
215. W.B. Yang, S.H. Fan, M. Ge, G.L. Zhang, Z.M. Shen and S.Z. Yang, *Acta Physica Sini.* **55**, 351 (2006).
216. P.R. Somani, S.P. Somani and M. Umeno, *Chem. Phys. Lett.* **430**, 56 (2006).
217. E. Fitzer, D. Hegen and H. Strohmeier, *J. Electrochem. Soc.* **126**, C353 (1979).
218. S. Kato, Y. Yamada and T. Taguchi, *J. Cryst. Growth* **190**, 223 (1998).
219. L.A. Dobrzanski and J. Mikula, *J. Mater. Processing Technol.* **167**, 438 (2005).

220. J. Berjonneau, F. Langlais and G. Chollon, *Surface & Coatings Technol.* **201**, 7273 (2007).
221. M.M. Alkaisi, R.J. Blaikie and S.J. McNab, *Advan. Mater.* **13**, 877 (2001).
222. H. Iwasaki, T. Yoshinobu and K. Sudoh, *Nanotechnology* **14**, R55 (2003).
223. M. Munz, B. Cappella, H. Sturm, M. Geuss and E. Schulz, in *Filler-Reinforced Elastomers Scanning Force Microscopy.* **87** (2003).
224. L.B. Zhang, J.X. Shi, J.L. Yuan, S.M. Ji and M. Chang. 2004.
225. R.P. Seisyan, *Technical Phys.* **50**, 535 (2005).
226. X.N. Xie, H.J. Chung, C.H. Sow and A.T.S. Wee, *Mater. Sci. Engin. R-Rep.* **54**, 1 (2006).
227. N. Stokes, A.M. McDonagh and M.B. Cortie, *Gold Bulletin* **40**, 310 (2007).
228. H. Mattoussi, J.M. Mauro, E.R. Goldman, T.M. Green, G.P. Anderson, V.C. Sundar and M.G. Bawendi, *Phys. Status Solidi B* **224**, 277 (2001).
229. E.R. Goldman, G.P. Anderson, P.T. Tran, H. Mattoussi, P.T. Charles and J.M. Mauro, *Anal. Chem.* **74**, 841 (2002).
230. A.R. Clapp, I.L. Medintz, H.T. Uyeda, B.R. Fisher, E.R. Goldman, M.G. Bawendi and H. Mattoussi, *J. Am. Chem. Soc.* **127**, 18212 (2005).
231. M.N. Rhyner, A.M. Smith, X.H. Gao, H. Mao, L.L. Yang and S.M. Nie, *Nanomed.* **1**, 209 (2006).
232. Y. Xing, Q. Chaudry, S. Chen, K.Y. Kong, H.E. Zhau, L.W. Chung, J.A. Petros, R.M. O'Regan, M.V. Yezhelyev, J.W. Simons, M.D. Wang and S. Nie, *Nature Protocols* **2**, 1152 (2007).
233. S. Yoshitake, Y. Yamada, E. Ishikawa and R. Masseyeff, *European J. Biochem.* **101**, 395 (1979).
234. A.M. Smith, S. Dave, S.M. Nie, L. True and X.H. Gao, *Expert Rev. Mol. Diagn.* **6**, 231 (2006).
235. N. Pradhan and X.G. Peng, *J. Am. Chem. Soc.* **129**, 3339 (2007).
236. J.Z. Zhang, Z.L. Wang, J. Liu, S. Chen and G.-y. Liu, *Self-assembled Nanostructures.* Nanoscale Science and Technology, New York: Kluwer Academic/Plenum Publishers. **316** (2003).
237. G. Cao, *Nanostructures & Nanomaterials: Synthesis, Properties & Applications*, London: Imperial College Press. **433** (2004).
238. J.D. Bryan and D.R. Gamelin, *Prog. Inorganic Chem.* **54**, 47 (2005).
239. C. Burda, X.B. Chen, R. Narayanan and M.A. El-Sayed, *Chem. Rev.* **105**, 1025 (2005).

Chapter 5

Optical Properties of Semiconductor Nanomaterials

Semiconductors possess interesting and important optical and electronic properties useful for diverse technologies including microelectronics, detectors, sensors, lasers and photovoltaics. Semiconductor nanomaterials have been extensively studied and exploited for various applications ranging from energy conversion to medicine. This chapter covers some of the basic concepts about semiconductors and then focuses on optical properties of semiconductor nanomaterials. Emphasis is placed on unique optical and related properties of nanomaterials pertaining to bulk semiconductors.

5.1. Some basic concepts about semiconductors

Before we discuss optical properties of semiconductor nanomaterials, it is useful to review briefly some basic concepts related to general semiconductor physics of bulk materials, including crystal structure, electronic band structure, bandgap, Fermi energy, trap states and charge carrier mobility. This is needed partly because most of the fundamental physics underlying nanomaterials are the same as for bulk materials. Another reason is that optical properties are intrinsically connected with electronic transitions and both depend on the crystal structure of the material.

5.1.1. *Crystal structure and phonons*

Crystal structure refers to the spatial arrangement of atoms and ions in a solid. The specific arrangement of atoms or ions in a solid is often referred to as *lattice* or *lattice structure*. Atoms can be completely ordered on the long range and through the entire crystal, and this is usually called a *single crystal*, the size of which can vary substantially for different materials. If the atomic order is limited to part of the crystal only and the solid is composed of many small crystals or domains, the solid is referred to as *polycrystalline*. If there is no order on the atomic scale in any part of the solid, then it is considered as *amorphous*. For crystalline solids, single crystals or polycrystals, their crystal structure can be determined experimentally and, in certain cases, calculated theoretically. Two of the most powerful experimental techniques for crystal structure determination are X-ray diffraction (XRD) or X-ray crystallography and NMR (nuclear magnetic resonance), discussed in Chapter 3.

The crystal structures of many commonly encountered solids have been determined and can be easily found in the literature or data base [1]. For the same material, reducing the size to nanoscale could change their crystal structure but in practice most materials retain their crystal structures as that of bulk, down to a few nanometers. Due to the large *surface-to-volume (S/V) ratio*, the surface plays a more and more important role as size decreases. This is a unique aspect to nanomaterials and will be discussed further later.

It should be emphasized that the crystal structure is intimately coupled to electronic and thereby optical properties of solids. It is thus important to determine and understand the crystal structure of the nanomaterials even if we are primarily interested in their electronic and/or optical properties in this book.

One important character of crystal structures is *phonons*. Phonons are collective vibrational modes of the atoms forming the solid. They are often classified as *optical phonons*, which have high frequencies as a result of atoms vibrating out of phase, and *acoustic phonons*, which have lower frequencies as a result of atoms vibrating in phase or in unison. Phonons are analogous to vibrational modes in molecules. They are essentially the normal modes of vibration involving a huge number of atoms simultaneously. Typical phonon frequencies are in the tens to hundreds of

cm^{-1} range. Generally, the phonon frequencies are lower for heavier atoms and weaker force constants of vibration. The force constants depend on the electronic structures of the atoms, ions and/or molecules involved. For example, bulk crystalline Si has a phonon mode at 520 cm^{-1} that can be easily detected as the most intense peak in Raman scattering, which shifts to lower frequency in Si nanowires (~505 cm^{-1}), attributed to quantum phonon confinement [2].

5.1.2. *Electronic energy bands and bandgap*

An *energy band* refers to a large number of electronic states within a certain energy range. In atoms and small molecules, there is a relatively small number of electronic states per unit energy. When atoms, ions or molecules interact and form solids, a huge number of electronic states become clustered within a relatively narrow energy range and an energy band is formed. The band formed from electronic states with valence electrons is called the *valence band* (VB) and is fully occupied if there are two electrons per electronic state originally, or half occupied if there is only one electron per electronic state in the original atom. The band formed from the lowest unoccupied electronic states is called the *conduction band* (CB), since it is primarily responsible for electronic and thermal conduction when electrons are promoted into it.

The solid with a half occupied valence band is a metal. A metal can also result when the valence and conduction bands overlap in energy, as illustrated in Fig. 5.1. If the VB and CB do not overlap in energy, the energy difference between the top of the VB and the bottom of the CB is called a *bandgap*, usually denoted as E_g, which determines the thermal, electrical and optical properties of the material.

For example, the bandgap corresponds to the lowest energy or longest wavelength for optical absorption of the solid. This is true certainly for so-called *direct bandgap* solids. Direct bandgap refers to electrical dipole allowed transitions. In contrast, *indirect bandgap* refers to electrical dipole forbidden transitions that require photon assistance for electronic transitions to occur. This is illustrated in Fig. 5.2.

Small bandgap solids are called *semiconductors* while large bandgap solids are named *insulators*. The distinction between semiconductor and

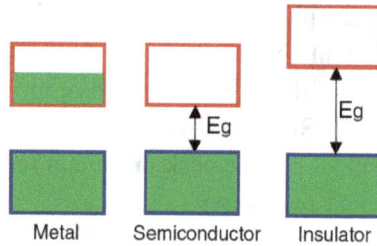

Fig. 5.1. Comparison of different electronic band structures of metal, semiconductor and insulator. E_g represents the bandgap energy. The boxes represent VB (blue) or CB (red) and green region represents electron occupied states and the white (uncolored) region represents unoccupied states.

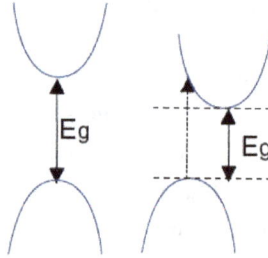

Fig. 5.2. Comparison of direct (left) and indirect (right) bandgap structures. E_g is the bandgap energy. The dotted vertical arrow indicates electronic transition that is not allowed by electrical dipole but can be allowed with phonon assistance. The band structure is plotted in terms of energy as a function of wave vector $k = 2\pi/\lambda$, namely reciprocal or momentum space ($p = \hbar k$).

insulator is only quantitative while the distinction between semiconductor and metal is more profound. Also, materials are traditionally classified as metal, semiconductor or insulator based on their conductivity [3]. In terms of bandgap, most materials with a bandgap of about 3.5 eV or less are generally considered as semiconductors while those with E_g above 3.5 eV are considered as insulators. However, there is no strict definition or anything clear cut about this.

In principle, when there are no defects or doping and thereby no electronic states in the bandgap, semiconductors are considered as intrinsic and are practically nonconductive unless the bandgap is very small, near the energy of kT. In reality, there are impurities, defects or intentional doping that introduce electronic states within the bandgap that alter the electrical and optical properties. These are considered as extrinsic semiconductors.

5.1.3. *Electron and hole effective masses*

The mass of an electron in a solid is usually different from that of a free electron or electron in vacuum, m_e. The idea of *effective mass*, designated as $m*$, is introduced to describe the mass of an electron (m_e*) or hole (m_h*) in a solid and it is mathematically expressed as:

$$m* = \frac{1}{\hbar^2} \frac{d^2E}{dk^2} \qquad (5.1)$$

where E is energy as a function of wave vector k in momentum space ($p = \hbar k$), $\hbar = h/2\pi$ and h is Plank's constant. For an electron in the CB, the CB energy as a function of k is used to calculate m_e*. Likewise, for the hole in the VB, the VB energy as a function of k is used to calculate m_h*. The CB and VB structures can be determined experimentally or calculated theoretically, depending on the complexity of the solid.

For an *electron-hole pair*, it is often needed to describe them as one entity (like an *exciton* to be discussed later) using an effective reduced mass, μ, defined as:

$$\mu = \frac{m_e^* m_h^*}{m_e^* + m_h^*} \qquad (5.2)$$

For example, for GaAs, $\mu = 0.059 m_e$, which is much smaller than the free electron mass, m_e [3].

5.1.4. *Density-of-states, Fermi energy, and carrier concentration*

Density-of-states (*DOS*) is defined as the number of states per unit energy. It is determined by the electronic or chemical nature as well as crystal

structure of the solid. It, in turn, critically affects the properties and performance of the material, including optical, electronic, thermal,
mechanical and magnetic characteristics.

The DOS for an electron in a parabolic CB can be shown to be [4]:

$$N(E) = \frac{1}{2\pi^2 \hbar^3} (2m_e^*)^{3/2} (E - E_g)^{1/2} \qquad (5.3)$$

where m_e^* is the electron effective mass, $N(E)$ is the DOS at energy E, E_g
is the bandgap energy. Similarly, the DOS for a hole in a parabolic VB is
given by:

$$N(E) = \frac{1}{2\pi^2 \hbar^3} (2m_h^*)^{3/2} E^{1/2} \qquad (5.4)$$

where m_h^* is the hole effective mass.

Electrons are fermions and their occupancy in electronic states
follows the Fermi–Dirac distribution function:

$$f(E) = \frac{1}{\exp\dfrac{(E - E_F)}{kT} + 1} \qquad (5.5)$$

where E_F is the so-called *Fermi energy or Fermi level*, k is the
Boltzmann's constant, T is the absolute temperature, $f(E)$ is the probability of finding the electron at energy E. *Fermi energy* can be defined as the
energy at which the probability for finding an electron is 50% or 1/2. The
density of electrons is a simple product of the DOS and the Fermi–Dirac
function.

Likewise, the density of holes is the product of the DOS and the
probability of the state being empty. This probability is given by:

$$1 - f(E) = \frac{1}{\exp\dfrac{(E_F - E)}{kT} + 1} \qquad (5.6)$$

5.1.5. *Charge carrier mobility and conductivity*

Electrical conduction is a fundamental property of semiconductors critical to the electronics industry. The *electrical conductivity* σ is the sum of contributions from the concentrations of electrons n and of holes p as expressed by:

$$\sigma = (ne\mu_e + pe\mu_h) \tag{5.7}$$

where e is the unit electronic charge, μ_e and μ_h are the electron and hole *mobilities*, which can be expressed as charge carrier drift velocity per unit electric field [3]. The charge carrier concentrations, n and p, are generally dependent on temperature, doping, or photoexcitation. For example, above bandgap photoexcitation increases electrical conductivity due to the promotion of electrons into the CB and the creation of holes in the VB. The electrical conductivity is important for any devices or applications that involve charge transport.

5.1.6. *Exciton, exciton binding energy, and exciton Bohr radius*

An *exciton* is an electron and hole pair with Columbic attraction. When treated quantum mechanically, similar to that of the hydrogen atom consisting of an electron and proton, the *exciton binding energy* is given by:

$$E_{ex} = -\frac{13.6\mu/m_e}{\left(\dfrac{\varepsilon}{\varepsilon_0}\right)^2 n^2}eV \tag{5.8}$$

where μ is the reduced effective mass of the electron-hole pair or effective mass of the exciton, as given by Eq. (5.2), m_e is the free electron mass, $\varepsilon/\varepsilon_0$ is the relative dielectric constant of the material with ε the dielectric constant of the material and ε_0 the dielectric constant or permittivity of free space (a known constant), n is the quantum number and a positive integer ($n = 1, 2, 3 \ldots \infty$) with $n = 1$ the lowest energy state or ground state.

When $n = \infty$, E_{ex} is zero and the exciton dissociates or is unbound and the electron and hole are free.

The exciton binding energy for most semiconductors is in the range of a few to a few tens of meV. For example, for GaAs, E_{ex} is about 4.6 meV for the ground state and for CdS around 28 meV [3, 5]. In comparison, the ground state of the hydrogen atom is located at 13.6 eV below the energy for the free electron and proton. Thus, the exciton binding energy is usually small, smaller than or comparable to kT at room temperature, which is about 40 meV. As a result, most excitons are not bound or dissociative at room temperature. Therefore, transition to and from the excitonic state is often considered the same as the CB edge. They differ strictly speaking by the exciton binding energy.

Another important concept related to the exciton is the *Bohr exciton radius* defined as:

$$\alpha_B = \varepsilon_0 \varepsilon h^2 / \pi \mu e^2 \qquad (5.9)$$

This is similar to the Bohr radius for the ground state of the hydrogen atom, which is usually denoted as $a_0 = 0.529$ nm. The Bohr exciton radius for solids are mostly much larger than a_0. For CdS, a_B is around 3 nm and for GaAs 11.8 nm. From Eq. (5.9), it is clear that one can calculate the exciton Bohr radius once the dielectric constant ε and exciton effective mass μ are known since the rest of the parameters are known constants. The exciton Bohr radius provides a picture on how delocalized the exciton or electron-hole pair is spatially.

When the electron and hole are weakly bound, the exciton formed is named *Mott–Wannier exciton*. A strongly or tightly bound electron and hole pair, similar to a long-lived excited state in an atom or molecule, is called a *Frenkel exciton*. Both types of excitons are mobile in the crystal lattice with the Mott–Wannie excitons more mobile than the Frenkel excitons in general. We will focus our discussion only on Mott–Wannier excitons in this chapter since they are most commonly encountered in semiconductors and in nanostructures.

5.1.7. *Fundamental optical absorption due to electronic transitions*

Absorption of light is a fundamental property of semiconductor materials. Light absorption can be expressed in terms of a coefficient $\alpha(hv)$ which is defined as the relative rate of decrease in light intensity, $I(hv)$, along its propagation path (x):

$$\alpha(hv) = \frac{1}{(hv)} \frac{dI(hv)}{dx} \qquad (5.10)$$

If the light radiation is represented as a plane wave with a frequency v propagating through a semiconductor with a complex index of refraction and the imaginary part represented as n_i, one can show that [4]:

$$\alpha(hv) = \frac{4\pi hvn_i}{c} \qquad (5.11)$$

where c is the speed of light. This equation shows that the absorption is mainly due to the imaginary part of the complex index of refraction of the material.

It can be further shown that for electrical dipole allowed, direct bandgap transitions,

$$\alpha(hv) = A*(hv - E_g)^{1/2} \qquad (5.12)$$

where $A*$ is a constant determined by the index of refraction, and electron and hole effective masses [4, 6].

For transitions that are only allowed at $k \neq 0$ but forbidden at $k = 0$, the absorption coefficient has the following expression:

$$\alpha(hv) = A'(hv - E_g)^{3/2} \qquad (5.13)$$

where, similar to $A*$ in Eq. (5.12), A' is related to the index of refraction and effective masses of the electron and hole.

For electric dipole forbidden or indirect bandgap transitions that require phonon assistance, the electronic absorption coefficient is given by:

$$\alpha_a(hv) = \frac{A(hv - E_g + E_p)^2}{\exp\dfrac{E_P}{kT} - 1}$$

(5.14)

For indirect bandgap transitions involving phonon emission, the absorption coefficient is given by:

$$\alpha_a(hv) = \frac{A(hv - E_g - E_p)^2}{1 - \exp\left(-\dfrac{E_p}{kT}\right)}$$

(5.15)

where A is a constant similar to A^* and A', E_p is the phonon energy. The above four equations show that the absorption coefficient dependence on the light frequency varies for different transitions. One can, therefore, use the absorption spectrum to determine what kind of transition is involved as well as to estimate the bandgap energy. It is an estimate since approximations have been made to derive the above equations, including the assumption of parabolic band structures for both the VB and CB. Nonetheless, these equations have been frequently used in practice to estimate bandgap energy.

5.1.8. *Trap states and large surface-to-volume ratio*

In comparison to bulk solids, nanomaterials are unique in their huge surface-to-volume (S/V) ratio, typically, million-fold increase compared to bulk materials. For a solid sphere with 1 cm diameter dispersed into 10 nm spherical particles, the surface area increases by a factor of one million. For a typical nanoparticle containing a few hundred atoms, there are a significant number of surface atoms. For instance, for a cube with 10 atoms in each direction, there are 1000 atoms total and 488 on the surface, which is nearly 50%. These surface atoms tend to be unsaturated chemically or have different bonding environment compared to atoms in the interior, and would thus have energy level different from those in the interior, e.g. introducing electronic states within the bandgap, often called *trap states*. These trap states will, in turn, alter the electronic band structure of the nanoparticles, if the band picture is still valid, and also associated properties and functionalities. At very small size, e.g. particles with a few or fen tens of atoms, the nanoparticles or nanoclusters behave more like molecules and the band picture developed for bulk solids do not really apply anymore. Trap states due to large S/V is an important reality of nanomaterials and need to be taken into account in explaining their properties.

5.2. Energy levels and density of states in reduced dimension systems

5.2.1. *Energy levels*

As discussed above, the electronic energy levels and density-of-states (DOS) determine the properties of materials, including optical and electronic properties as well as their functionalities. For nanoscale materials, the energy levels and DOS vary as a function of size, resulting in dramatic changes in the materials properties. The energy level spacing increases with decreasing dimension and this is known as the *quantum size confinement effect*. It can be understood using the classic particle-in-a-box model. With decreasing particle size, the energy level spacing increases quadratically as the length of the box decreases. The quantum confinement effect can be qualitatively explained using the effective mass approximation [7–11]. For a spherical particle with radius R, the effective bandgap, $E_{g,\,\text{eff}}(R)$, is given by:

$$E_{g,\,\text{eff}}(R) = E_g(\infty) + \frac{\hbar^2 \pi^2}{2R^2}\left(\frac{1}{m_e^*} + \frac{1}{m_h^*}\right) - \frac{1.8e^2}{\varepsilon R} \quad (5.16)$$

where $E_g(\infty)$ is the bulk bandgap, m_e^* and m_h^* are the effective masses of the electron and hole, and ε is the bulk optical dielectric constant or relative permittivity. The second term on the right-hand side shows that the effective bandgap is inversely proportional to R^2 and increases as size decreases. On the other hand, the third term shows that the bandgap energy decreases with decreasing R due to increased Coulombic interaction. However, since the second term becomes dominant at small R, the effective bandgap is expected to increase with decreasing R, especially when R is small. Figure 5.3 illustrates how the bandgap and energy level spacing change with size.

The quantum size confinement effect becomes significant particularly when the particle size becomes comparable to or smaller than the Bohr exciton radius, α_B, as given by Eq. (5.9). The quantum mechanical treatment of this problem involving an electron and a hole is similar to that of the hydrogen atom containing an electron and a proton. As an example, the exciton Bohr radius for GaAs is around 11.8 nm with an exciton binding energy of 4.6 meV [3]. As another example, the Bohr radius of CdS is around 2.4 nm [5] and particles with radius smaller or comparable to 2.4 nm show strong quantum confinement effects, as indicated by a significant blue-shift of their optical absorption relative to that of bulk [12–14]. Likewise, the absorption spectra of CdSe nanoparticles (NPs) show a dramatic blue-shift with decreasing particle size, as shown in Fig. 5.4. The emission spectra usually show a similar blue-shift with decreasing size.

There are some interesting differences between semiconductors and metals in terms of quantum confinement. A major difference is in their

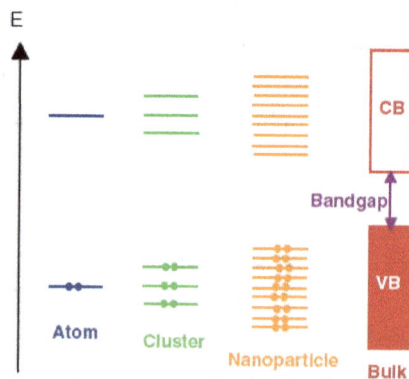

Fig. 5.3. Schematic illustration of the quantum size confinement effect. As the size increases the energy level spacing decreases.

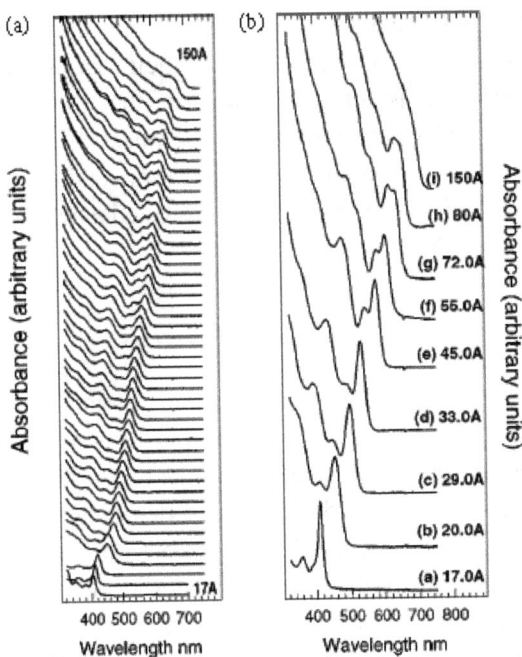

Fig. 5.4. Decreasing the nanocrystal (NC) diameter increases the separation between states. (a) Blue-shift in the absorption edge and a larger separation between electronic transitions for a homologous size series of CdSe NC dispersions, collected at room temperature. (b) Observation of discrete electronic transitions in optical absorption as a measure of spectroscopic information that can be uncovered in monodisperse NC samples ($\sigma \leq 5\%$). Adapted with permission from (Fig. 6) of Ref. 15.

electronic band structure. The bandgap that plays a central role in semiconductors is much less pronounced in metals, since electrons are free to move even at fairly low temperature in their half-filled conduction band. As the particle size increases, the center of the band develops first and the edge last [16]. In metals with Fermi level lying in the band center, the relevant energy level spacing is very small even in relatively small particles, whose properties resemble those of the bulk [17]. For semiconductors, in contrast, the Fermi level lies between two bands and the edges of the bands dominate the low energy optical and electronic properties. Bandgap optical excitation depends strongly on particle size. As a consequence, the quantum size confinement effect is expected to become important at much smaller sizes for metal than for semiconductor nanoparticles. This difference has fundamental consequences on their optical, electronic and dynamic properties. It is interesting to note that for very small metal nanoparticles, a significant bandgap can be observed [18]. In principle, metal nanoparticles can become essentially semiconductors or even insulators at very small size, experiencing the so-called *metal to semiconductor and insulator transition*.

5.2.2. *Density of states (DOS) in nanomaterials*

The density-of-states of a reduced dimension system also changes significantly with decreasing size. For example, for a three-dimensional bulk material, the DOS is proportional to the square root of energy E. For a system with confinement in one dimension (a two-dimensional material or quantum well), the DOS is a step function. For systems with confinement in two dimensions (a one-dimensional material or quantum wire), the DOS has a peculiarity. For systems with confinement in three dimensions (zero-dimensional material or so-called quantum dots, QDs), the DOS has the shape of a δ-function [19].

Theoretically, the DOS can be described by the following equations. For a 3D (bulk) system, the DOS has the form [19]:

$$\frac{dN}{dE} \propto \frac{d}{dE} E^{3/2} \propto E^{1/2} \tag{5.17}$$

For a 2D system (quantum well), the DOS is a step function:

$$\frac{dN}{dE} \propto \frac{d}{dE} \sum_{\varepsilon_i < E} (E - \varepsilon_i) \propto \sum_{\varepsilon_i < E} (1) \tag{5.18}$$

For a 1D system (e.g. quantum wire), the DOS is given by:

$$\frac{dN}{dE} \propto \frac{d}{dE} \sum_{\varepsilon_i < E} (E - \varepsilon_i)^{1/2} \propto \sum_{\varepsilon_i < E} (E - \varepsilon_i)^{-1/2} \tag{5.19}$$

For a 0-D system, i.e. quantum dot, the DOS can be described by:

$$\frac{dN}{dE} \propto \frac{d}{dE} \sum_{\varepsilon_i < E} \theta(E - \varepsilon_i) \propto \sum_{\varepsilon_i < E} \delta(E - \varepsilon_i) \qquad (5.20)$$

The change in DOS in reduced dimension systems is illustrated in Fig. 5.5. The DOS strongly influences the electronic and optical properties of semiconductors as well as metals. The changes in DOS, along with changes in electronic and vibrational energy levels, alter the properties of the nanomaterials compared to bulk. This results in changes in both their dynamic and equilibrium properties. For example, changes in the DOS of the electrons and phonons affect the electron–phonon coupling and thereby the electronic relaxation due to electron–phonon interaction. A direct measure of this effect is usually difficult since the surface often plays a dominant role in electronic relaxation through trapping and the effect of electron–phonon interaction on relaxation cannot be determined independently.

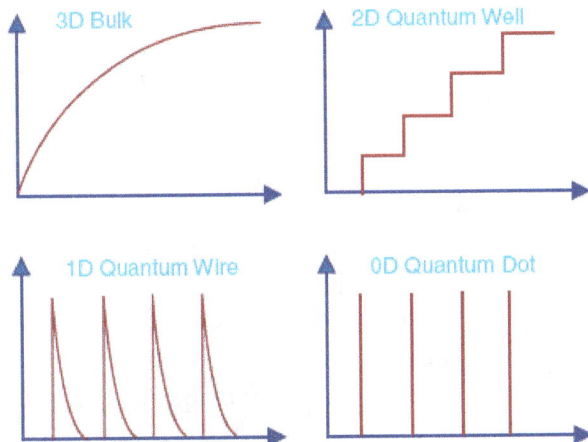

Density of States (DOS) in Reduced Dimension Systems

Fig. 5.5. Illustration of change in DOE as a function of physical dimension of the system, from 3D (bulk) to 2D, 1D, and 0D.

5.2.3. *Size dependence of absorption coefficient, oscillator strength, and exciton lifetime*

Size dependence of properties such as *absorption coefficient* or *oscillator strength* and *exciton lifetime* in nanoparticles is a fundamental issue of interest. Both the absorption coefficient and oscillator strength are proportional to and determined by the square of the electrical transition dipole moment. In a theoretical study, the oscillator strength per unit volume has been found to have a different size dependence from the oscillator strength per particle [20]. With the assumption that the excitation is coherent throughout the nanoparticle, the *oscillator strength per volume* (normalized to volume) was predicted to increase with decreasing particle radius. This is not intuitive but has been attributed to the quantum confinement effect on the spatial correlation between the electron and hole. On the contrary, the *oscillator strength per nanoparticle* was predicted to increase with increasing particle radius. This appears to make sense since one may expect the oscillator strength per particle to increase with increasing volume or radius of the particle. However, in a different theoretical study, the oscillator strength per particle has been predicted to be a constant with varying particle size in the strongly confined regime, where the particle radius is smaller than the Bohr radius [8].

Experimental studies of this issue have been somewhat limited and more complicated than one may anticipate. The main reason is that experimental studies are often complicated by defect or surface trap states and inhomogeneous size distribution, making it challenging to extract a simple quantitative size dependence of the absorption coefficient or oscillator strength. In particular, as size changes, the surface area, volume, and surface-to-volume ratio all change, which could have competing effects on the size dependence of interest. In addition, it is often challenging to determine the number of nanoparticles in an accurate way [21].

Even given these issues, there have been some experimental studies that show qualitative agreement with theory. For example, an experimental study of CdTe nanoparticles has found that, in the size range of 20–40 Å, the molar absorption coefficient, based on CdTe concentration, follows an inverse cube dependence on particle diameter ($1/R^3$), while the extinction coefficient per particle remained constant [22]. In a similar experimental study of CdS nanoparticles, it has been found that the oscillator strength, normalized to that of bulk CdS, increases with decreasing particle size proportional to $1/R^3$, in the size range of about radius $R = 6$ to 48 Å [21], consistent with previous theoretical prediction that attributed the dependence to strong overlap of the wave functions of the confined electron and hole [23]. The oscillator strength was determined based on the analytical amount of cadmium since the number of nanoparticles cannot be easily determined. The results fit well with a theoretical model predicting that the oscillator strength per cluster is independent of the cluster size within the strong confinement region [8].

In a recent experimental study of InAs quantum dots, it has been found that while the per-dot absorption cross-section increases as the cube of the dot radius at high photon energy, at the first-exciton peak it increases approximately linearly with dot radius [24]. It was also found that the first-exciton oscillator strength remains constant with dot size in the range studied ($R = 1.5$ to 3.5 nm), which is consistent with the theory for strongly confined excitons, and with previous experimental studies [8]. Consequently, the radiative lifetime of the first-exciton transition calculated from the absorption properties is predicted to increase with increasing dot size. This is also qualitatively consistent with theoretical predictions that attributed the size dependence to electron-hole correlation or the excitonic effect [20]. Further experimental and theoretical studies are needed to better understand this fundamental issue at a more quantitative level and in a boarder particle size range.

5.3. Electronic structure and electronic properties

5.3.1. *Electronic structure of nanomaterials*

The electronic band structure of semiconductors and metals determines their optical and electronic properties. The electronic band structure is a description of the energy levels of electronic states as a function of lattice phonon wave vector k. If the band maximum for the valence band (VB) matches the band minimum of the conduction band (CB) for the same *k* value, electronic transition from VB to CB is allowed by the electrical dipole and this is called an *allowed transition* or *direction bandgap transition*, as shown in Fig. 5.2. Otherwise, the transition is electrical dipole forbidden. In this case, transitions can be made possible with phonons involved. These transitions are called *phonon-assisted transitions* or *indirect bandgap transitions,* and are usually much weaker than electric dipole allowed transitions (Fig. 5.2).

Optical absorption and emission as well as electronic transport properties of direct and indirect bandgap semiconductors can be very different. For direct bandgap semiconductors with electrical dipole allowed transitions from VB to CB, their electronic absorption as well as emission is usually strong with a distinct excitonic feature in the absorption spectrum. For indirect bandgap semiconductors with electrical dipole forbidden transitions from VB to CB, both energy and momentum of the electron-hole pair are changed in the transition with phonon assistance [4]. Both their absorption and emission are weaker compared to those of direct bandgap semiconductors, and no obvious excitonic features are observed.

For nanomaterials, the electronic band structure could be altered as compared to bulk. As a result, the properties of the materials can change with a reduction of dimensions of the system. The band structure of nanomaterials is generally not as well characterized as with bulk materials. One example is silicon nanoparticles for which there have been intensive debates over the observed luminescence on if it originates from surface effect or fundamental changes in electronic band structure, from indirect in bulk to direct on the nanoscale [25–31]. It is most likely that the luminescence from nanostructured silicon was partly due to surface effect, especially the red emission, and partly due to quantum confinement (electronic structure change), especially the bluer emission. This is similar to the red trap state emission due to surface states versus bandedge emission determined by the band structure observed typically for CdS [32]. In principle, the band structure could be altered in nanostructured materials due to the confinement. However, clear demonstration of this effect is rare due to the complication of surface effect and requires further theoretical and experimental research.

5.3.2. *Electron–phonon interaction*

Electron–phonon interaction plays a critical role in the materials proper-ties and functionalities of bulk as well as nanoscale materials. With reduced dimensions in nanomaterials, the electron–phonon interaction is expected to change. From the standpoint of DOS, one may expect that the interaction becomes weaker with decreasing size, since the DOS of both the electrons and phonons decreases with decreasing size and their spec-tral overlap will decrease. However, other factors, such as surface, could play a competing role by introducing electronic trap states and should be considered in practice. The change in electron–phonon interaction and electron–surface interaction with particle size will, in turn, affect the charger carrier lifetime, as to be discussed in more detail later.

Compared to bulk materials, another unique aspect of nanomaterials, in terms of electron–phonon interaction, is the involvement of surface phonons or vibrations. These include phonons due to surface atoms of the nanocrystals, which tend to have a lower frequency than bulk phonons, and vibrations of molecules or atoms adsorbed on the surface of the nanocrystal that can have frequencies higher or lower compared to bulk phonon frequencies. The combination of changes of DOS and surface phonon frequencies could significantly affect the electron–phonon inter-action in nanostructured materials [33].

5.4. Optical properties of semiconductor nanomaterials

5.4.1. *Absorption: direct and indirect bandgap transitions*

Optical absorption of semiconductors, determined by their electronic structure, is often used as a way to probe their optical and electronic properties. Optical absorption is a result of interaction between the mate-rial and light. When the frequency of light is on resonance with the energy difference between electronic states and the transition is allowed or partly allowed by selection rules, a photon is absorbed by the material. This is reflected as a decrease of transmission or an increase in absorbance of the light passing through the sample. By measuring the transmission or absorbance of samples as a function of the frequency of light, one obtains an absorption spectrum of the sample. The spectrum is characteristic of a given material. For direct bandgap semiconductor, the excitonic peak is usually well-defined and blue-shifts with decreasing particle size.

Blue-shift of the absorption spectrum due to quantum size confinement is one of the most prominent features of nanoparticles compared to bulk semiconductors. This has been observed for many semiconductor nanoparticles with CdSe receiving the most attention. As shown in Fig. 5.4, the electronic absorption spectra of CdSe nanoparticles show a significant blue-shift with decreasing size. Many studies have been conducted on high quality samples of CdSe to measure and assign their size-dependent optical spectrum [15, 34–36]. The measurements of the absorption and emission spectra were often carried out at low temperature (~10 K) to reduce inhomogeneous spectral broadening due to thermal effect. These studies have shown that the size dependence of up to ten excited states in the absorption spectra of CdSe nanocrystals can be successfully described by uncoupled multiband effective mass (MBEM) theory that includes valence band degeneracy but not coupling between the conduction and valence bands. The assignment of the states provides a foundation for the discussion of electronic structure of semiconductor nanoparticles [37]. Similar to CdSe, the absorption properties of CdS have also been extensively studied and similar blue-shift in the spectra has been observed [12, 14]. The absorption spectra of CdS nanoparticles are blue-shifted compared to those of CdSe mainly due to the lighter mass and lower electron density of S than Se.

Other metal sulfide nanoparticles such as PbS, Cu_xS and Ag_2S have also been studied with respect to their optical properties. PbS NPs are interesting in that their particle shapes can be readily varied by controlling synthetic conditions [38–42]. Also, since the Bohr radius of PbS is relatively large, 18 nm, and its bulk bandgap is small, 0.41 eV [43], it is easy to prepare particles with size smaller than the Bohr radius that show strong quantum confinement effects and are still absorbed in the visible part of the spectrum. It was observed that the ground state electronic absorption spectrum significantly changes when particle shapes are changed from mostly spherical to needle and cube shaped by changes in the surface capping polymers [39].

Copper sulfides (Cu_xS, $x = 1 - 2$) are interesting due to their ability to form with various stoichiometries. The copper–sulfur system ranges between the chalcocite (Cu_2S) and covellite (CuS) phases with several stable and metastable phases of varying stoichiometry in between. Their complex structures and valence states result in some unique properties [44–54]. In a study of CuS nanoparticles by Drummond *et al.* a near IR band was found to disappear following reduction to Cu_2S via viologen and was therefore attributed to the presence of Cu(II) [44, 45]. The IR band was assigned to a state in the bandgap due to surface oxidation, which lies 1 eV below the conduction band. It was believed that this new middle-gap state was occupied by electrons and thus has electron donor character. However, the model appeared to have some inconsistencies and a recent study has suggested instead that the state has electron acceptor character [55].

Ag_2S is potentially useful for photoimaging and photodetection in the IR region due to its strong absorption in the IR [56]. Ag_2S nanoparticles are usually difficult to synthesize due to their tendency to aggregate into bulk. One approach was to synthesize them in reverse micelles [57]. A new synthetic method for preparing Ag_2S nanoparticles has recently been developed using cysteine (Cys) and glutathione (GSH) as capping molecules [58]. The ground state electronic absorption spectra of the Ag_2S nanoparticles show a simple continuous increase in absorption cross-section towards shorter wavelengths starting from the red (600–800 nm). There is no apparent excitonic feature in the visible and UV regions of the spectrum. The absorption spectrum of Au_2S nanoparticles also shows featureless absorption that increases with decreasing wavelength [59], similar to that of Ag_2S.

Silver halide nanoparticles, e.g. AgI and AgBr, play an important role in photography [60, 61] and their synthesis is relatively simple [62, 63]. The optical absorption spectrum of AgI features a sharp excitonic peak at 416 nm, while the absorption spectrum of AgBr nanoparticles lacks such a excitonic feature and the spectrum is blue-shifted with respect to that of AgI [64].

Other common and important semiconductors include GaN, Si and Ge. GaN nanostructures, including nanoparticles, thin films, quantum wells, nanowires and nanorods, have received considerable attention as a potential blue emitting laser or LED material [65–76]. Bulk GaN is direct bandgap semiconductor with a bandgap energy of 3.4 eV at room temperature. For 3 nm particles, the electronic absorption spectrum features an excitonic band around 330 nm and the emission around 350–550 nm was broad and featureless [77]. Recent effort has shifted towards 1D nanostructures. Figure 5.6 shows TEM images of GaN nanorods synthesized at high temperature [76]. Absorption spectrum is usually hard to measure for 1D nanostructures due to substantial scattering, especially when the length of the nanostructures is on the order of or longer than that of the wavelength of light. In this case, photoluminescence excitation (PLE) spectrum can be obtained in lieu of the absorption spectrum. For example, PL properties of such GaN nanorods will be discussed in the next section.

Silicon NPs have attracted considerable attention recently because of their photoluminescence properties. Since bulk silicon is an indirect bandgap semiconductor with a bandgap of 1.1 eV, it is very weakly luminescent. For opto-electronics applications, it is highly desirable to develop luminescent materials that are compatible with the current existing silicon technology developed and matured for the electronics industry. The weak luminescence of bulk silicon presents a major obstacle to its use for the fast-growing opto-electronics industry. The discovery that porous and

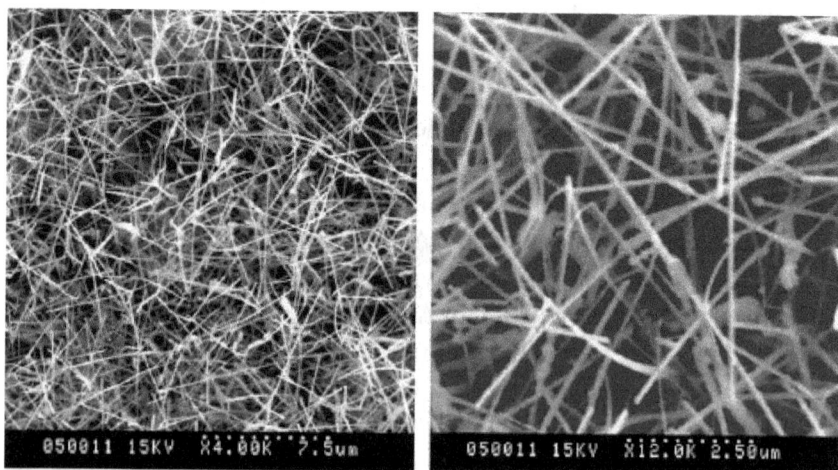

Fig. 5.6. TEM images of GaN nanorods synthesized on Si(111) substrates through annealing sputtered Ga$_2$O$_3$/Nb films under flowing ammonia at 950°C in a quartz tube. Reproduced with permission from Ref. 76.

nanocrystalline Si emit visible light with high quantum yield in 1990 [78] has raised hopes for new photonic devices based on silicon and stimulated strong research interest in porous silicon and Si nanoparticles [25–31, 79, 80]. Various methods have been used to make Si nanoparticles, including slow combustion of silane [29], reduction of $SiCl_4$ by Na [81], separation from porous Si following HF acid electrochemical etching [82–84], microwave discharge [85], laser vaporization/controlled condensation [25], high pressure aerosol reaction [26], decomposition of alkyl silanes [86], laser-induced chemical vapor deposition [28], ball milling [87], chemical vapor deposition [27], and various other solution-based methods [88–91]. Si NPs are usually difficult to make using conventional wet colloidal chemistry techniques. The absorption spectrum of Si nanoparticles shows no excitonic features in the near IR to near UV region and the absorption cross-section increases with decreasing wavelength. An example is shown in Fig. 5.7 [90]. Interestingly, Si nanowires

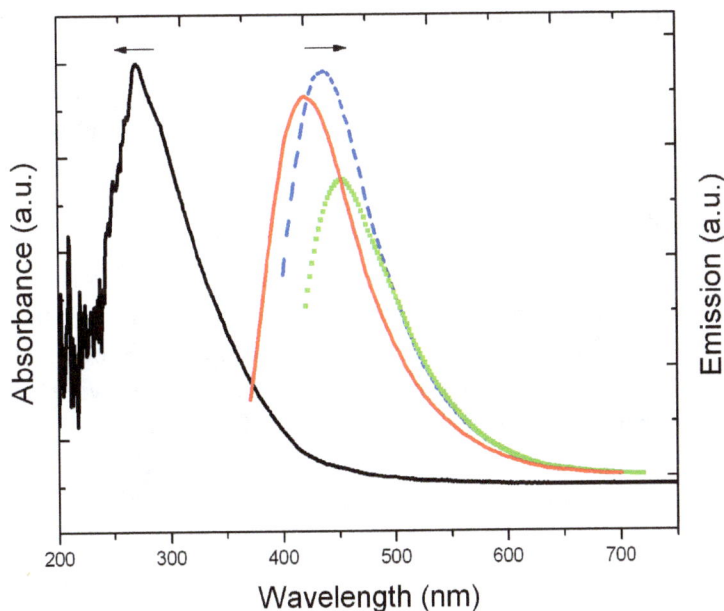

Fig. 5.7. UV-vis electronic absorption spectra (black) and room temperature PL spectra of silicon nanoparticles at different excitation wavelengths: 360 (solid, red); 380 (dashed, blue); and 400 (dotted, green). Reproduced with permission from Ref. 90.

synthesized with a supercritical fluid solution approach was found to show sharp discrete absorption features in the electronic absorption spectrum and relatively strong "bandedge" photoluminescence [92]. These optical properties have been suggested to be due to quantum confinement effects, even though surface effects cannot be ruled out. Si nanowires and nanowire arrays have also been recently fabricated using different techniques [93–99].

Nanoparticles of Ge, an element with similar properties to those of Si, have also been synthesized and studied spectroscopically [100–102]. The absorption spectrum starting in the near IR was dominated by direct bandgap transitions for the largest dots while being dominated by indirect transitions for the smallest dots.

Another interesting class of semiconductors with some unique properties are layered semiconductors such as PbI_2 Bi_2I_3, and MoS_2 [103]. Because of their layered structure, nanoparticles of these materials can be considered as quantum dots composed of quantum wells. Nanoparticles of layered semiconductors can be prepared using techniques similar to those used for other semiconductors. Some can also be made by simply dissolving bulk crystals in suitable solvents. For PbI_2 nanoparticles, there have been some controversies over the nature of the optical absorption spectrum, whether it is from the nanoparticles or from some kind of iodine complexes, and if the three major absorption peaks are due to different sized "magic" numbered particles [103–105]. These two questions have been addressed in detail by Sengupta *et al.* and no evidence was found for "magic" numbered particles of different size correlating with the three absorption peaks and the optical absorption seemed to be dominated by PbI_2 NPs [106]. Figure 5.8 shows some representative UV-vis electronic absorption spectra of PbI_2 nanoparticles and their dependence on aging under light [106]. Using a particle-in-a-rectangular box model, the peak positions and the observed blue-shift of the peaks with simultaneous decrease in particle size upon aging under light have been satisfactorily explained [107]. TEM images and optical studies have led to the proposal that in the photodecomposition process the initially formed large, single-layered particles break down into smaller, multilayered particles, resulting in significant increase in the optical absorption intensity in the visible region and slight blue-shift of the absorption peaks [107].

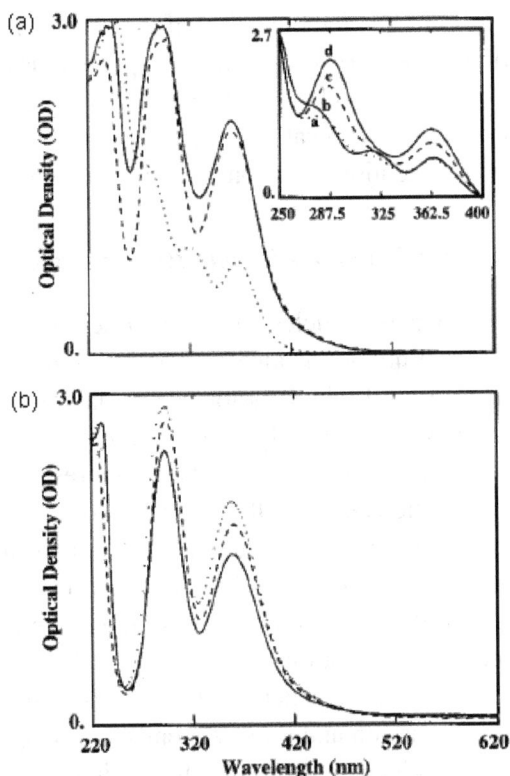

Fig. 5.8. (a) Electronic absorption spectra of PbI₂ nanoparticles in acetonitrile: aged in dark for one week (dotted), aged under light for one week (solid), and aged under light for three weeks (dashed). Inset: (a) fresh; (b) aged in dark for one day; (c) aged under light for one day; (d) aged under light for two days. (b) Electronic absorption spectra of PbI₂ nanoparticles in propanol aged under light for a few days (solid line) and a few weeks (dashed line), and in butanol aged under light for few weeks (dotted line). The samples aged for three weeks under light were diluted by a factor of 3 for measuring the spectra in both (a) and (b). Reproduced with permission from Ref. 106.

Similar studies conducted on two related layered semiconductors, bismuth iodide, BiI_3, and bismuth sulfide, Bi_2S_3, support the model developed for PbI_2 [107, 108]. BiI_3 is promising for nonsilver based and thermally controlled photographic applications [109]. The bandgap of bulk BiI_3 has been reported to be 2.1 eV [110]. Colloidal BiI_3 nanoparticles and BiI_3 clusters in zeolite LTA have been synthesized [111–113]. Similar to

PbI_2, the peak positions and the blue-shift of the peaks with simultaneous decrease in particle size in BiI_3 NPs can be explained using the particle-in-a-rectangular-box model [107]. In contrast to PbI_2 and BiI_3, Bi_2S_3 nanoparticles show no sharp peaks in their absorption spectrum and no evidence of photodegradation based on TEM measurements [108].

5.4.2. *Emission: photoluminescence and Raman scattering*

Light emission from nanoparticles serves as a sensitive probe of their electronic properties. Emission is also the basis of applications in lasers, optical sensors and LEDs. Light emission can be photoinduced or electrically induced, commonly referred to as photoluminescence (PL) and electroluminescence (EL). In PL, the emission is a result of photoexcitation of the material. The energy of the emitted photons is usually lower than the energy of the incident excitation photons and this emission is called Stokes emission. In EL, the light emission is a result of electron-hole recombination following electrical injection of the electron and hole, as to be discussed in more detail in Sec. 5.4.3.

Photoluminescence can be generally divided into bandedge emission, including excitonic emission and trap state emission. Trap state emission is usually red-shifted compared to bandedge emission. For instance, in Fig. 5.9, the emission of PL spectra of CdS nanoparticles clearly shows both trap state and bandedge emission with the latter at a shorter wavelength [114]. Two features are worth noting here. First, the overall PL intensity for the surface passivated sample with less trap states is much higher than the unpassivated sample. Second, the intensity ratio between the bandedge PL, peaked around 450 nm, and trap state PL, peaked around 550 nm, is much higher for the passivated sample than for the unpassivated sample. The ratio between the bandedge and trap state PL is determined by the density and distribution of trap states. A high trap state emission, relative to bandedge PL, indicates a high density of trap states and efficient trapping.

It is possible to prepare high quality samples that have mostly bandedge emission when the surface is well capped. For example, TOPO capped CdSe show mostly bandedge emission and little trap state emission, which is an indication of the high quality of the sample [115–117].

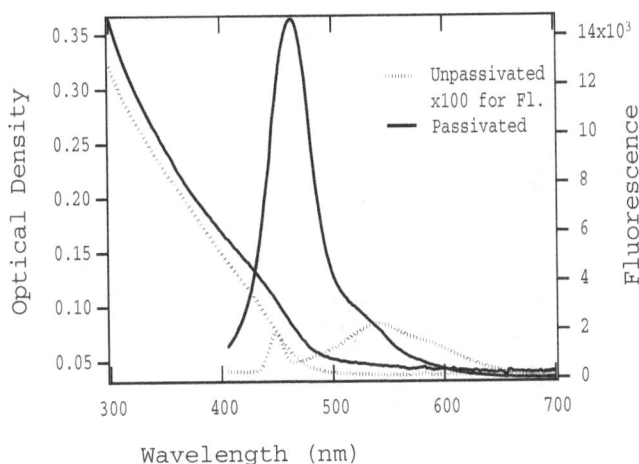

Fig. 5.9. Electronic absorption and fluorescence ($\lambda_{ex} = 390$ nm) spectra of unpassivated (UP) (dotted line) and surface passivated (SP) (solid line) CdS nanoparticles. For comparison, the UP fluorescence signal is multiplied by 100. Reproduced with permission from Ref. 114.

Luminescence can be enhanced by surface modification [32, 118–122] or using core/shell structures [123–126]. Nanoparticles that have been found to show strong photoluminescence include CdSe, CdS, ZnS [127]. Other nanoparticles have generally been found to be weakly luminescent or non-luminescent at room temperature, e.g. PbS [39], PbI$_2$ [106], CuS [55], Ag$_2$S [58]. The weak luminescence can be due to either indirect nature of the semiconductor or a high density of internal and/or surface trap states that quench the luminescence. Controlling the surface by removing surface trap states can lead to significant enhancement of luminescence as well as of the ratio of bandedge over trap state emission [32, 118–122]. The surface modification often involves capping of the particle surface with organic, inorganic or biological molecules or ions that can result in reduction of trap states that fall within the bandgap and quench the luminescence. This scheme of surface states reduction and luminescence enhancement is important for many applications that require high luminescence yield of nanoparticles, e.g. laser, LEDs, fluorescence imaging and optical sensing. PL usually increases with decreasing temperature, from both bandedge and trap states, due to suppression of electron–phonon interaction and thereby lengthened excited electronic state lifetime.

In general, compared to QDs with spatial confinement in 3D, quantum wells with confinement in 1D and nanorods/nanowires with confinement in 2D typically show less dramatic changes in their optical properties due to weaker quantum confinement. Of course, the specific changes depend on the chemical nature of the material and the exact dimensions involved. As an example, Fig. 5.10 shows a PL spectrum of single crystalline GaN nanorods [76]. The main PL peak is clearly due to bandedge emission, with little trap state emission, and quantum confinement effect, reflected in blue-shift in PL peak compared to bulk, is not significant due to the fact that the dimensions of the nanorods (200 nm in average diameter and several microns in length) are much larger than the Bohr exciton radius of 11 nm.

In contrast to Stokes emission for which the emission is red-shifted compared to the excitation wavelength, the energy of emitted photons can be

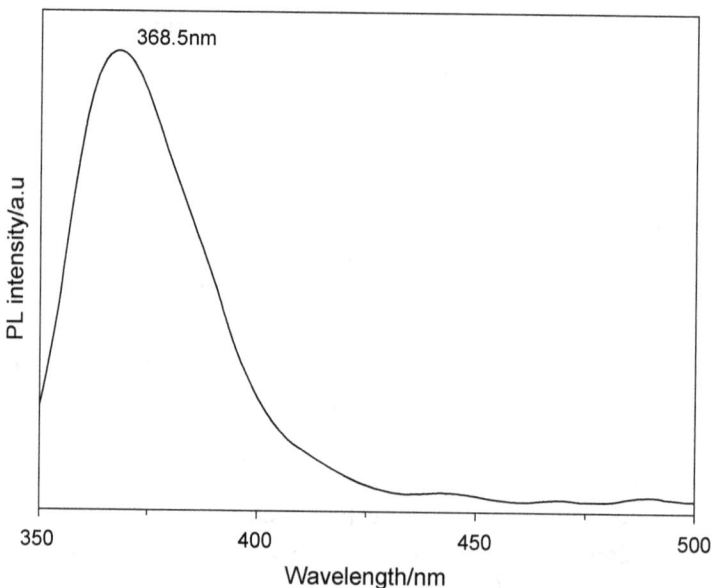

Fig. 5.10. PL spectrum of GaN nanorods ammoniated at 950°C. TEM images of these nanorods are shown in Fig. 5.6. Reproduced with permission from Ref. 76.

higher than that of the incident photons, i.e. blue-shifted, and this is called *anti-Stokes emission* or *up-conversion*, similar to anti-Stokes Raman scattering. This has been observed in a number of nanoparticle systems, including CdS [128], InP [129, 130], CdSe [129], InAs/GaAs [131], Er^{3+}-doped $BaTiO_3$ [132], and Mn-doped ZnS [133]. A more detailed discussion of luminescence up-conversion (LUC) will be given in the next section on "Non-linear Optical Properties".

Another interesting observation in PL studies is photoenhanced luminescence, as observed in several nanoparticles, including CdSe [134–136], porous Si [137], as well as Mn^{2+}-doped ZnS [138, 139]. Several tentative explanations have been provided, including decreased trapping rates due to trap state filling [140], surface transformation [135], change in density of dangling bonds [137] for CdSe, and increasing energy transfer rate from ZnS to Mn^{2+} [141], photooxidation of the surface [142], or photoinduced adsorption of oxygen [138] in the case of ZnS:Mn. All these explanations seem to indicate that the surface plays a critical role in the photoenhancement. However, there still lacks a molecular level model for the observed photoinduced PL enhancement. It should be pointed out that photoinduced reduction of PL has also been reported [134]. The difference between photoenhanced and photoreduced PL could be due to both the mechanisms involved and/or the original PL yield of the samples. If the original PL yield is low, photoexcitation of the sample could cause changes to the surface of the nanoparticles so as to induce an increase in PL intensity. On the contrary, if the initial PL yield is high, photoexcitation could lead to changes to the surface in a way that decreases the PL intensity. Photoexcitation could alter surface characteristics that are sensitive to the surrounding medium and have strong influence on PL [134]. In any case, more studies are needed to better understand the molecular mechanism behind photoenhancement or photoreduction of PL in semiconductor nanomaterials.

A special type of light emission is Raman scattering. Raman spectroscopy is a powerful technique for studying vibrational or phonon modes, electron–phonon coupling, as well as symmetries of excited electronic states of nanoparticles. Raman scattering is discussed here since it is similar to PL in terms of light scattering. Even theoretically, PL and

Raman can be treated using the same framework with the main difference that Raman scattering is a much faster process than PL is dynamically [143]. Raman spectra of NPs have been studied in a number of cases, including CdS [144–149], CdSe [150–152], ZnS [147], PbS [153], InP [154], Si [155–159], and Ge [100, 160–164]. Resonance Raman spectra of nanostructures of GaAs [165], Ge [166], and CdZnSe/ZnSe [167] have also been determined. For CdS nanocrystals, resonance Raman spectrum reveals that the lowest electronic excited state is coupled strongly to the lattice and the coupling decreases with decreasing nanocrystal size [146]. For 4.5 nm nanocrystals of CdSe, the coupling between the lowest electronic excited state and the LO phonons is found to be 20 times weaker than in the bulk solid [150]. For CdZnSe/ZnSe quantum wires, resonance Raman spectroscopy revealed that the ZnSe-like LO phonon position depends on the Cd content as well as excitation wavelength due to relative intensity changes of the peak contributions of the wire edges and of the wire center [167, 168].

Figure 5.11 shows examples of Raman spectra taken from $Cd_{0.2}Zn_{0.8}Se$ nanowires with different wire sizes [167]. The signal in the ZnSe 1LO

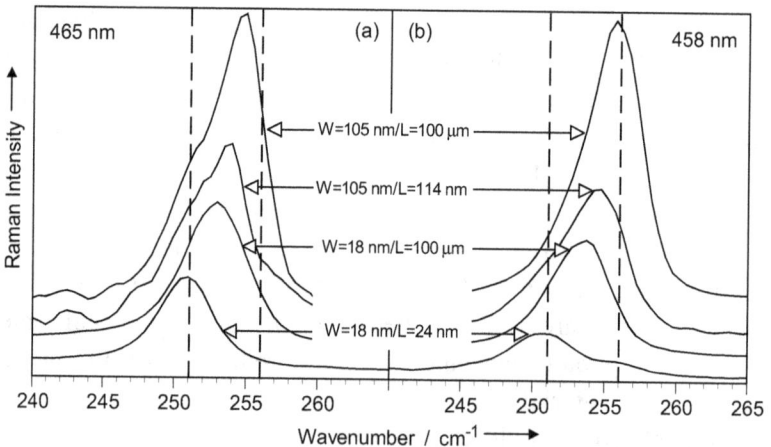

Fig. 5.11. Raman spectra of $Cd_{0.2}Zn_{0.8}Se/ZnSe$ quantum wires with different sizes recorded with excitation wavelengths of (a) 465 and (b) 458 nm. The dashed lines mark the LO phonon positions of the ZnSe barrier and the unstrained CdZnSe nanowire. Reproduced with permission from Ref. 167.

phonon region shows at least two peaks at low temperatures. The high wave number peak at 256 cm^{-1} is assigned to the ZnSe LO from the bottom and top ZnSe layer of the quantum wire and the low wave number peaks are assigned to the ZnSe-like LO phonon of the Cd$_x$Zn$_{1-x}$Se quantum well layer. The spectra are apparently sensitive to the nanowire width and length.

Most optical emission studies have been conducted on isolated particles, i.e. particles with weak or no interparticle interaction. A few studies have been done on nanoparticles assemblies. Both optical absorption and emission properties can be altered for assembled particles relative to isolated particles. Interparticle interaction due to dipole-dipole or electrostatic couplings usually results in spectral line broadening and red-shift [169]. More discussion on optical properties of assembled systems will be given later (Sec. 5.4.4).

5.4.3. *Emission: chemiluminescence and electroluminescence*

Chemiluminescence (CL) refers to luminescence generated from energy released from a chemical reaction, and electrochemiluminescence (ECL) involves luminescence generated from an excited species produced from electron-transfer reactions at electrode surfaces. They are both of interest for potential analysis and other applications due to their sensitivity and selectivity [170]. Nanoparticles have been used in CL in several different ways, including as catalysts, labels, and, more directly, as luminophors. As catalysts, they aid directly the chemical reactions involved in CL [171]. As labels, nanoparticles, such as gold, have been used to bind or support specific biological molecules for chemiluminescent bioassays [172, 173]. As luminophors, bandgap CL is produced due to chemical reactions involving nanoparticles directly, e.g. CdSe/CdS core/shell [174] or CdTe [175]. This resembles PL except that the energy comes from chemical reactions in CL instead of photoexcitation in PL. The bandgap CL and PL spectra are actually very similar, likely due to the same or similar electron-hole recombination mechanism, despite the fact that the electrons and holes are generated differently in PL and CL.

As discussed in Chapter 2, electroluminescence (EL) is based on electrical pumping or excitation for light generation. The basic mechanism is electron-hole recombination following electrical injection of electrons

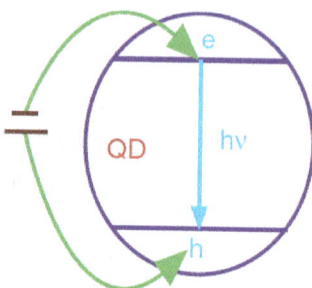

Fig. 5.12.   Schematic illustration of the basic idea behind electronluminescence (EL) based on electrical injection of electron and hole into a semiconductor quantum dot (QD).

and holes into a semiconductor [176–181]. Figure 5.12 illustrates the basic idea behind EL in a semiconductor QD. Applications of EL in devices such as LEDs are discussed in Chapter 10.

5.4.4. *Optical properties of assembled nanostructures: interaction between nanoparticles*

Optical properties of isolated nanostructures is relatively simple in comparison to that of assembled systems in which there is strong interparticle interaction. Assemblies of nanoparticles often exhibit properties modified from those of isolated particles [16, 182, 183]. Most samples studied exist in the form of a collection of a large number of particles with weak or no interaction among them. The optical properties observed are an ensemble average of that of all particles. Since each particle has different size, shape and surface properties, the ensemble spectrum observed is inhomogeneously broadened. It would be ideal to study the properties of a single particle to avoid the problem of inhomogeneity. There have been recent attempts to use single particle spectroscopy to study optical properties of single particles, as to be discussed in Sec. 5.7 later.

Even though there are a large number of particles for a given sample in most cases, the interaction among particles can usually be ignored if they are far apart, e.g. in relatively dilute colloidal solutions. However, when the particles are close in distance such as in films or assembled systems, their interaction becomes important and can be reflected in changes in properties of the particles, e.g. optical absorption and emission. Their spectral features often become broader and red-shifted compared to that of isolated particles, as demonstrated in the case of CdSe nanoparticles [184]. This has been considered to be a result of delocalization and formation of collective electronic states in the close-packed assembled particles. It was also found that the collective states can collapse and primary localization of electronic states can be restored under strong electric fields. So observed large electro-absorption response was thought to be potentially useful for the development of large area electro-optic devices [184].

The attractive potential for the dispersion interaction, $V(D)$, between two spheres of finite volume as a function of distance between them can be explained using the theory derived by Hamaker [185]:

$$V(D) = \frac{-A_H}{12} \left\{ \frac{R_{12}}{D[1 + D/2(R_1 + R_2)]} + \frac{1}{1 + D/R_{12} + D^2/4R_1R_2} \right. $$
$$\left. + 2\ln\left(\frac{D[1 + D/2(R_1 + R_2)]}{R_{12}[1 + D/R_{12} + D^2/4R_1R_2]} \right) \right\}$$

(5.21)

where A_H is Hamaker constant, which is material dependent, R_1 and R_2 are the radii of two particles with the separation between them as D, and R_{12} is the reduced radius of particles 1 and 2. Equation (5.21) indicates very different behavior for two extremes. When $D \gg R_{12}$, $V(D)$ simply becomes the Van der Waals potential (D^{-6} dependence), and when $D \ll R_{12}$, $V(D)$ shows D^{-1} dependence. Another aspect of the theory is that $V(D)$ has to be comparable to kT at room temperature, otherwise the driving force towards ordering becomes negligible, even though the entropy term is still present [186]. If $V(D)$ is much greater than kT, the interparticle attraction is strong and the particles will form nonequlibrium aggregate structures [183].

Assembled structures can be ordered or random in terms of spatial arrangement of the nanoparticles with respective to each other. Aggregates or agglomerates are considered as random structures with no long-range order among nanoparticles. In ordered, assembled nanostructures, there is long-range spatial order among nanoparticles. These are often referred to as *superlattices* in which individual nanoparticles are ordered, similar to atoms in an atomic crystalline solid lattice [187–190].

Potentially, superlattices formed from assembled nanocrystals could have enhanced charge transport properties compared to isolated particles. The modification of electrical conductivity of a solid will depend on various parameters including spatial arrangement of lattice sites with respect to each other, charging energy of the individual lattice site, and coupling between various sites in a unit cell [183]. The properties of nanocrystal superlattices are expected to depend on factors such as size, stoichiometry of nanoparticles, inter-particle separation and symmetry of the nanocrystals. Murray *et al.* did extensive work on developing techniques to produce narrow size distribution of II-VI semiconductor nanoparticles and on understanding inter-QD interactions [115]. They crystallized long range ordered structure of CdSe QDs with extremely narrow size distribution (±3%) [187]. Long range resonance energy transfer (LRRT) between CdSe nanoparticles in the superlattice structure has been observed and Forster theory has been used to determine the distance of LRRT from experimental data [191, 192]. It was shown that the site charging energy, the coupling interaction between sites, which is mostly dipole–dipole in origin, and the lattice symmetry can be controlled in a semiconductor superlattice. However, due to the presence of surface states in semiconductor nanoparticles, the individual particle dipole may not be aligned in a crystallographic order with respect to each other. Therefore, true quantum mechanical wavefunction overlap may not be possible. For this reason, the role of the superlattice symmetry is not clear [183]. The coupling between nanoparticles in a superlattice has important implications in QD lasers, resonant tunneling devices and quantum cellular automata [193, 194]. Close-packed nanoparticle superlattices have also been used to develop various photonics devices, such as light emitting diodes (LEDs) [195, 196] and photovoltaics [197].

Optical properties of superlattice structures can be expected to be different from those of randomly assembled nanostructures. The order in superlattice is likely to result in narrower spectral and DOS distribution, as illustrated in Fig. 5.13.

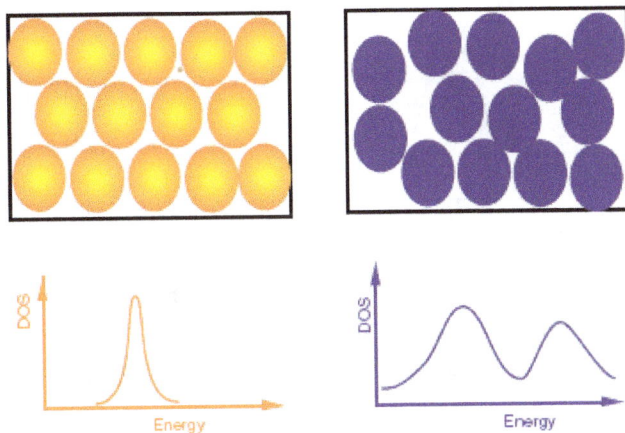

Fig. 5.13. Schematic illustration of a 2D superlattice (top left) and a random structure (top right), as well as their associated DOS as a function of energy (bottom). The superlattice is expected to have narrower dispersion or distribution of DOS.

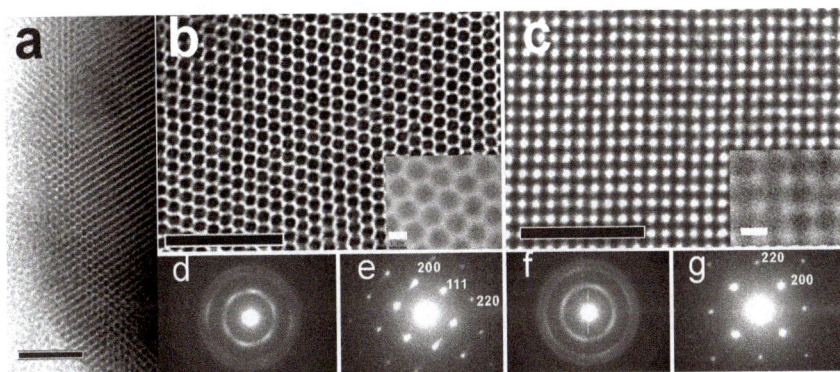

Fig. 5.14. TEM images obtained from CdSe faceted superlattices formed from 5.3 nm NCs: (a) typical image of a crystal edge; (b and c) <110> and <100> projections obtained in the thin areas close to the edges; inserts are respective HRTEM images; segmented rings of the corresponding wide-angle electron diffraction patterns (d and f) show partial alignment of NCs within an fcc superlattice identified by the small-angle electron diffraction (e and g). Bars are 50 nm in (a)–(c), and 5 nm in the inserts. Reproduced with permission from Ref. 198.

This notion is supported by recent experimental studies of CdSe QD superlattices compared to random structures [198]. Figure 5.14 shows TEM images of the superlattice structures, and Fig. 5.15 shows their UV-vis absorption and PL spectra in comparison to QD in solution and in randomly

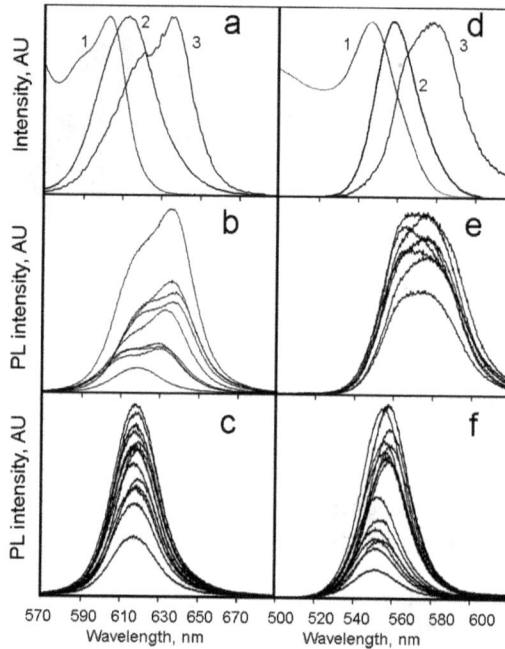

Fig. 5.15. Optical spectra of 5.3 (a–c) and 3.5 nm (d–f) CdSe NC: (a and d) normalized spectra of the initial solutions in nonanoic acid; 1 and 2, absorption and PL spectra of diluted solutions; 3, PL of saturated solutions; (b and e) PL spectra measured at random spots in the solid amorphous layers deposited on a glass slide from saturated solutions; (c and f) PL spectra measured in individual faceted superlattices. Reproduced with permission from Ref. 198.

assembled structures. It is very clear that the PL spectrum is much narrower than that of random structures. These results have important implications in applications of superlattice structures as compared to random structures.

The self-assembly and self-organization of nanomaterials in solution are important in terms of unique chemistry and material applications. However, to date, progress in this area has been limited. Covalent assembly of nanocrystals in solution to build superlattices or heterosupramolecules is another area related to the assembly of nanoparticles to yield well-defined functionalities. The method to make covalently linked nanoparticles has its advantages and disadvantages. On one hand, when they form irre-versible cross-linkages, the superlattices are more stable than those formed based on noncovalent interparticle interaction. This has been tested to produce devices like single electron tunnel junctions and nano-electrodes [183]. On the other hand, long-range order is harder to achieve by covalent linkages.

5.4.5. *Shape dependent optical properties*

Most optical studies of nanoparticles have focused on the effects of size and surface. Another very important factor to consider is the shape dependence of the absorption as well as emission properties. A limited number of studies have been carried out to address the issue of shape dependence on absorption and luminescence properties of semiconductor nanoparticles. The extreme case is nanowires (NW) that can be considered as a limiting case of nanoparticle in the general sense. For instance, the absorption spectra of InAs/InP self-assembled quantum wires show a dependence on polarization of the excitation light as well as polarization anisotropy in the emission spectrum [199]. Interestingly, it was also found that the nonradiative decay mechanism limiting the emission intensity at room temperature is related to the thermal escape of carriers out of the wire. More work needs to done to explore the dependence of optical properties on the shape of semiconductor nanoparticles and this is expected to be an interesting area of research in the future. Due to the extended dimension in one direction for NW or nanorod (NR) that is comparable or even longer than the wavelength of light, scattering becomes more significant than for nanoparticles with all three dimensions much smaller than the wavelength of light.

5.5. Doped semiconductors: absorption and luminescence

An interesting class of luminescent semiconductor nanoparticles is of those doped with transition metal or rare earth metal ions. Doped bulk semiconductor materials play a critical role in various technologies, including the semiconductor industry [200]. Compared to undoped semiconductors, doped materials offer the possibility of using the dopant to tune their electronic, magnetic and optical properties. Therefore, in addition to the existing advantages that nanomaterials offer in terms of controllable parameters such as size, shape and surface, dopants offer the additional flexibility to design new nanomaterials and to alter their properties.

Doped luminescent semiconductor nanoparticles are of strong interest for possible use in opto-electronics such as LEDs and lasers or as novel phosphors because of their interesting magnetic [201–204] and electro-optical properties [205–209]. One example is Mn^{2+}-doped ZnS nanoparticles, commonly denoted as ZnS:Mn that has received considerable attention. Bulk or powdered (micron sized) ZnS:Mn have already been used as phosphors and in electroluminescence [200]. In these materials, a small amount of transition metal ions, such as Mn^{2+}, is incorporated into the nanocrystalline lattice of the ZnS host semiconductor. The host semiconductor usually absorbs light and transfers energy to the dopant metal ion that emits photons with energies characteristic of the metal ion. Luminescence can also result from direct photoexcitation through transitions of the metal ions [133].

Figure 5.16 shows a schematic diagram of energy levels associated with semiconductor nanoparticles, including CB, VB and shallow trap (ST) and deep trap (DT) states, as well as dopant excited state (DE)

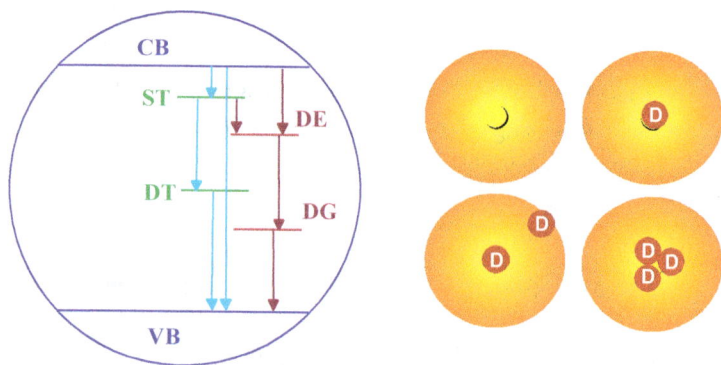

Fig. 5.16. (left) Schematic illustration of energy levels of shallow trap (ST), deep trap (DT), dopant excited state (DE) and dopant ground state (DG) in a doped semiconductor nanoparticle with respect to the band edges of the valence band (VB) and conduction band (CB). (right) Illustration of nanoparticles with different numbers of dopant ions per particles as well as different locations of the dopant ions in the nanoparticles. Reproduced from (Fig. 6) of Ref. 211.

and dopant ground state (DG). The right-hand side of the figure illustrates the issues related to the number and location of the dopant ions in the nanoparticles. In the ideal case, each doped nanoparticle should have the same number of dopant ions in the same location in order for them to have the same or very similar properties, since the properties will likely depend on the number and location of the dopants [210]. This is an important issue particularly for single nanostructure devices. For instance, for a nanoparticle with 100 atoms, a variation from 1 to 2 dopant ions can make a major difference in the property of the nanoparticles. Other possible complications involve aggregation or clustering of the dopants if more than one is present per nanostructure. The dopant ions usually exist in a substitutional site or interstitial site. Substitutional doping of a crystalline host material is often desired. Difference in the location of the dopant will affect its energy levels and spectroscopic signatures. In practice, most samples contain NPs with a distribution in the number and location of the dopants in them. It is very challenging to control the number of dopants per NP and the location of the dopants in each NP. This is an area that clearly needs further research.

Primary interest in doped semiconductor nanomaterials has been for their luminescence properties. For example, the PL and EL of Mn^{2+}, Cu^{2+}, and Er^{3+} doped ZnS nanoparticles as free colloids [208, 209, 212] and in polymer matrices and thin films [207, 213–216] have been extensively studied. As in bulk Mn^{2+}-doped ZnS, the Mn^{2+} ion acts as a luminescence color center, emitting near 585 nm as a result of $^4T_1$ to $^6A_1$ transition [203, 208, 212, 217, 218]. Mn^{2+}-doped ZnS NP was first reported by Becker and Bard in 1983 [138]. It was found that the Mn^{2+} emission at 538 nm with a quantum yield of about 8% was sensitive to chemical species on the particle surface and that the emission can be enhanced with photoirradition in the presence of oxygen, which was attributed to photoinduced adsorption of O_2. In 1994, Bhargava *et al.* made the claim that the luminescence yield is much higher and the emission lifetime is much shorter in ZnS:Mn NPs than in bulk [205]. There were later some debates cast over the issue on if the emission quantum yield is indeed higher in ZnS:Mn NPs relative to bulk. Several subsequent studies [139, 219–223], including a theoretical study [224], made claims of enhancement that seem to support the original claim of enhanced luminescence by Bhargava *et al.* [205]. However, most of these studies failed to provide a calibrated, quantitative measure of the luminescence quantum yield in comparison to bulk ZnS:Mn. Enhancement has been observed mostly with respect to different NPs, instead of a true calibrated measure against the quantum yield of the corresponding bulk. Such measurement is critically needed to establish if there is a true enhancement relative to bulk, which is an issue to be resolved. However, several recent time-resolved studies, to be discussed later (Chapter 9), have found that the emission lifetime in ZnS:Mn NPs is the same as in bulk [225–227]. These lifetime studies seem to suggest that the luminescence yield in NPs should not be higher than that of bulk.

Besides ZnS:Mn, Cu-doped ZnS is another important phosphor material with strong emission in the blue region of the visible spectrum. Extensive studies on bulk and powdered (micron sized) ZnS:Cu have been conducted over the years. Bulk ZnS doped with copper is known to have three PL emission bands: blue, green and red [200, 228]. Polarization experiments showed that the blue and red copper luminescent center had lower symmetry than the host lattice, indicating that they must be associated centers [229, 230]. The green peak was found not to have lower symmetry than the lattice, hence not being spatially associated with the co-activator such as Cl^{-1} [231]. The appearance of the three peaks was dependent on the ratio of activator to co-activator, e.g. Cu^{+1}/Cl^{-1} [232, 233]. In particular, the blue peak was present when the concentration of Cu^+ was greater than the concentration of Cl^-. Recently, interest in nano-sized ZnS:Cu has been on the rise due to the anticipation of potentially improved optical properties [215, 216, 234–238]. There seems to be some inconsistency in previous literature in assigning the oxidation state of copper in ZnS:Cu, +1 versus +2. A very recent study based on combined structural (EXAFS) and optical (PL) studies has found that copper exists primarily as Cu^{+1} in ZnS nanocrystals and is located in the interior but near the surface of the NCs [239]. Figure 5.17 shows a representative PL spectra of undoped and Cu-doped ZnS NCs. The PL peak in undoped ZnS NCs is due to electron-hole recombination from trapped states while the PL for Cu-doped ZnS NCs with high doping level is mainly from shallow electron trap states to Cu^{+1} energy levels. For moderate doping level, the

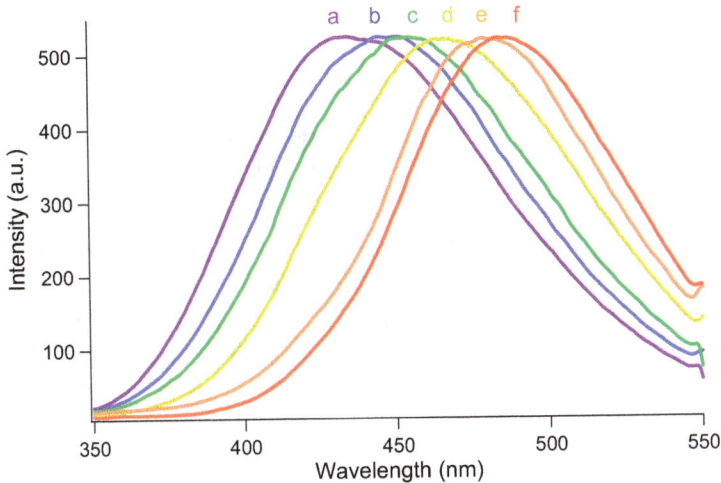

Fig. 5.17. PL spectra of ZnS NCs with different Cu dopant concentrations ($\lambda_{ex} = 280$ nm): (a) 0%, (b) 0.2%, (c) 0.5%, (d) 1%, (e) 2%, (f) 3% (from left to right). Reproduced with permission from Ref. 239.

PL spectrum is clearly composed of both emission bands, one from ZnS and another associated with the Cu^{+1} dopant.

5.6. Nonlinear optical properties

5.6.1. *Absorption saturation and harmonic generation*

Nanoparticles have interesting nonlinear optical properties at high excitation intensities, including absorption saturation, shift of transient bleach, second and third harmonic generation, and up-conversion luminescence. The most commonly observed nonlinear effect in semiconductor nanoparticles is absorption saturation and transient bleach shift at high intensities [21, 32, 119, 240–247]. Similar nonlinear absorption have been observed for quantum wires of GaAs [248, 249] and porous Si [250, 251]. These nonlinear optical properties have been considered potentially useful for optical limiting and switching applications [252]. Further discussion of nonlinear optical properties will be given in Sec. 9.3.7 in relation to charge carrier dynamics.

Another nonlinear optical phenomenon is harmonic generation, mostly based on the third-order nonlinear optical properties of semiconductor nanoparticles [42, 253–255]. The third order nonlinearity is also responsible for phenomena such as the Kerr effect and degenerate four wave mixing (DFWM) [256]. For instance, the third-order nonlinear susceptibility, $\chi^{(3)}$, for PbS nanoparticles (~5.6 × 10^{-12} esu) has been determined using time-resolved optical Kerr effect spectroscopy and it was found to be dependent on surface modification [42]. Third-order nonlinearity of porous silicon has been measured with the Z-scan technique and found to be significantly enhanced over crystalline silicon [252]. DFWM studies of thin films containing CdS nanoparticles found a large $\chi^{(3)}$ value, ~10^{-7} esu, around the excitonic resonance at room temperature [257].

Only a few studies have been carried out on second-order nonlinear optical properties since it is usually believed that the centrosymmetry or near centrosymmetry of the spherical nanoparticles reduces their first-order hyperpolarizability (β) to zero or near zero. Using hyper-Rayleigh scattering, second harmonic generation in CdSe nanocrystals has been observed [258]. The first hyperpolarizibility β per nanocrystal was found to be dependent on particle size, decreasing with size down to about 1.3 nm in radius and then increasing with further size reduction. The results are explained in terms of surface and bulk-like contributions. A similar technique has been used for CdS nanoparticles for which the β-value per particle (4 nm mean diameter) was found to be on the order of 10^{-27} esu, which is quite high for solution samples [259]. Second harmonic generation has also been observed for magnetic cobalt ferrite ($CoFe_2O_4$) colloidal particles when oriented with a magnetic field [260]. The nonlinear optical properties of nanoparticles are found to be strongly influenced by the surface.

As discussed earlier, the optical properties of isolated nanoparticles can be very different from those of assembled nanoparticle films. This is true for both linear and nonlinear optical properties. Theoretical calculations on nonlinear optical properties of nanoparticle superlattice solids have shown that an ideal resonant state for a nonlinear optical process is the one that has large volume and narrow line width [261–263]. The calculations also showed that nonlinear optical responses could be enhanced greatly with a decrease in interparticle separation distance.

5.6.2. *Luminescence up-conversion*

Anti-Stokes photoluminescence or photoluminescence up-conversion is another interesting nonlinear optical phenomenon. In contrast to Stokes emission that is red-shifted with respect to the excitation wavelength, the photon energy of the luminescence output is higher than the excitation photon energy or is blue-shifted. This effect has been previously reported for both doped [264, 265] and high purity bulk semiconductors [266, 267]. For bulk semiconductors, the energy up-conversion is usually achieved by (i) an Auger recombination process, (ii) anti-Stokes Raman scattering mediated by thermally populated phonons, or (iii) two-photon absorption [129, 268].

Luminescence up-conversion (LUC) has recently been observed in semiconductor heterojunctions and quantum wells [268–284] and has been explained based on either Auger recombination [272, 277, 285] or two-photon absorption [278]. Long-lived intermediate states have been suggested as essential for LUC in some heterostructures such as GaAs/Al$_x$Ga$_{1-x}$As [277]. For semiconductor nanoparticles or quantum dots with confinement in three dimensions, LUC has been reported for CdS [128], InP [129, 130], CdSe [129], InAs/GaAs [131], ZnS:Mn [286, 287], and CdSe/ZnS [288, 289]. Surface states have been proposed to play an important role in the up-conversion in nanoparticles such as InP and CdSe [129]. LUC is so efficient that it can be observed in single QDs such as InAs/GaAs [290, 291]. Figure 5.18 shows normal PL (a) and LUC (b) of single InAs/GaAs QDs [290]. For the LUC, the excitation energy is 1.345 eV and the PL is at 1.42 eV.

In the case of ZnS:Mn nanoparticles, when 767 nm excitation was used, Mn^{2+} emission near 620 nm was observed with intensity increasing almost quadratically with excitation intensity [133]. Comparison with 383.5 nm excitation showed similar luminescence spectrum and decay kinetics, indicating that the up-converted luminescence with 767 nm excitation is due to a two-photon excitation process. Some intriguing temperature dependence of the up-converted luminescence was observed for ZnS:Mn [287]. It was found that LUC of ZnS:Mn NPs first decreases and then increases with increasing temperature. This is in contrast to bulk ZnS:Mn in which the luminescence intensity decreases monotonically

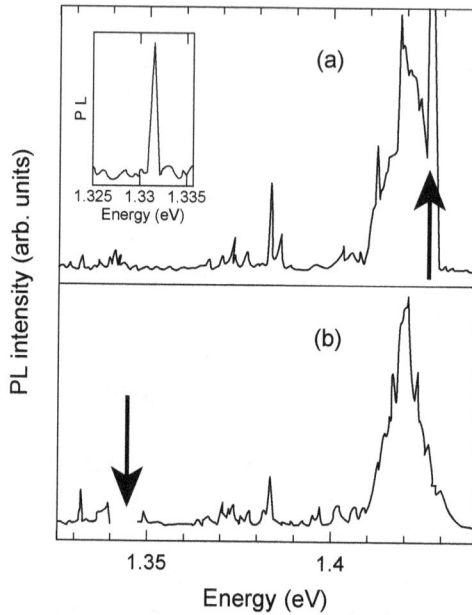

Fig. 5.18. Micro PL spectra of single InAs/GaAs QDs recorded at 10 K under CW excitation at two different energies (indicated by the arrows). (a) Excitation at 1.428 eV with $P_{ex} \sim 300$ W/cm². (b) Excitation at 1.345 eV with $P_{ex} \sim 10$ kW/cm². The inset is an expanded view of the PL spectrum around 1.33 eV where a single QD line is observed. The PL band with a peak at 1.42 eV when excited at 1.345 eV is an up-converted signal. Reproduced with permission from Ref. 290.

with increasing temperature as a result of increased electron–phonon interaction. The increase in luminescence intensity with increasing temperature for nanoparticles was attributed tentatively to the involvement of surface trap states. With increasing temperature, surface trap states can be thermally activated, resulting in increased energy transfer to the excited state of Mn^{2+} and thereby increased luminescence. This factor apparently is significant enough to overcome the increased electron–phonon coupling with increasing temperature that usually results in decreased luminescence [287].

5.7. Optical properties of single particles

As mentioned earlier, most spectroscopy studies of nanoparticles have been carried out on ensembles of a large number of particles. The properties measured are thus ensemble averages of the properties of individual particles. Due to heterogeneous distributions in size, shape, environment and surface properties, the spectrum measured is thus inhomogeneously broadened. This results in loss of spectral information [292]. For instance, it has been predicted by theory that nanocrystallites should have a spectrum of discrete, atom-like energy states [7, 8]. However, transition line widths observed experimentally appear significantly broader than expected, even though the discrete nature of the excited states has been verified [36, 293]. This is true even when size-selective optical techniques are used to extract homogeneous line widths [36, 292–297].

One way to solve the above problem is to make particles with truly uniform size, surface, environment and shape. However, this is almost impossible or at least very difficult. Another, perhaps simpler, approach to remove the heterogeneity is to conduct the measurement on one single particle at a time. This approach is similar to that used in the field of single-molecular spectroscopy [298, 299]. A number of single nanoparticle studies have been reported on semiconductor nanoparticles, including CdS [300] and CdSe [301–307]. Compared to ensemble averaged samples, single particle spectroscopy studies of CdSe nanoparticles revealed several new features, including fluorescence blinking, ultranarrow transition line width, a range of phonon couplings between individual particles, and spectral diffusion of the emission spectrum over a wide range of energies [307, 308]. Furthermore, electrical field studies showed both polar and polarizable character of the emitting state. The polar component has been attributed to an induced excited state dipole resulting from the presence of local electrical fields. The fields could be the result of trapped charge carriers on the particle surface. These fields seemed to change over time and result in spectral diffusion [306, 308]. It was noted that the spectral line widths are strongly dependent on experimental conditions such as integration time and excitation intensity. Therefore, it has been concluded that the line widths are primarily the result of spectral shifting and not the intrinsic properties of the nanoparticle [306].

The emission spectra of a single 5.65 nm overcoated CdSe NC with different integration time exhibit two interesting features. First, the line shape of a single particle contains information about changes in the surrounding environment and not the intrinsic physics of the NC. Second, changes in line shape of a single NC spectrum resulting from different experimental conditions are likely to be caused by changes in spectral diffusion. The line width was found to become broader with increasing excitation intensity [308].

With low-temperature (20 K) confocal microscopy, fluorescence spectra of isolated single CdS nanoparticles have been recorded [300]. The narrowest measured line width of the main fluorescence band has been determined to be about 5 meV at the lowest excitation intensities. At higher intensities, the main fluorescence band broadened and two new peaks appeared on each side of the main band, one of which was attributed to the coupling to a longitudinal optical phonon. In a single CdS/HgS/CdS quantum well quantum dots structure, the spectral and intensities fluctuations were strongly reduced, due to charge carrier localization in the HgS region of the nanocrystal [309]. Similarly, for CdSe nanoparticles, the elimination of spectral inhomogeneities reveals resolution limited spectral line widths of <120 meV at 10K, more than 50 times narrower than expected from ensemble measurements [302].

One interesting observation in measuring the emission of a single nanoparticle is an intermittent on–off behavior in emission intensity under CW light excitation, first discovered by Nirmal *et al.* (Fig. 5.19) [303]. The intermittency was analyzed in terms of random telegraph nose and the on–off times were extracted. The off times were found to be power independent while the on times, over a narrow intensity range, revealed a linear dependence on power. The intermittency observed was attributed to an Auger photoionization mechanism that leads to ejection of one charge (electron or hole) outside the particle. A "dart exciton" state was assigned to such ionized nanocrystal in which the emission is quenched because of the excess charge. The feasibility of this mechanism was further demonstrated in theoretical calculations which also considered the possibility of thermal ionization [310]. Similar studies on single CdSe/CdS core/shell nanocrystals as a function of temperature and excitation intensity and the observations are consistent with a darkening mechanism that is a combination of Auger photoionization and thermal trapping of charge [311].

Interestingly, single particle spectroscopy also reveals nonlinear optical properties of single nanoparticles. For example, in the low temperature

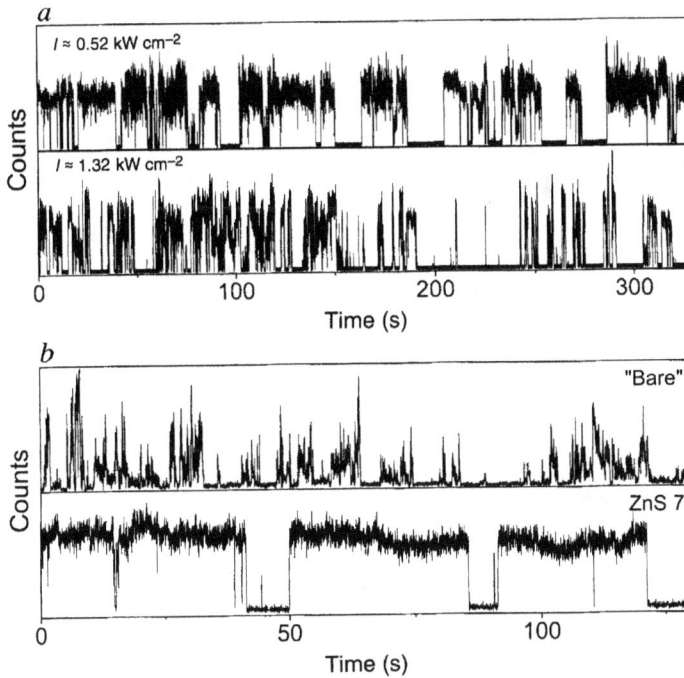

Fig. 5.19. (a) Comparison of fluorescence-intensity versus time traces at excitation intensities (I) of ~ 52 and ~1.32 kW cm^{-2} with a sampling interval of 10 ms. The average on/off times at the two excitation intensities are: ($I \approx 0.52$ kWcm^{-2}; $<\tau_{on}> \sim 0.97$ s, $<\tau_{off}> \sim 0.44$s: $I \approx 1.32$ kWcm^{-2}; $<\tau_{on}> \sim 0.32$ s, $<\tau_{off}> \sim 0.43$ s). Intensity-dependent studies were carried out on CdSe nanocrystals with ~4 monolayers of ZnS on their surface. (b) Fluorescence-intensity versus time trace of a "bare" nanocrystal compared with that of an overcoated one with a shell thickness of ~7 monolayers of ZnS (ZnS 7) at the same excitation intensity ($I \approx 0.70$ kW cm^{-2}) and a sampling interval of 20 ms. Reproduced with permission from (Fig. 2) of Ref. 303.

near-field absorption spectroscopy study of InGaAs single quantum dots, the absorption change was found to depend nonlinearly on the excitation intensity [312]. This nonlinearity was suggested to originate from state filling of the ground state.

Single nanoparticle spectroscopy has been applied recently to study single CdSe nanorods [313]. It has been found that the on and off blinking statistics in single CdSe nanorods are influenced by the excitation wavelength and embedding environment (Fig. 5.20). The results are explained

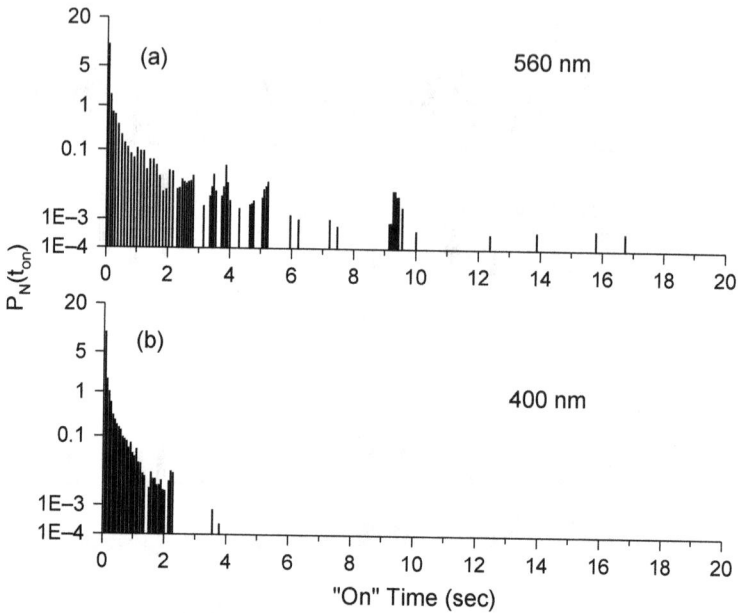

Fig. 5.20. Wavelength-dependent "on"-time probability distribution for CdSe/TOPO nanorods excited with (a) 560 nm and (b) 400 nm. The probabilities are plotted on a log-arithmic scale as the dependent term. The "on"-time is plotted as the independent term on a linear scale. These plots highlight the wavelength-dependent effect, showing the decreased probability of continuous fluorescence emission beyond a few seconds when the nanoparticle is excited with the more energetic wavelength. Reproduced with permission from Ref. 313.

by a threshold to access nonemissive trap states, attributed to self-trapping of hot charge carriers at the higher-photon excitation energies.

  Single nanoparticle spectroscopy should also be useful for studying doped nanoparticles, especially for potentially probing the dopant level of single doped nanoparticles. However, to date, such study has been very limited with one example reported on Te-doped CdSe nanoparticles passivated by ZnS [314]. While the dopant did not alter the host CdSe emission noticeably, the dopant emission showed blue-shift with repeated measurements, which was attributed to instability of the dopant in the nanocrystal upon photoexcitation. As nanostructure features become smaller and smaller, the importance of doping becomes

more and more significant. It thus becomes more important to understand the effect of doping on the host nanoparticle, as well as the
location and local environment of the dopant. Single nanoparticle spectroscopy is potentially powerful, in conjunction with other experimental techniques with structural sensitivity, for probing the properties
of doped nanoparticles.

Time-resolved studies of single nanoparticles or quantum dots have
also been carried out to probe the dynamic properties of single particles.
For example, time-resolved emission from self-assembled InGaAs/GaAs
quantum dots have been measured at low temperature by the time-corre-
lated single photon counting method using near-field microscopy [315].
The decay time of the emission from discrete levels of a single dot was
found to increase with the decrease in emission energy and with the
increase in excitation intensity. The results are explained with a model that
includes initial filling of the states, cascade relaxation, state filling and
carrier feeding from a wetting layer. Room temperature photoluminescence was also acquired and the homogeneous line widths were found to
be around 9.8–14.5 meV [316]. A more detailed coverage of charge carrier dynamics studied using time-resolved laser techniques will be given
in Chapter 9.

5.8. Summary

Semiconductor nanomaterials represent a significant part of nanomaterials and have optical properties that are both fundamentally fascinating
and technologically important. This chapter has highlighted some of
the interesting optical properties of semiconductor nanomaterials with
emphasis on their absorption and emission properties as well as some
nonlinear optical characteristics. Single nanoparticle spectroscopy was
also covered to show its usefulness for probing properties of individual
nanostructures. Applications of optical properties of semiconductor
nanomaterials will be given in Chapter 10. The next chapter will focus
on optical properties of metal oxides, many of which are also semiconductors. Given the unique properties and importance of metal oxides,
we have decided to discuss them in a separate chapter.

References

1. N.W. Ashcroft and N.D. Mermin, *Solid State Physics*, Philadelphia: Saunders College. 826 (1976).
2. J.X. Liu, J.J. Niu, D.R. Yang, M. Yan and H. Sha, *Physica E* **23**, 221 (2004).
3. J. Poole, C.P. and F.J. Owens, *Introduction to Nanotechnology*, Hoboken: John Wiley & Sons. 388 (2003).
4. J.I. Pankove, *Optical Properties in Semiconductors*, New York: Dover Publications, Inc. 422 (1971).
5. M. Gratzel, *Heterogeneous Photochemical Electron Transfer*, Boca Raton: CRC Press, 176 (1989).
6. J. Bardeen, F.J. Blatt and L.H. Hall, in *Atlantic City Photoconductivity Conference.* 1954 (pub 1956). Atlantic City: J. Wiley and Chapman and Hall.
7. A.L. Efros and A.L. Efros, *Fizika i Tekhnika Poluprovodnikov* **16**, 1209 (1982).
8. L.E. Brus, *J. Chem. Phys.* **80**, 4403 (1984).
9. A.I. Ekimov, A.L. Efros, M.G. Ivanov, A.A. Onushchenko and S.K. Shumilov, *Solid State Commun.* **56**, 921 (1985).
10. L. Brus, *J. Phys. Chem.* **90**, 2555 (1986).
11. M.G. Bawendi, W.L. Wilson, L. Rothberg, P.J. Carroll, T.M. Jedju, M.L. Steigerwald and L.E. Brus, *Phys. Rev. Lett.* **65**, 1623 (1990).
12. V.L. Colvin, A.N. Goldstein and A.P. Alivisatos, *J. Am. Chem. Soc.* **114**, 5221 (1992).
13. D. Duonghong, J.J. Ramsden and M. Gratzel, *J. Amer. Chem. Soc.* **104**, 2977 (1982).
14. J.Z. Zhang, R.H. O'Neil and T.W. Roberti, *J. Phys. Chem.* **98**, 3859 (1994).
15. C.B. Murray, C.R. Kagan and M.G. Bawendi, *Ann. Rev. Mater. Sci.* **30**, 545 (2000).
16. A.P. Alivisatos, *J. Phys. Chem.* **100**, 13226 (1996).
17. W.A. de Heer, *Rev. Mod. Phys.* **65**, 611 (1993).
18. S.W. Chen, R.S. Ingram, M.J. Hostetler, J.J. Pietron, R.W. Murray, T.G. Schaaff, J.T. Khoury, M.M. Alvarez and R.L. Whetten, *Science* **280**, 2098 (1998).
19. L. Jacak, A. Wójs and P. Hawrylak, *Quantum Dots*, Berlin; New York: Springer. 176 (1998).
20. T. Takagahara, *Phys. Rev. B* **36**, 9293 (1987).
21. T. Vossmeyer, L. Katsikas, M. Giersig, I.G. Popovic, K. Diesner, A. Chemseddine, A. Eychmuller and H. Weller, *J. Phys. Chem.* **98**, 7665 (1994).

22. T. Rajh, O.I. Micic and A.J. Nozik, *J. Phys. Chem.* **97**, 11999 (1993).
23. Y. Kayanuma, *Phys. Rev. B* **38**, 9797 (1988).
24. P.R. Yu, M.C. Beard, R.J. Ellingson, S. Ferrere, C. Curtis, J. Drexler, F. Luiszer and A.J. Nozik, *J. Phys. Chem. B* **109**, 7084 (2005).
25. S.T. Li, S.J. Silvers and M.S. ElShall, *J. Phys. Chem. B* **101**, 1794 (1997).
26. K.A. Littau, P.J. Szajowski, A.J. Muller, A.R. Kortan and L.E. Brus, *J. Phys. Chem.* **97**, 1224 (1993).
27. W.Q. Cao and A.J. Hunt, *Appl. Phys. Lett.* **64**, 2376 (1994).
28. W.X. Wang, S.H. Liu, Y. Zhang, Y.B. Mei and K.X. Chen, *Physica B* **225**, 137 (1996).
29. A. Fojtik and A. Henglein, *Chem. Phys. Lett.* **221**, 363 (1994).
30. L. Brus, *J. Phys. Chem.* **98**, 3575 (1994).
31. L.B. Zhang, J.L. Coffer, W. Xu and T.W. Zerda, *Chem. Mater.* **9**, 2249 (1997).
32. T.W. Roberti, N.J. Cherepy and J.Z. Zhang, *J. Chem. Phys.* **108**, 2143 (1998).
33. J.Z. Zhang, *J. Phys. Chem. B* **104**, 7239 (2000).
34. A.I. Ekimov, F. Hache, M.C. Schanneklein, D. Ricard, C. Flytzanis, I.A. Kudryavtsev, T.V. Yazeva, A.V. Rodina and A.L. Efros, *J. Opt. Soc. Am. B* **10**, 100 (1993).
35. D.J. Norris, A. Sacra, C.B. Murray and M.G. Bawendi, *Phys. Rev. Lett.* **72**, 2612 (1994).
36. D.J. Norris and M.G. Bawendi, *Phys. Rev. B-Condensed Matter* **53**, 16338 (1996).
37. L.E. Brus, A.L. Efros and T. Itoh, *J. Luminescence*, **70**, R7 (1996).
38. M.T. Nenadovic, M.I. Comor, V. Vasic and O.I. Micic, *J. Phys. Chem.* **94**, 6390 (1990).
39. A.A. Patel, F.X. Wu, J.Z. Zhang, C.L. Torres-Martinez, R.K. Mehra, Y. Yang and S.H. Risbud, *J. Phys. Chem. B* **104**, 11598 (2000).
40. M.Y. Gao, Y. Yang, B. Yang, J.C. Shen and X.C. Ai, *J. Chem. Soc. Faraday T* **91**, 4121 (1995).
41. T. Schneider, M. Haase, A. Kornowski, S. Naused, H. Weller, S. Forster and M. Antonietti, *Berichte Der Bunsen-Gesellschaft-Physical Chemistry Chemical Physics* **101**, 1654 (1997).
42. X.C. Ai, L. Guo, Y.H. Zou, Q.S. Li and H.S. Zhu, *Mater. Lett.* **38**, 131 (1999).
43. J.L. Machol, F.W. Wise, R.C. Patel and D.B. Tanner, *Phys. Rev. B-Condensed Matter* **48**, 2819 (1993).
44. E.J. Silvester, F. Grieser, B.A. Sexton and T.W. Healy, *Langmuir* **7**, 2917 (1991).

45. K.M. Drummond, F. Grieser, T.W. Healy, E.J. Silvester and M. Giersig, *Langmuir* **15**, 6637 (1999).
46. H.J. Gotsis, A.C. Barnes and P. Strange, *J. Phys. Condens. Matter* **4**, 10461 (1992).
47. H. Grijalva, M. Inoue, S. Boggavarapu and P. Calvert, *J. Mater. Chem.* **6**, 1157 (1996).
48. H. Nozaki, K. Shibata and N. Ohhashi, *J. Solid State Chem.* **91**, 306 (1991).
49. S. Saito, H. Kishi, K. Nie, H. Nakamaru, F. Wagatsuma and T. Shinohara, *Phys. Rev. B-Condensed Matter* **55**, 14527 (1997).
50. C. Sugiura, H. Yamasaki and T. Shoji, *J. Phys. Soc. Japan* **63**, 1172 (1994).
51. K.V. Yumashev, P.V. Prokoshin, A.M. Malyarevich, V.P. Mikhailov, M.V. Artemyev and V.S. Gurin, *Appl. Phys. B-Lasers O* **64**, 73 (1997).
52. V. Klimov, P.H. Bolivar, H. Kurz, V. Karavanskii, V. Krasovskii and Y. Korkishko, *Appl. Phys. Lett.* **67**, 653 (1995).
53. I. Grozdanov and M. Najdoski, *J. Solid State Chem.* **114**, 469 (1995).
54. M.V. Artemyev, V.S. Gurin, K.V. Yumashev, P.V. Prokoshin and A.M. Maljarevich, *J. Appl. Phys.* **80**, 7028 (1996).
55. M.C. Brelle, C.L. Torres-Martinez, J.C. McNulty, R.K. Mehra and J.Z. Zhang, *Pure and Applied Chemistry* **72**, 101 (2000).
56. S. Kitova, J. Eneva, A. Panov and H. Haefke, *J. Imag. Sci. Technol.* **38**, 484 (1994).
57. L. Motte, F. Billoudet and M.P. Pileni, *J. Mater. Sci.* **31**, 38 (1996).
58. M.C. Brelle, J.Z. Zhang, L. Nguyen and R.K. Mehra, *J. Phys. Chem. A* **103**, 10194 (1999).
59. C.D. Grant, J. Norman, T.J., T. Morris, G. Szulczewski and J.Z. Zhang, *SPIE Proc.* **4807**, 216 (2002).
60. B.H. Carroll, G.C. Higgins and T.H. James, *Introduction to Photographic Theory: the Silver Halide Process*, New York: J. Wiley. 355 (1980).
61. J.F. Hamilton, *Adv. Phys.* **37**, 359 (1988).
62. H. Saijo, M. Iwasaki, T. Tanaka and T. Matsubara, in *Photographic Science and Engineering.* 1982.
63. T. Tanaka, H. Saijo and T. Matsubara, *J. Photographic Sci.* **27**, 60 (1979).
64. M.C. Brelle and J.Z. Zhang, *J. Chem. Phys.* **108**, 3119 (1998).
65. J.I. Pankove and T.D. Moustakas, *Gallium Nitride (GaN).* Semiconductors and Semimetals; v. 50, 57, San Diego, CA: Academic Press. 2 v. (1998).
66. S.M. Zhou, Y.S. Feng and L.D. Zhang, *Physics of Low-Dimensional Structures* **1–2**, 87 (2003).
67. Y.D. Wang, K.Y. Zang, S.J. Chua, S. Tripathy, H.L. Zhou and C.G. Fonstad, *Appl. Phys. Lett.* **88**, 211908 (2006).

68. K.Y. Zang, Y.D. Wang, H.F. Liu and S.J. Chua, *Appl. Phys. Lett.* **89**, 171921 (2006).
69. L.H. Robins, K.A. Bertness, J.M. Barker, N.A. Sanford and J.B. Schlager, *J. Appl. Phys.* **101**, 113506 (2007).
70. X.T. Zhang, Z.A. Liu, C.C. Wong and S.K. Hark, *Solid State Commun.* **139**, 387 (2006).
71. F.K. Yam, Z. Hassan, L.S. Chuah and Y.P. Ali, *Appl. Surf. Sci.* **253**, 7429 (2007).
72. S.V. Bhat, K. Biswas and C.N.R. Rao, *Solid State Commun.* **141**, 325 (2007).
73. P.W. Wang, X.J. Zhang, B.Q. Wang, X.Z. Zhang and D.P. Yu, *Chinese Phys. Lett.* **25**, 3040 (2008).
74. Z.X. Zhang, X.J. Pan, T. Wang, L. Jia, L.X. Liu, W.B. Wang and E.Q. Xie, *J. Electron. Mater.* **37**, 1049 (2008).
75. Q.T. Zhou, Y.Q. Chen, Y. Su, C. Jia, B. Peng, S. Yin, S. Li and W.H. Kong, *Mater. Res. Bull.* **43**, 2207 (2008).
76. H.Z. Zhuang, B.L. Li, C.S. Xue, S.Y. Zhang, D.X. Wang and J.B. Shen, *Vacuum* **82**, 1224 (2008).
77. O.I. Micic, S.P. Ahrenkiel, D. Bertram and A.J. Nozik, *Appl. Phys. Lett.* **75**, 478 (1999).
78. L.T. Canham, *Appl. Phys. Lett.* **57**, 1046 (1990).
79. Y.H. Tang, X.H. Sun, F.C.K. Au, L.S. Liao, H.Y. Peng, C.S. Lee, S.T. Lee and T.K. Sham, *Appl. Phys. Lett.* **79**, 1673 (2001).
80. Q.S. Li, R.Q. Zhang, S.T. Lee, T.A. Niehaus and T. Frauenheim, *Appl. Phys. Lett.* **91** (2007).
81. J.R. Heath, *Science* **258**, 1131 (1992).
82. J.L. Heinrich, C.L. Curtis, G.M. Credo, K.L. Kavanagh and M.J. Sailor, *Science* **255**, 66 (1992).
83. R.A. Bley, S.M. Kauzlarich, J.E. Davis and H.W.H. Lee, *Chem. Mater.* **8**, 1881 (1996).
84. L.B. Zhang, J.L. Coffer and T.W. Zerda, *J. Sol-Gel. Sci. Techn.* **11**, 267 (1998).
85. H. Takagi, H. Ogawa, Y. Yamazaki, A. Ishizaki and T. Nakagiri, *Appl. Phys. Lett.* **56**, 2379 (1990).
86. N. Zaitseva, S. Hamel, Z.R. Dai, C. Saw, A. Williamson and G. Galli, *J. Phys. Chem. C* **112**, 3585 (2008).
87. C. Lam, Y.F. Zhang, Y.H. Tang, C.S. Lee, I. Bello and S.T. Lee, *J. Cryst. Growth* **220**, 466 (2000).
88. R.K. Baldwin, K.A. Pettigrew, J.C. Garno, P.P. Power, G.Y. Liu and S.M. Kauzlarich, *J. Am. Chem. Soc.* **124**, 1150 (2002).

89. J. Zou, P. Sanelle, K.A. Pettigrew and S.M. Kauzlarich, *J. Clust. Sci.* **17**, 565 (2006).
90. X.M. Zhang, D. Neiner, S.Z. Wang, A.Y. Louie and S.M. Kauzlarich, *Nanotechnology* **18**, 095601 (2007).
91. J. Zou and S.M. Kauzlarich, *J. Clust. Sci.* **19**, 341 (2008).
92. J.D. Holmes, K.P. Johnston, R.C. Doty and B.A. Korgel, *Science* **287**, 1471 (2000).
93. Y. Cui and C.M. Lieber, *Science* **291**, 851 (2001).
94. Y. Wu, Y. Cui, L. Huynh, C.J. Barrelet, D.C. Bell and C.M. Lieber, *Nano Lett.* **4**, 433 (2004).
95. A. Mao, H.T. Ng, P. Nguyen, M. McNeil and M. Meyyappan, *J. Nanosci. Nanotechnol.* **5**, 831 (2005).
96. Z.H. Zhong, Y. Fang, C. Yang, W. Lu and C.M. Lieber, *Abstr. Pap. Am. Chem. S* **229**, U703 (2005).
97. Y.W. Wang, V. Schmidt, S. Senz and U. Gosele, *Nature Nanotechnology* **1**, 186 (2006).
98. C.Y. Meng, B.L. Shih and S.C. Lee, *J. Nanoparticle Res.* **9**, 657 (2007).
99. A. San Paulo, N. Arellano, J.A. Plaza, R.R. He, C. Carraro, R. Maboudian, R.T. Howe, J. Bokor and P.D. Yang, *Nano Lett.* **7**, 1100 (2007).
100. J.R. Heath, J.J. Shiang and A.P. Alivisatos, *J. Chem. Phys.* **101**, 1607 (1994).
101. H.W. Chiu, C.N. Chervin and S.M. Kauzlarich, *Chem. Mater.* **17**, 4858 (2005).
102. X. Ma, F. Wu and S.M. Kauzlarich, *J. Solid State Chem.* **181**, 1631 (2008).
103. M.W. Peterson and A.J. Nozik, in *Photoelectrochemistry and Photovoltaics of Layered Semiconductors*, A. Aruchamy, Editor., Kluwer Academic Publishers: Dordrecht. 297(1992).
104. C.J. Sandroff, D.M. Hwang and W.M. Chung, *Phys. Rev. B: Cond. Matt.* **33**, 5953 (1986).
105. O.I. Micic, L. Zongguan, G. Mills, J.C. Sullivan and D. Meisel, *J. Phys. Chem.* **91**, 6221 (1987).
106. A. Sengupta, B. Jiang, K.C. Mandal and J.Z. Zhang, *J. Phys. Chem. B* **103**, 3128 (1999).
107. A. Sengupta, K.C. Mandal and J.Z. Zhang, *J. Phys. Chem. B* **104**, 9396 (2000).
108. A. Sengupta and J.Z. Zhang, in *First International Chinese Workshop on Nanoscience and Nanotechnology.* 2001. Beijing, China: Tsinghua University Press.
109. T.K. Chaudhuri, A.B. Patra, P.K. Basu, R.S. Saraswat and H.N. Acharya, *Mater. Lett.* **8**, 361 (1989).

110. T. Karasawa, K. Miyata, T. Komatsu and Y. Kaifu, *J. Phys. Soc. Japan* **52**, 2592 (1983).

111. C.J. Sandroff, S.P. Kelty and D.M. Hwang, *J. Chem. Phys.* **85**, 5337 (1986).

112. T. Kamatsu, T. Karasawa, I. Akai and T. Iida, *J. Luminescence* **70**, 448 (1996).

113. Z.K. Tang, Y. Nozue and T. Goto, *J. Phys. Soc. Japan* **61**, 2943 (1992).

114. F. Wu, J.Z. Zhang, R. Kho and R.K. Mehra, *Chem. Phys. Lett.* **330**, 237 (2000).

115. C.B. Murray, D.J. Norris and M.G. Bawendi, *J. Am. Chem. Soc.* **115**, 8706 (1993).

116. L.R. Becerra, C.B. Murray, R.G. Griffin and M.G. Bawendi, *J. Chem. Phys.* **100**, 3297 (1994).

117. J.E.B. Katari, V.L. Colvin and A.P. Alivisatos, *J. Phys. Chem.* **98**, 4109 (1994).

118. L. Spanhel, M. Haase, H. Weller and A. Henglein, *J. Am. Chem. Soc.* **109**, 5649 (1987).

119. P.V. Kamat and N.M. Dimitrijevic, *J. Phys. Chem.* **93**, 4259 (1989).

120. P.V. Kamat, M.D. Vanwijngaarden and S. Hotchandani, Israel *J. Chem.* **33**, 47 (1993).

121. C. Luangdilok and D. Meisel, Israel *J. Chem.* **33**, 53 (1993).

122. M. Gao, S. Kirstein, H. Mohwald, A. Rogach, A. Kornowski, A. Eychmuller and H. Weller, *J. Phys. Chem. B* **102**, 8360 (1998).

123. M.A. Hines and P. Guyot-Sionnest, *J. Phys. Chem.* **100**, 468 (1996).

124. M.C. Schlamp, X.G. Peng and A.P. Alivisatos, *J. Appl. Phys.* **82**, 5837 (1997).

125. X.G. Peng, M.C. Schlamp, A.V. Kadavanich and A.P. Alivisatos, *J. Am. Chem. Soc.* **119**, 7019 (1997).

126. P. Palinginis and W. Hailin, *Appl. Phys. Lett.* **78**, 1541 (2001).

127. M.A. Hines and P. Guyot-Sionnest, *J. Phys. Chem. B* **102**, 3655 (1998).

128. S.A. Blanton, M.A. Hines, M.E. Schmidt and P. Guyotsionnest, *J. Luminescence* **70**, 253 (1996).

129. E. Poles, D.C. Selmarten, O.I. Micic and A.J. Nozik, *Appl. Phys. Lett.* **75**, 971 (1999).

130. I.V. Ignatiev, I.E. Kozin, H.W. Ren, S. Sugou and Y. Matsumoto, *Phys. Rev. B-Condensed Matter* **60**, R14001 (1999).

131. P.P. Paskov, P.O. Holtz, B. Monemar, J.M. Garcia, W.V. Schoenfeld and P.M. Petroff, *Appl. Phys. Lett.* **77**, 812 (2000).

132. H.X. Zhang, C.H. Kam, Y. Zhou, X.Q. Han, S. Buddhudu and Y.L. Lam, *Opt. Mater.* **15**, 47 (2000).

133. W. Chen, A.G. Joly and J.Z. Zhang, *Phys. Rev. B* **64**, 41202 (2001).

134. S.R. Cordero, P.J. Carson, R.A. Estabrook, G.F. Strouse and S.K. Buratto, *J. Phys. Chem. B* **104**, 12137 (2000).
135. B.C. Hess, I.G. Okhrimenko, R.C. Davis, B.C. Stevens, Q.A. Schulzke, K.C. Wright, C.D. Bass, C.D. Evans and S.L. Summers, *Phys. Rev. Lett.* **86**, 3132 (2001).
136. D.F. Underwood, T. Kippeny and S.J. Rosenthal, *J. Phys. Chem. B* **105**, 436 (2001).
137. S. Shih, K.H. Jung, J. Yan, D.L. Kwong, M. Kovar, J.M. White, T. George and S. Kim, *Appl. Phys. Lett.* **63**, 3306 (1993).
138. W.G. Becker and A.J. Bard, *J. Phys. Chem.* **87**, 4888 (1983).
139. J.Q. Yu, H.M. Liu, Y.Y. Wang, F.E. Fernandez and W.Y. Jia, *J. Luminescence* **76–7**, 252 (1998).
140. R.W. Meulenberg, H.W. Offen and G.F. Strouse, *Materials Research Society Symposium Proceedings* **636**, D9.46, 1 (2001).
141. J.Q. Yu, H.M. Liu, Y.Y. Wang, F.E. Fernandez, W.Y. Jia, L.D. Sun, C.M. Jin, D. Li, J.Y. Liu and S.H. Huang, *Opt. Lett.* **22**, 913 (1997).
142. A.A. Bol and A. Meijerink, *Phys. Status Solidi. B* **224**, 291 (2001).
143. S.O. Williams and D.G. Imre, *J. Phys. Chem.* **92**, 3363 (1988).
144. J.J. Shiang, A.N. Goldstein and A.P. Alivisatos, *J. Chem. Phys.* **92**, 3232 (1990).
145. G. Scamarcio, M. Lugara and D. Manno, *Phys. Rev. B-Condensed Matter* **45**, 13792 (1992).
146. J.J. Shiang, S.H. Risbud and A.P. Alivisatos, *J. Chem. Phys.* **98**, 8432 (1993).
147. M. Abdulkhadar and B. Thomas, *Nanostruct. Mater.* **5**, 289 (1995).
148. A.A. Sirenko, V.I. Belitsky, T. Ruf, M. Cardona, A.I. Ekimov and C. TralleroGiner, *Phys. Rev. B-Condensed Matter* **58**, 2077 (1998).
149. V.G. Melehin and V.D. Petrikov, *Physics of Low-Dimensional Structures* **9–10**, 73 (1999).
150. A.P. Alivisatos, T.D. Harris, P.J. Carroll, M.L. Steigerwald and L.E. Brus, *J. Chem. Phys.* **90**, 3463 (1989).
151. J.J. Shiang, I.M. Craig and A.P. Alivisatos, *Z Phys. D Atom. Mol. Cl.* **26**, 358 (1993).
152. V. Spagnolo, G. Scamarcio, M. Lugara and G.C. Righini, *Superlattice Microst.* **16**, 51 (1994).
153. J.P. Ge, J. Wang, H.X. Zhang, X. Wang, Q. Peng and Y.D. Li, *Chem.-Eur. J.* **11**, 1889 (2005).
154. J.J. Shiang, R.H. Wolters and J.R. Heath, *J. Chem. Phys.* **106**, 8981 (1997).

155. V.A. Volodin, M.D. Efremov, V.A. Gritsenko and S.A. Kochubei, *Appl. Phys. Lett.* **73**, 1212 (1998).

156. M.D. Efremov, V.V. Bolotov, V.A. Volodin and S.A. Kochubei, *Solid State Commun.* **108**, 645 (1998).

157. W.F.A. Besling, A. Goossens and J. Schoonman, *J. Phys. Iv.* **9**, 545 (1999).

158. G.H. Li, K. Ding, Y. Chen, H.X. Han and Z.P. Wang, *J. Appl. Phys.* **88**, 1439 (2000).

159. J.D. Prades, J. Arbiol, A. Cirera, J.R. Morante and A.F.I. Morral, *Appl. Phys. Lett.* **91**, 123107 (2007).

160. Y.Y. Wang, Y.H. Yang, Y.P. Guo, J.S. Yue and R.J. Gan, *Mater. Lett.* **29**, 159 (1996).

161. W.K. Choi, V. Ng, S.P. Ng, H.H. Thio, Z.X. Shen and W.S. Li, *J. Appl. Phys.* **86**, 1398 (1999).

162. A.V. Kolobov, Y. Maeda and K. Tanaka, *J. Appl. Phys.* **88**, 3285 (2000).

163. Y.W. Ho, V. Ng, W.K. Choi, S.P. Ng, T. Osipowicz, H.L. Seng, W.W. Tjui and K. Li, *Scripta Materialia* **44**, 1291 (2001).

164. U. Pal and J.G. Serrano, *Appl. Surf. Sci.* **246**, 23 (2005).

165. R. Rinaldi, R. Cingolani, M. Ferrara, A.C. Maciel, J. Ryan, U. Marti, D. Martin, F. Moriergemoud and F.K. Reinhart, *Appl. Phys. Lett.* **64**, 3587 (1994).

166. K.L. Teo, S.H. Kwok, P.Y. Yu and S. Guha, *Phys. Rev. B* **62**, 1584 (2000).

167. B. Schreder, A. Materny, W. Kiefer, G. Bacher, A. Forchel and G. Landwehr, *J. Raman Spectroscopy* **31**, 959 (2000).

168. B. Schreder, T. Kummell, G. Bacher, A. Forchel, G. Landwehr, A. Materny and W. Kiefer, *J. Cryst. Growth* **214**, 792 (2000).

169. H. Dollefeld, H. Weller and A. Eychmuller, *Nano Lett.* **1**, 267 (2001).

170. Z. Wang and J.H. Li, in *Annual Review of Nano Research* G. Cao and C.J. Brinker, Editors., World Scientific Singapore. 63(2008).

171. Z.F. Zhang, H. Cui, C.Z. Lai and L.J. Liu, *Anal. Chem.* **77**, 3324 (2005).

172. A.P. Fan, C.W. Lau and J.Z. Lu, *Anal. Chem.* **77**, 3238 (2005).

173. Z.P. Li, Y.C. Wang, C.H. Liu and Y.K. Li, *Analytica Chimica Acta* **551**, 85 (2005).

174. S.K. Poznyak, D.V. Talapin, E.V. Shevchenko and H. Weller, *Nano Lett.* **4**, 693 (2004).

175. Z.P. Wang, J. Li, B. Liu, J.Q. Hu, X. Yao and J.H. Li, *J. Phys. Chem. B* **109**, 23304 (2005).

176. S. Nakamura, K. Kitamura, H. Umeya, A. Jia, M. Kobayashi, A. Yoshikawa, M. Shimotomai and K. Takahashi, *Electron. Lett.* **34**, 2435 (1998).

177. M.Y. Gao, B. Richter, S. Kirstein and H. Mohwald, *J. Phys. Chem. B* **102**, 4096 (1998).

178. T. Brunhes, P. Boucaud, S. Sauvage, F. Aniel, J.M. Lourtioz, C. Hernandez, Y. Campidelli, O. Kermarrec, D. Bensahel, G. Faini and I. Sagnes, *Appl. Phys. Lett.* **77**, 1822 (2000).

179. I.E. Itskevich, S.T. Stoddart, S.I. Rybchenko, Tartakovskii, II, L. Eaves, P.C. Main, M. Henini and S. Parnell, *Phys. Status Solidi. A* **178**, 307 (2000).

180. R. Schmidt, M. Vitzethum, R. Fix, U. Scholz, S. Malzer, C. Metzner, P. Kailuweit, D. Reuter, A. Wieck, M.C. Hubner, S. Stufler, A. Zrenner and G.H. Dohler, *Physica E* **26**, 110 (2005).

181. X.L. Xu, A. Andreev, D.A. Williams and J.R.A. Cleaver, *Appl. Phys. Lett.* **89**, 091120 (2006).

182. J.H. Fendler and F.C. Meldrum, *Advan. Mater.* **7**, 607 (1995).

183. C.P. Collier, T. Vossmeyer and J.R. Heath, *Ann. Rev. Phys. Chem.* **49**, 371 (1998).

184. M.V. Artemyev, U. Woggon, H. Jaschinski, L.I. Gurinovich and S.V. Gaponenko, *J. Phys. Chem. B* **104**, 11617 (2000).

185. H.C. Hamaker, *Physica* **4**, 1058 (1937).

186. W. vanMegen and S.M. Underwood, *Phys. Rev. Lett.* **70**, 2766 (1993).

187. C.B. Murray, C.R. Kagan and M.G. Bawendi, *Science* **270**, 1335 (1995).

188. Z.L. Wang, *Australian J. Chem.* **54**, 153 (2001).

189. J. Park, E. Kang, S.U. Son, H.M. Park, M.K. Lee, J. Kim, K.W. Kim, H.J. Noh, J.H. Park, C.J. Bae, J.G. Park and T. Hyeon, *Advan. Mater.* **17**, 429 (2005).

190. Q. Song, Y. Ding, Z.L. Wang and Z.J. Zhang, *J. Phys. Chem. B* **110**, 25547 (2006).

191. C.R. Kagan, C.B. Murray, M. Nirmal and M.G. Bawendi, *Phys. Rev. Lett.* **76**, 1517 (1996).

192. C.R. Kagan, C.B. Murray and M.G. Bawendi, *Phys. Rev. B-Condensed Matter* **54**, 8633 (1996).

193. R. Ugajin, *J. Appl. Phys.* **76**, 2833 (1994).

194. S. Lloyd, *Science* **261**, 1569 (1993).

195. V.L. Colvin, M.C. Schlamp and A.P. Alivisatos, *Nature* **370**, 354 (1994).

196. B.O. Dabbousi, M.G. Bawendi, O. Onitsuka and M.F. Rubner, *Appl. Phys. Lett.* **66**, 1316 (1995).

197. N.C. Greenham, X.G. Peng and A.P. Alivisatos, *Synthet. Metal* **84**, 545 (1997).

198. N. Zaitseva, Z.R. Dai, F.R. Leon and D. Krol, *J. Am. Chem. Soc.* **127**, 10221 (2005).

199. B. Alen, J. Martinez-Pastor, A. Garcia-Cristobal, L. Gonzalez and J.M. Garcia, *Appl. Phys. Lett.* **78**, 4025 (2001).
200. S. Shionoya and W.M. Yen, eds., CRC Press: New York (1999).
201. T.A. Kennedy, E.R. Glaser, P.B. Klein and R.N. Bhargava, *Phys. Rev. B-Condensed Matter* **52**, 14356 (1995).
202. G. Counio, S. Esnouf, T. Gacoin and J.P. Boilot, *J. Phys. Chem.* **100**, 20021 (1996).
203. T. Igarashi, T. Isobe and M. Senna, *Phys. Rev. B-Condensed Matter* **56**, 6444 (1997).
204. N. Feltin, L. Levy, D. Ingert and M.P. Pileni, *J. Phys. Chem. B* **103**, 4 (1999).
205. R.N. Bhargava, D. Gallagher and T. Welker, *J. Luminescence* **60–1**, 275 (1994).
206. U.W. Pohl and H.E. Gumlich, *Phys. Rev. B-Condensed Matter* **40**, 1194 (1989).
207. P. Devisschere, K. Neyts, D. Corlatan, J. Vandenbossche, C. Barthou, P. Benalloul and J. Benoit, *J. Luminescence* **65**, 211 (1995).
208. R.N. Bhargava, *J. Luminescence* **70**, 85 (1996).
209. J.Q. Yu, H.M. Liu, Y.Y. Wang and W.Y. Jia, *J. Luminescence* **79**, 191 (1998).
210. T.J. Norman, D. Magana, T. Wilson, C. Burns, J.Z. Zhang, D. Cao and F. Bridges, *J. Phys. Chem. B* **107**, 6309 (2003).
211. C.D. Grant and J.Z. Zhang, in *Annual Review of Nano Research*, G. Cao and C.J. Brinker, Editors., World Scientific Publisher: Singapore. 1(2008).
212. K. Sooklal, B.S. Cullum, S.M. Angel and C.J. Murphy, *J. Phys. Chem.* **100**, 4551 (1996).
213. L.D. Sun, C.H. Yan, C.H. Liu, C.S. Liao, D. Li and J.Q. Yu, *J. Alloy Compd.* **277**, 234 (1998).
214. D.D. Papakonstantinou, J. Huang and P. Lianos, *J. Mater. Sci. Lett.* **17**, 1571 (1998).
215. A.A. Khosravi, M. Kundu, L. Jatwa, S.K. Deshpande, U.A. Bhagwat, M. Sastry and S.K. Kulkarni, *Appl. Phys. Lett.* **67**, 2702 (1995).
216. J.M. Huang, Y. Yang, S.H. Xue, B. Yang, S.Y. Liu and J.C. Shen, *Appl. Phys. Lett.* **70**, 2335 (1997).
217. C.M. Jin, J.Q. Yu, L.D. Sun, K. Dou, S.G. Hou, J.L. Zhao, Y.M. Chen and S.H. Huang, *J. Luminescence* **66–7**, 315 (1995).
218. R.N. Bhargava, D. Gallagher, X. Hong and A. Nurmikko, *Phys. Rev. Lett.* **72**, 416 (1994).
219. G. Counio, T. Gacoin and J.P. Boilot, *J. Phys. Chem. B* **102**, 5257 (1998).
220. A.D. Dinsmore, D.S. Hsu, H.F. Gray, S.B. Qadri, Y. Tian and B.R. Ratna, *Appl. Phys. Lett.* **75**, 802 (1999).

221. W. Chen, R. Sammynaiken and Y.N. Huang, *J. Appl. Phys.* **88**, 5188 (2000).
222. W. Chen, R. Sammynaiken, Y.N. Huang, J.O. Malm, R. Wallenberg, J.O. Bovin, V. Zwiller and N.A. Kotov, *J. Appl. Phys.* **89**, 1120 (2001).
223. M. Konishi, T. Isobe and M. Senna, *J. Luminescence* **93**, 1 (2001).
224. K. Yan, C.K. Duan, Y. Ma, S.D. Xia and J.C. Krupa, *Phys. Rev. B-Condensed Matter* **58**, 13585 (1998).
225. A.A. Bol and A. Meijerink, *Phys. Rev. B-Condensed Matter* **58**, R15997 (1998).
226. B.A. Smith, J.Z. Zhang, A. Joly and J. Liu, *Phys. Rev. B* **62**, 2021 (2000).
227. J.H. Chung, C.S. Ah and D.-J. Jang, *J. Phys. Chem. B* **105**, 4128 (2001).
228. R. Bowers and N.T. Melamed, *Phys. Rev.* **99**, 1781 (1955).
229. K. Urabe and S. Shionoya, *J. Phys. Soc. Japan* **24**, 543 (1968).
230. A. Suzuki and S. Shionoya, *J. Phys. Soc. Japan* **31**, 1462 (1971).
231. A. Suzuki and S. Shionoya, *J. Phys. Soc. Japan* **31**, 1719 (1971).
232. C.S. Kang, P. Beverley, P. Phipps and R.H. Bube, *Phys. Rev.* **156**, 998 (1967).
233. W. Van Gool, in *Philips Res. Rept., Suppl.*; *Vol: No. 3*; *Thesis*; *Amsterdam, Universiteit, 1961*: Not Available. Pages: 122(1961).
234. M. Wang, L. Sun, X. Fu, C. Liao and C. Yan, *Solid State Commun.* **115**, 493 (2000).
235. W. Sang, Y. Qian, J. Min, D. Li, L. Wang, W. Shi and L. Yinfeng, *Solid State Commun.* **121**, 475 (2002).
236. A.A. Bol, *J. Luminescence* **99**, 325 (2002).
237. K. Manzoor, S.R. Vadera, N. Kumar and T.R.N. Kutty, *Mater. Chem. Phys.* **82**, 718 (2003).
238. W.Q. Peng, G.W. Cong, S.C. Qu and Z.G. Wang, *Opt. Mater.* **29**, 313 (2006).
239. C. Corrado, Y. Jiang, F. Oba, M. Kozina, F. Bridges and J.Z. Zhang, *J. Phys. Chem. C*, in press (2009).
240. N.M. Dimitrijevic and P.V. Kamat, *J. Phys. Chem.* **91**, 2096 (1987).
241. M. Haase, H. Weller and A. Henglein, *J. Phys. Chem.* **92**, 4706 (1988).
242. E.F. Hilinski, P.A. Lucas and W. Ying, *J. Chem. Phys.* **89**, 3534 (1988).
243. P.V. Kamat, N.M. Dimitrijevic and A.J. Nozik, *J. Phys. Chem.* **93**, 2873 (1989).
244. Y. Wang, A. Suna, J. McHugh, E.F. Hilinski, P.A. Lucas and R.D. Johnson, *J. Chem. Phys.* **92**, 6927 (1990).
245. A. Henglein, A. Kumar, E. Janata and H. Weller, *Chem. Phys. Lett.* **132**, 133 (1986).
246. K.I. Kang, A.D. Kepner, S.V. Gaponenko, S.W. Koch, Y.Z. Hu and N. Peyghambarian, *Phys. Rev. B-Condensed Matter* **48**, 15449 (1993).

247. V. Klimov, S. Hunsche and H. Kurz, *Phys. Rev. B-Condensed Matter* **50**, 8110 (1994).

248. V. Dneprovskii, N. Gushina, O. Pavlov, V. Poborchii, I. Salamatina and E. Zhukov, *Phys. Lett. A* **204**, 59 (1995).

249. N.V. Gushchina, V.S. Dneprovskii, E.A. Zhukov, O.V. Pavlov, V.V. Poborchii and I.A. Salamatina, *Jetp. Lett.-Engl. Tr.* **61**, 507 (1995).

250. V. Dneprovskii, A. Eev, N. Gushina, D. Okorokov, V. Panov, V. Karavanskii, A. Maslov, V. Sokolov and E. Dovidenko, *Phys. Status Solidi. B* **188**, 297 (1995).

251. V. Dneprovskii, N. Gushina, D. Okorokov, V. Karavanskii and E. Dovidenko, *Superlattice Microst.* **17**, 41 (1995).

252. F.Z. Henari, K. Morgenstern, W.J. Blau, V.A. Karavanskii and V.S. Dneprovskii, *Appl. Phys. Lett.* **67**, 323 (1995).

253. B.L. Yu, C.S. Zhu and F.X. Gan, *J. Appl. Phys.* **82**, 4532 (1997).

254. Y. Wang, *Acc. Chem. Res.* **24**, 133 (1991).

255. T. Dannhauser, M. O'Neil, K. Johanseon, D. Whitter and G. McLendon, *J. Phys. Chem.* **90**, 6074 (1986).

256. Y.R. Shen, *The Principles of Nonlinear Optics*, New York: J. Wiley. **563** (1984).

257. T. Yamaki, K. Asai, K. Ishigure, K. Sano and K. Ema, *Synthet. Metal* **103**, 2690 (1999).

258. M. Jacobsohn and U. Banin, *J. Phys. Chem. B* **104**, 1 (2000).

259. Z. Yu, F. Dgang, W. Xin, L. Juzheng and L. Zuhong, *Colloids and Surfaces A: Phyiscochemical and Enginnering Aspects* **181**, 145 (2001).

260. J. Lenglet, A. Bourdon, J.C. Bacri, R. Perzynski and G. Demouchy, *Phys. Rev. B-Condensed Matter* **53**, 14941 (1996).

261. T. Takagahara, *Solid State Commun.* **78**, 279 (1991).

262. T. Takagahara, *Surface Sci.* **267**, 310 (1992).

263. Y. Kayanuma, *J. Phys. Soc. Japan* **62**, 346 (1993).

264. B. Clerjaud, F. Gendron and C. Porte, *Appl. Phys. Lett.* **38**, 212 (1981).

265. Y. Mita, in *Phosphor Handbook*, S.a.Y. Shionoya, W.M., Editor, CRC Press: New York. 643 (1999).

266. L.G. Quagliano and H. Nather, *Appl. Phys. Lett.* **45**, 555 (1984).

267. E.J. Johnson, J. Kafalas, R.W. Davies and W.A. Dyes, *Appl. Phys. Lett.* **40**, 993 (1982).

268. Y.H. Cho, D.S. Kim, B.D. Choe, H. Lim, J.I. Lee and D. Kim, *Phys. Rev. B-Condensed Matter* **56**, R4375 (1997).

269. M. Potemski, R. Stepniewski, J.C. Maan, G. Martinez, P. Wyder and B. Etienne, *Phys. Rev. Lett.* **66**, 2239 (1991).

270. P. Vagos, P. Boucaud, F.H. Julien, J.M. Lourtioz and R. Planel, *Phys. Rev. Lett.* **70**, 1018 (1993).

271. W. Seidel, A. Titkov, J.P. Andre, P. Voisin and M. Voos, *Phys. Rev. Lett.* **73**, 2356 (1994).

272. F. Driessen, H.M. Cheong, A. Mascarenhas, S.K. Deb, P.R. Hageman, G.J. Bauhuis and L.J. Giling, *Phys. Rev. B-Condensed Matter* **54**, R5263 (1996).

273. R. Hellmann, A. Euteneuer, S.G. Hense, J. Feldmann, P. Thomas, E.O. Gobel, D.R. Yakovlev, A. Waag and G. Landwehr, *Phys. Rev. B-Condensed Matter* **51**, 18053 (1995).

274. Z.P. Su, K.L. Teo, P.Y. Yu and K. Uchida, *Solid State Commun.* **99**, 933 (1996).

275. J. Zeman, G. Martinez, P.Y. Yu and K. Uchida, *Phys. Rev. B-Condensed Matter* **55**, 13428 (1997).

276. L. Schrottke, H.T. Grahn and K. Fujiwara, *Phys. Rev. B-Condensed Matter* **56**, 15553 (1997).

277. H.M. Cheong, B. Fluegel, M.C. Hanna and A. Mascarenhas, *Phys. Rev. B-Condensed Matter* **58**, R4254 (1998).

278. Z. Chine, B. Piriou, M. Oueslati, T. Boufaden and B. El Jani, *J. Luminescence* **82**, 81 (1999).

279. T. Kita, T. Nishino, C. Geng, F. Scholz and H. Schweizer, *Phys. Rev. B-Condensed Matter* **59**, 15358 (1999).

280. S.C. Hohng and D.S. Kim, *Appl. Phys. Lett.* **75**, 3620 (1999).

281. L. Schrottke, R. Hey and H.T. Grahn, *Phys. Rev. B-Condensed Matter* **60**, 16635 (1999).

282. W. Heimbrodt, M. Happ and F. Henneberger, *Phys. Rev. B-Condensed Matter* **60**, R16326 (1999).

283. T. Kita, T. Nishino, C. Geng, F. Scholz and H. Schweizer, *J. Luminescence* **87–9**, 269 (2000).

284. A. Satake, Y. Masumoto, T. Miyajima, T. Asatsuma and T. Hino, *Phys. Rev. B* **61**, 12654 (2000).

285. G.G. Zegrya and V.A. Kharchenko, *Zhurnal Eksperimentalnoi I Teoreticheskoi Fiziki* **101**, 327 (1992).

286. W. Chen, A.G. Joly and J.Z. Zhang, *Phys. Rev. B* **6404**, 1202 (2001).

287. A.G. Joly, W. Chen, J. Roark and J.Z. Zhang, *J. Nanosci. & Nanotechnol.* **1**, 295 (2001).

288. Y.P. Rakovich, J.F. Donegan, S.A. Filonovich, M.J.M. Gomes, D.V. Talapin, A.L. Rogach and A. Eychmuller, *Physica E* **17**, 99 (2003).

289. K.I. Rusakov, A.A. Gladyshchuk, Y.P. Rakovich, J.F. Donegan, S.A. Filonovich, M.J.M. Gomes, D.V. Talapin, A.L. Rogach and A. Eychmuller, *Opt. Spectroscopy* **94**, 859 (2003).

290. C. Kammerer, G. Cassabois, C. Voisin, C. Delalande, P. Roussignol and J.M. Gerard, *Phys. Rev. Lett.* **87**, 207401 (2001).

291. G. Cassabois, C. Kammerer, C. Voisin, C. Delalande, P. Roussignol and J.M. Gerard, *Physica E* **13**, 105 (2002).

292. A.P. Alivisatos, A.L. Harris, N.J. Levinos, M.L. Steigerwald and L.E. Brus, *J. Chem. Phys.* **89**, 4001 (1988).

293. D.J. Norris and M.G. Bawendi, *J. Chem. Phys.* **103**, 5260 (1995).

294. D.M. Mittleman, R.W. Schoenlein, J.J. Shiang, V.L. Colvin, A.P. Alivisatos and C.V. Shank, *Phys. Rev. B-Condensed Matter* **49**, 14435 (1994).

295. U. Woggon, S. Gaponenko, W. Langbein, A. Uhrig and C. Klingshirn, *Phys. Rev. B-Condensed Matter* **47**, 3684 (1993).

296. H. Giessen, B. Fluegel, G. Mohs, N. Peyghambarian, J.R. Sprague, O.I. Micic and A.J. Nozik, *Appl. Phys. Lett.* **68**, 304 (1996).

297. V. Jungnickel and F. Henneberger, *J. Luminescence* **70**, 238 (1996).

298. W.E. Moerner, *Science* **265**, 46 (1994).

299. T. Basche, W.E. Moerner, M. Orrit and U.P. Wild, eds., VCH: Weinheim; Cambridge. 250 (1997).

300. J. Tittel, W. Gohde, F. Koberling, T. Basche, A. Kornowski, H. Weller and A. Eychmuller, *J. Phys. Chem. B* **101**, 3013 (1997).

301. S.A. Blanton, A. Dehestani, P.C. Lin and P. Guyot-Sionnest, *Chem. Phys. Lett.* **229**, 317 (1994).

302. S.A. Empedocles, D.J. Norris and M.G. Bawendi, *Phys. Rev. Lett.* **77**, 3873 (1996).

303. M. Nirmal, B.O. Dabbousi, M.G. Bawendi, J.J. Macklin, J.K. Trautman, T.D. Harris and L.E. Brus, *Nature* **383**, 802 (1996).

304. S.A. Blanton, M.A. Hines and P. Guyot-Sionnest, *Appl. Phys. Lett.* **69**, 3905 (1996).

305. S.A. Empedocles and M.G. Bawendi, *Science* **278**, 2114 (1997).

306. S.A. Empedocles, R. Neuhauser, K. Shimizu and M.G. Bawendi, *Advan. Mater.* **11**, 1243 (1999).

307. S.A. Empedocles and M.G. Bawendi, *J. Phys. Chem. B* **103**, 1826 (1999).

308. S. Empedocles and M. Bawendi, *Acc. Chem. Res.* **32**, 389 (1999).

309. F. Koberling, A. Mews and T. Basche, *Phys. Rev. B (Condensed Matter)* **60**, 1921 (1999).

310. A.L. Efros, M. Rosen, M. Kuno, M. Nirmal, D.J. Norris and M. Bawendi, *Phys. Rev. B-Condensed Matter* **54**, 4843 (1996).
311. U. Banin, M. Bruchez, A.P. Alivisatos, T. Ha, S. Weiss and D.S. Chemla, *J. Chem. Phys.* **110**, 1195 (1999).
312. T. Matsumoto, M. Ohtsu, K. Matsuda, T. Saiki, H. Saito and K. Nishi, *Appl. Phys. Lett.* **75**, 3246 (1999).
313. J. Knappenberger, K.L., D.B. Wong, W. Xu, A.M. Schwartzberg, A. Wolcott, J.Z. Zhang and S.R. Leone, *ACS Nano* **2**, 2143 (2008).
314. N. Murase, *Chem. Phys. Lett.* **368**, 76 (2003).
315. M. Ono, K. Matsuda, T. Saiki, K. Nishi, T. Mukaiyama and M. Kuwata-Gonokami, *Jpn. J. Appl. Phys.* **38**, L1460 (1999).
316. K. Matsuda, T. Saiki, H. Saito and K. Nishi, *Appl. Phys. Lett.* **76**, 73 (2000).

Chapter 6

Optical Properties of Metal Oxide Nanomaterials

Metal oxides (MO) are an important class of materials with applications in many areas of science and technology. Depending on their bandgaps, some metal oxides are considered as semiconductors while others with larger bandgaps are considered as insulators. Optically, semiconducting MO with a bandgap below 3.5 eV have visible or near UV absorption, while true insulating MO have no visible or near UV absorption, but only absorption in the UV or shorter wavelength. Insulators are poor electrical and thermal conductors, and appear to be colorless or white. But they can have color when impurities or dopants are present. Many gem stones, e.g. ruby and sapphire, acquire their color due to metal ions such as Cr and Ti in the insulators such as Al_2O_3 and SiO_2.

The bandgap of MO can vary significantly depending on the nature of the metal. For examples, Fe_2O_3 has rich color due to its small bandgap while ZnO that has no color because of its relatively large bandgap. However, in practice, ZnO as well as TiO_2 are also often considered as semiconductors. The distinction between semiconductor and insulators is not totally well-defined sometimes. For convenience, we discuss metal oxides in this chapter, and separate them from other semiconductors presented in the previous chapter.

Since insulators usually have little visible or near UV absorption, their optical properties only become important in the UV or VUV region, we have decided not to discuss insulators in detail in this book. The fundamental photophysical and photochemical properties of insulators are similar to semiconductors, with the quantitative difference that insulators absorb in the UV and VUV regions, instead of in the visible and near UV for most semiconductors. Doped insulators are exception and have interesting optical properties in the visible, as discussed at the end of this chapter.

The exciton Bohr radius for most metal oxides are very small, e.g. 3 Å for TiO_2 [1], thus the quantum size effect is not significant for typical nanoparticles of MO in the few to few tens of nm region, which is in contrast to II–VI or III–V semiconductors that tend to have large Bohr exciton radius and thereby strong quantum confinement effect for nanoparticles with a few nm in radius. Even though the quantum confinement effect is not strong in the size range typically encountered, MO do show some confinement effects, such as blue-shift in absorption and PL spectra with decreasing size.

6.1. Optical absorption

Similar to semiconductors, the optical absorption of insulators depends strongly on the electronic band structure, particularly the lowest energy transition being a direct or an indirect band gap transition. Like in semiconductors, direct bandgap transitions are usually strong and feature a sharp excitonic peak, while indirect bandgap transitions are characterized by weaker and featureless absorption. Due to the large bandgap, even the lowest energy electronic transitions usually occur in the UV–near UV region. Due to lack of or weak visible absorption, the samples are often colorless or appear white. The sample could appear cloudy if the feature sizes are larger than or comparable to the wavelength of light. This is particularly true if the size distribution is broad.

As mentioned above, color can arise from impurities or dopants. Sometimes, color can also result from "color centers" or trapped electrons or, in yet other cases, from ordered porous structures like opal. Again, similar to doped semiconductors, doping can be used as an effective way to alter or control the optical and electronic properties of insulators. This is achieved by introducing bandgap states that act as electron donor or acceptor states. Electronic transitions involving these bandgap states and/or the intrinsic valence and conduction bands are responsible for the optical absorption and emission properties. The optical properties and nature of transitions depend strongly on the chemical nature and doping level. Sometimes, co-dopants are introduced to enhance the flexibility of manipulating the bandgap states and associated transitions and optical properties.

Certain metal oxides have relatively small bandgaps, and thus have strong visible absorption or color. Examples include Fe_2O_3, Fe_3O_4, and related iron oxide and/or hydroxide nanoparticles. Fe_2O_3 NPs show a red color due to absorption in the blue and green region. Fe_2O_3 can exist in the γ phase (maghemite) or α phase (hematite). In its α phase it can be used as a photocatalyst and in its γ phase it can be used as a component in magnetic recording media because of its magnetic properties [2–8]. The optical properties of the two forms of Fe_2O_3 NPs are very similar despite their very different magnetic properties [9]. Figure 6.1 shows UV-vis absorption spectra of three different Fe_2O_3 nanorod samples [8]. The spectra

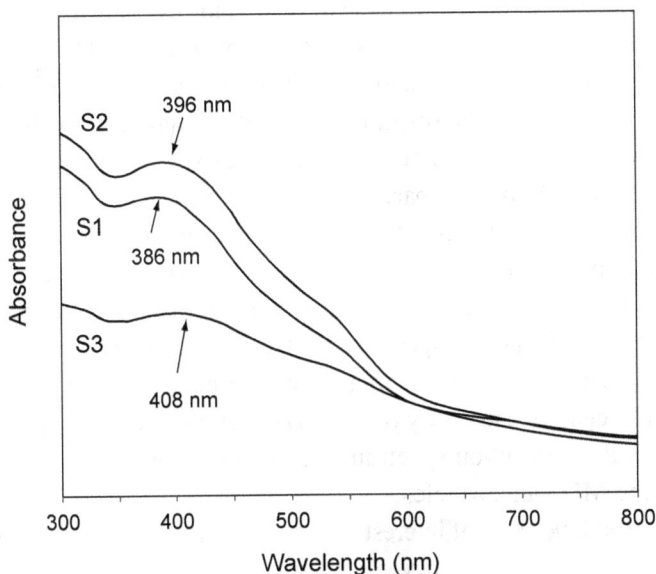

Fig. 6.1. UV-vis spectra of three different Fe_2O_3 nanorod samples (S1, S2 and S3). The diameter/length of nanorods are (in nm): 20–30/40–50 for S1, 20–30/400–500 for S2, and 30–40/700–800 for S3, respectively. Reproduced with permission from Ref. 8.

are overall similar to those of α-Fe_2O_3 NPs and blue-shifted with respect to that of bulk. There is also a noticeable red-shift with increasing dimensions (diameter and/or length) of the nanorods, likely due to spatial confinement effect.

It should be mentioned that iron oxides and iron oxyhydroxides on the nano and larger scales are important for various applications including pigments, catalysis, sensing, magnetic recording and magnets. These materials are very rich in color that varies with the chemical nature, composition, crystal structure, and physical dimensions such as size and shape. There are books dedicated to these materials and their applications [6, 10].

Other common metal oxide nanoparticles, e.g. TiO_2, SnO_2, WO_3, and ZnO, have very little visible absorption in their pristine form. They may exhibit visible absorption due to impurities or dopants. These metal oxides play an important role in catalysis and photocatalysis and as paint pigments. Their nanoparticles can usually be prepared by hydrolysis. Among the different metal oxide nanoparticles, TiO_2 and ZnO have received the most attention because of their stability, easy availability, and promise for applications, e.g. solar energy conversion [11].

As an example, Fig. 6.2 shows the UV-vis spectra of ZnO nanoparticles and nanorods. The absorption onsets are around 3.6 eV for the NPs and 3.2 eV for the nanorods, respectively. The excitonic peak clearly blue-shifts for the NPs as compared to that of the nanorods, attributed to stronger quantum confinement in the NPs than in the nanorods [12].

As another example, Fig. 6.3 shows the UV-vis spectra of TiO_2 nanoparticles dispersed in ethanol with a small amount of water added [13]. It has been found that the spectrum red-shifts with increasing water concentration. This has been attributed to interaction between water and surface Ti^{3+} centers caused by oxygen vacancy defect sites. This example shows that the absorption spectrum can be sensitive to surface characteristics of the MO nanoparticles.

WO_3 is another MO of interest for various applications including photocatalysts and photochromic devices. Figure 6.4 shows an UV-vis absorption spectrum of nanostructured WO_3 films with different thickness [14]. The absorption edge of the films is at ~480 nm, corresponding to a bandgap of 2.60 eV, which is in good agreement with previous report for WO_3 colloidal

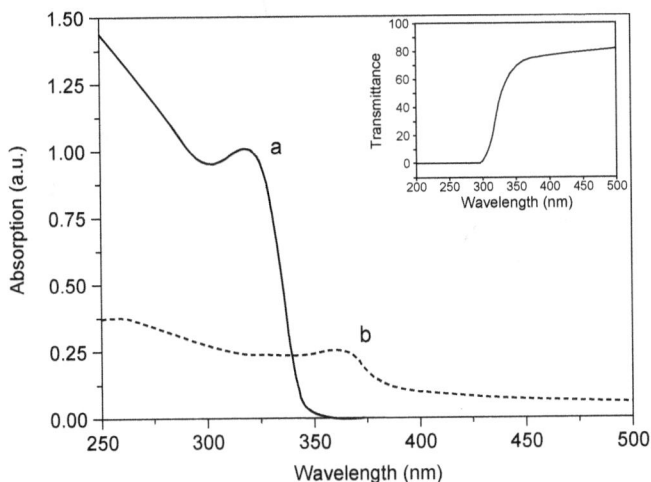

Fig. 6.2. UV-vis absorption spectra of ZnO quantum dots (curve a) and nanorods (curve b). Inset: transmission spectrum of ITO (indium tin oxide), a common semiconductor substrate for different devices. Reproduced with permission from Ref. 12.

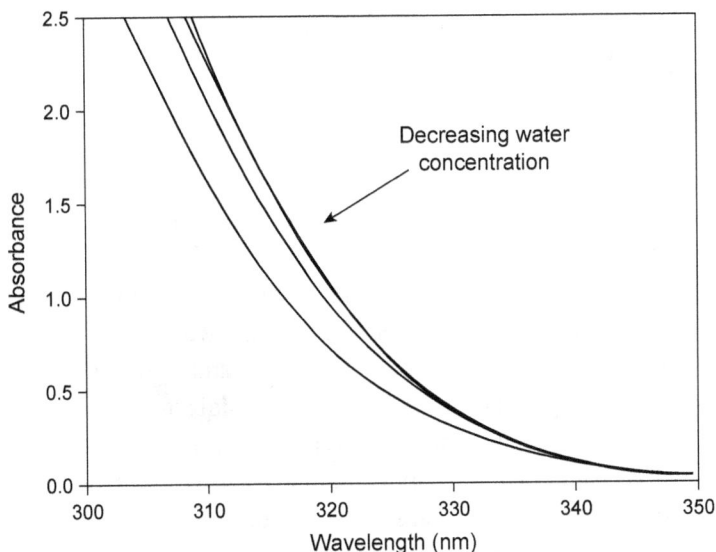

Fig. 6.3. Absorption spectra of TiO_2 nanoparticles in ethanol with different water concentrations. From bottom to top the concentrations of water are 0%, 0.33%, 0.66% and 1%. Reproduced with permission from Ref. 13.

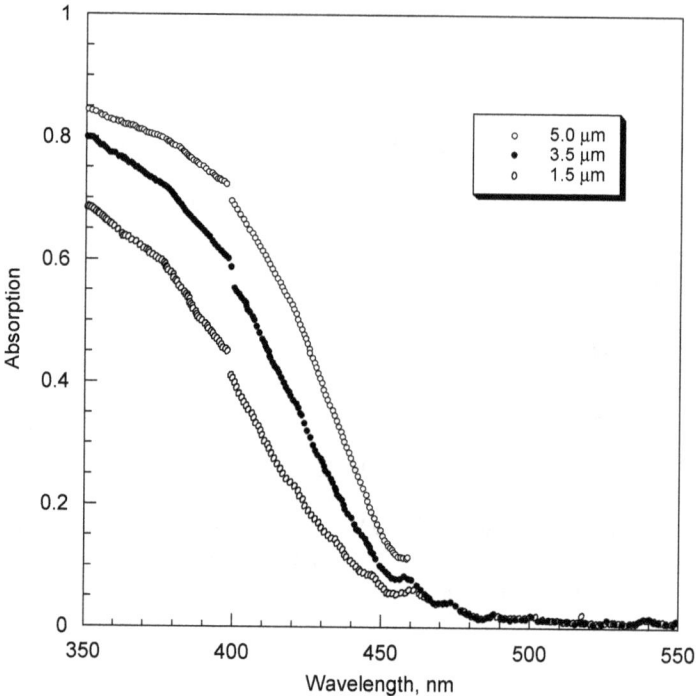

Fig. 6.4.   UV-vis electronic absorption spectra of nanostructured WO$_3$ films with different thickness. Reproduced with permission from Ref. 14.

nanoparticles [15]. The film thickness and inter-particle interaction does not seem to noticeably affect the bandgap.

A final example of MO is SnO$_2$ that is often used in thin film form as a semiconducting substrate for various device applications, usually used in combination with a significant portion of indium and/or fluorine. Figure 6.5 shows the UV-vis-NIR spectra of aqueous colloidal solutions of pure SnO$_2$ after being heated at 270°C under different atmospheres in an autoclave [16]. Under a reducing atmosphere, i.e. under a 1:1 mixture of Ar and H$_2$, the colloid has a blue color caused by a broad absorption in the red to IR region, which has been explained by the formation of oxygen vacancies in the crystal lattice. This broad absorption band becomes weaker when the colloid is exposed to oxygen or air and disappears completely in about one day. The strong bandgap absorption starts around 400 nm.

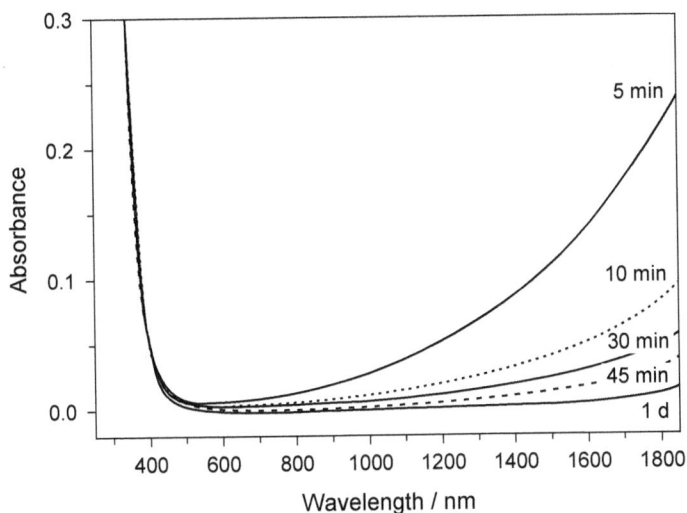

Fig. 6.5. UV-vis-NIR absorption spectra of a SnO_2 colloid after autoclavation for 3 h at 270°C under Ar/H_2 and subsequent exposure to air for the indicated periods of time. The spectra have been acquired in 1-mm cuvettes. Reproduced with permission from Ref. 16.

The above examples have shown that the absorption spectrum of most MO are active mainly in the near UV and UV regions with little or weak absorption in the visible region. There are some exceptions that do show strong visible absorption. For applications that require visible absorption, e.g. solar cells and photoelectrochemistry, the MO with weak visible absorption need to be sensitized using strategies such as dye sensitization, doping, or composite or hybridized structures to enhance visible absorption. This will be discussed in more detail in Sec. 6.3.

6.2. Optical emission

With above bandgap excitation, photoemission or photoluminescence (PL) is usually observed for MO nanomaterials, similar to other nonmetal oxide semiconductors. However, compared to QDs of semiconductors such as II–VI or III–V, the photoluminescence yield is often lower for MO nanomaterials. This is likely due to a higher density of defect or surface trap states that cause trapping of charge carriers and quenching of bandedge

PL. In addition, since most MOs do not have strong visible absorption, above bandgap excitation usually requires UV light. As discussed in Chapter 2, PL is more sensitive than absorption measurement due to the zero-background nature of PL measurement. Because of its sensitivity, one needs to be careful about contributions from impurities of other artifacts.

The use of UV light for excitation of MO requires special caution in PL measurement since some solvents that do not have PL with visible light excitation can become luminescent due to UV or near UV excitation. Thus, one needs to make sure that the solvent or solid matrix used for hosting the MO nanomaterials does not generate PL under the same excitation conditions. In addition, one should be careful with Rayleigh scattering that can appear as higher orders of wavelength due to the grating in the spectrometer, as discussed in Chapter 2. For example, a peak detected at 500 nm could be due to second order grating effect of the 250 nm light used for PL measurement (and similarly other multiples of 250 nm). Finally, Raman scattering due to solvent or solid matrix could appear, especially due to short wavelength excitation (resonance effect) and when the PL from the nanomaterial of interest is low. Higher orders of the Raman scattering due to the grating effect mentioned above could also appear. For example, if the excitation wavelength is at 200 nm and a Raman peak appears at 240 nm, and, due to second order grating effect, the same Raman peak could appear at 480 nm even though there is no real light at 480 nm. Long or short wavelength pass filters can be used to check if the peak at 480 nm is real or not. If it is real PL, a filter that does not allow 480 nm to pass through would block it. If the 480 nm peak is not real, e.g. due to second order grating peak of a Raman signal, the same filter would have no effect. Some of the potential artifacts in PL measurement have already been discussed in Chapter 2. Such artifacts could be more pronounced due to UV or near UV excitation.

It should also be mentioned that PL spectrum is sensitive to factors such as temperature, pressure, and excitation intensity and wavelength. In general, PL spectrum is narrower and appears to have higher energy at low temperature. Figure 6.6 shows temperature dependent PL spectra of ZnO nanorods fabricated on sapphire substrates using MOCVD techniques [17]. The sample is considered as high quality due to low impurity

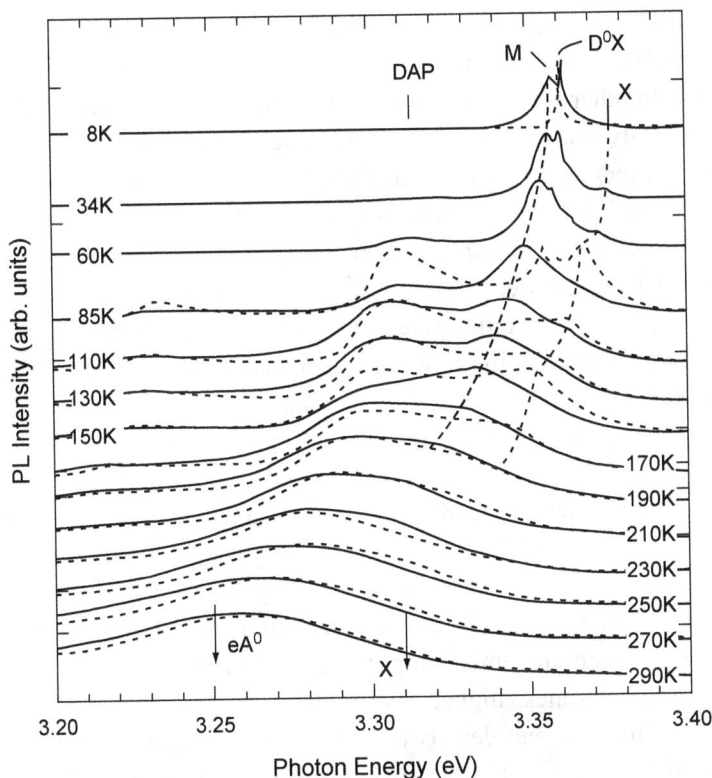

Fig. 6.6. Temperature dependence of PL spectra of ZnO nanorods fabricated by MOCVD on sapphire substrates with 325 nm excitation at two different excitation densities (54 and 1 kW/cm² for solid and dotted spectra, respectively). Each spectrum was normalized by its maximum peak intensity. Downward arrows indicate positions of eA^0 and X at 300 K. DAP, M, D^0X and X represent donor–acceptor–pair, biexcitons, neutral donor-bound excitons and free excitons, respectively. Reproduced with permission from Ref. 17.

and low density of defects. The PL spectra are characterized by free excitons, denoted by X, at 3.376 eV, neutral donor-bound excitons, denoted by D^0X, at 3.3647 and 3.3604 eV, and biexcitons, denoted by M, at 3.3571 eV. At the lowest excitation, the line width of D^0X at 3.3604 eV is 0.7 meV, which is comparable to that of bulk ~0.7 meV and indicates high optical qualities of the nanorods. In addition, there is a peak at 3.313 eV, which is attributed to donor–acceptor-pair emission, denoted as DAP. The biexciton emission was distinct until 190 K. The thermal energy at this temperature (~16.3 meV) corresponds well to the biexciton binding energy (17.3 meV, ~200 K). In other words, at temperatures higher than this value, the biexciton is easily ionized to free excitons due to thermal scattering, which results in a reduced biexciton population and then the emission intensity of biexciton. At lower excitation intensities, the intensity of D^0X decreases, whereas the free exciton emission becomes stronger with increasing temperature. This can be explained by thermal dissociation of D^0X. At higher excitation intensities, however, the biexciton emission M is predominant over the free exciton emission, indicating higher emission efficiency for the former. The relative intensity of the eA^0 and X peaks at high temperature depends on the impurity of defect level of the sample [17].

For samples with high density of trap states, especially deep traps, the PL spectrum could show relatively strong deep trap state emission in the low energy part of the spectrum [18, 19]. As discussed for semiconductor QDs in Chapter 5, the relative intensity of trap state versus bandedge or excitonic emission provides an indication of density of trap states or defect states, higher relative trap state/bandedge PL intensity corresponding to higher density of trap states and overall lower PL yield. For the same sample, this PL intensity ratio is expected to be lower at lower temperature due to hindered activation to populate the trap states from the bandedge states.

Figure 6.7 shows the PL spectra of three different-size anatase TiO_2 nanoparticle samples [20]. The PL spectra are different for the three samples, possibly due to different surface properties. Based on the analysis of absorption spectra, the low energy absorption band is attributed to indirect bandgap transitions with transition energy around 2.97 eV and 3.21 eV. The next stronger transition is due to direct bandgap transition with bandgap energy of around 4.03 eV. The absorption spectra shown in Fig. 6.7 does not show the low energy absorption band from the indirect bandgap transition due to very low TiO_2 loading or concentration. In the PL spectra shown, one can clearly observe several PL bands for each sample. These bands have been analyzed and

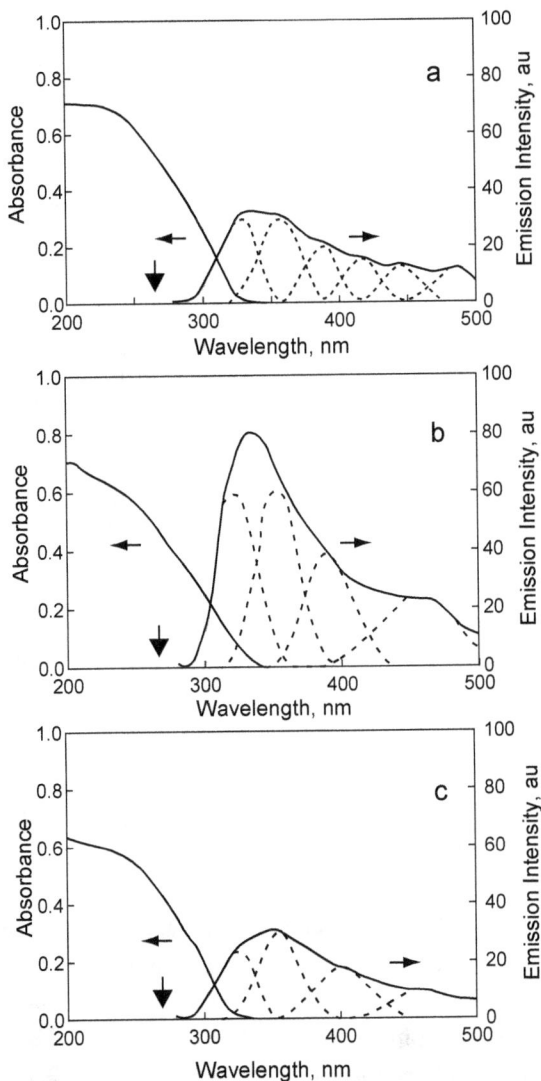

Fig. 6.7. Absorption and fluorescence spectra of different-size TiO$_2$ nanoparticles: (a) specimen A (average particle diameter $2R = 2.1$ nm), (b) specimen B ($2R = 13.3$ nm), and (c) specimen C ($2R = 26.7$ nm). TiO$_2$ loading, 0.015 g L^{-1} for (a, b) and 0.3 g L^{-1} for (c); pH 2.7 in all cases; excitation wavelength 270 nm (indicated by downward pointing arrow). Reproduced from Ref. 20.

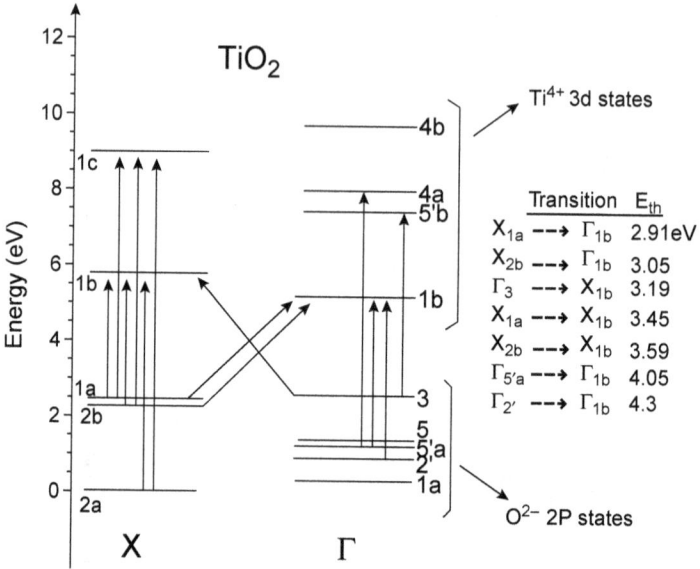

Fig. 6.8. Short energy level diagram illustrating the relative energy levels in TiO_2 as calculated by Daude *et al.* [21] The arrows indicate a few of the allowed direct and indirect transitions. The level X_{2a} positioned at zero energy for the sake of simplicity. X denotes the edge and Γ the center of the Brillouin zone (BZ). Reproduced from Ref. 20.

assigned based on theoretical studies of the electronic band structure of TiO_2 [21].

Figure 6.8 shows a summary of some of the states and transitions involved [20]. Generally, the agreement between the calculated transition energies and those determined experimentally is very good. The bandgap energies have been found to be independent of particle size down to about 2 nm [20], contrary to some earlier studies [22].

Similar to nonoxide semiconductors, Raman is a useful light scattering technique for characterizing structural properties of MO nanostructures. As discussed in Chapters 2 and 5, photoluminescence and Raman share some common features, and differ mainly in the time scale of the scattering process. Raman scattering has been applied to the study of many MO nanostructures, including TiO_2 [23, 24], ZnO [25, 26], and SnO_2 [27]. For example, Fig. 6.9 shows the Raman spectra of ZnO

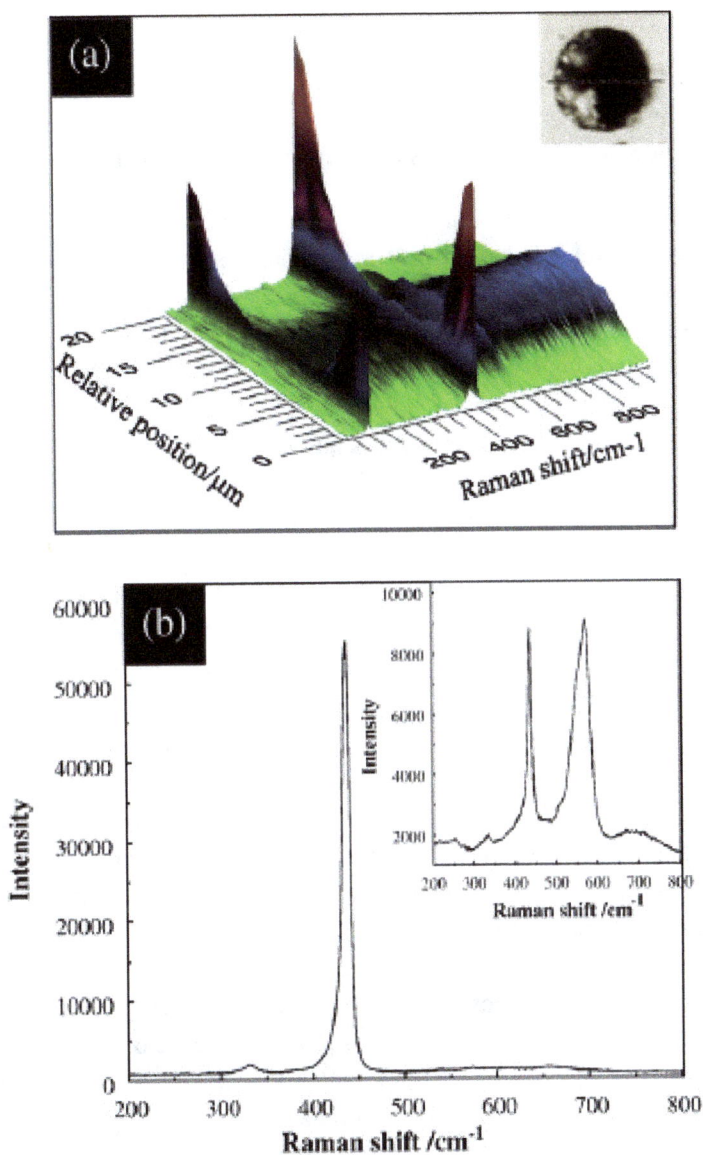

Fig. 6.9. (a) Raman spectra along the diameter of the irradiated area of ZnO nanorods every 0.5 μm; the insert shows the origin of the relative position of the ablated area. (b) Raman spectra of the side and center (inset) of the sample. Reproduced with permission from Ref. 26.

nanorods that demonstrate spatial dependence of the Raman spectrum, reflecting local structural variations [26].

6.3. Other optical properties: doped and sensitized metal oxides

Since most MOs have weak or no visible absorption, it is highly desired to enhance their visible absorption in applications such as photovoltaic (PV) solar cells, photocatalysis and photoelectrochemistry (PEC). To achieve this, different strategies have been developed, including doping and sensitization using visible absorbing chromophores such as dye molecules or small bandgap semiconductor QDs.

As mentioned before, doping is an effective way to alter the properties of a host insulator or semiconductor material [28–31]. The principle behind this is to introduce electronic states within the bandgap so as to create additional transitions and influence the property and functionality of the host material. Doping has been used for bulk materials for a long time, e.g. p- or n-doped Si critical to the electronics industry. For instance, MOs such as TiO_2 have been doped with elements such as C, S and N [32–36]. N-doping, in particular, has been found to be very effective in extending the visible absorption of TiO_2, which is highly desirable for PV and PEC applications [32, 37–46]. Figure 6.10 shows a visual comparison and the UV-vis reflectance spectra of N-doped TiO_2 and related nanoparticles [47]. As can be seen, successful N-doping (curve d) substantially shifts the absorption (less reflection) into the visible, giving a bright yellow color of the sample instead of the white TiO_2 without doping.

Similarly, ZnO nanostructures have been doped with different elements such as N, Ni, Ag, Mn and Co [48, 49]. Figure 6.11 shows UV-vis and magnetic circular dichroism (MCD) spectra of Co^{2+}-doped ZnO nanoparticles and photocurrent internal conversion efficiency data for photovoltaic cells fabricated based on the $ZnO:Co^{2+}$ nanoparticle films [49]. The enhancement of visible absorption by Co^{2+} doping is very obvious. Such systems are of interest for potential magnetic applications due to their strong ferromagnetism [50].

Sensitization of MO using dye molecules has attracted substantial attention due to its promise in solar energy conversion and other

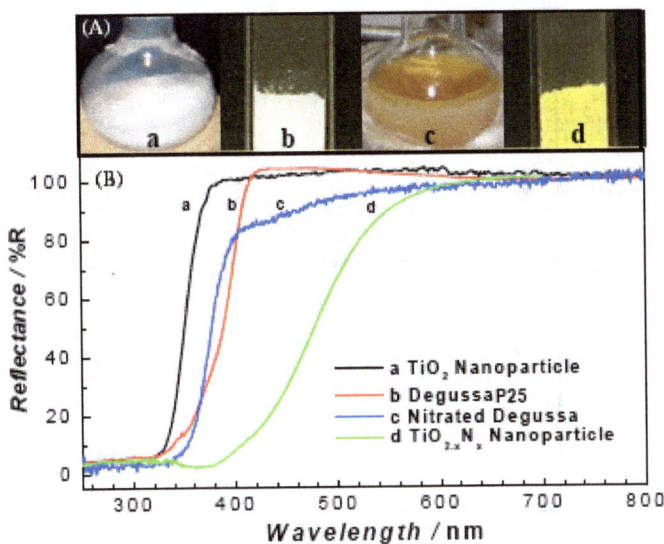

Fig. 6.10. Visual comparison (A) and UV-visible reflectance spectra (B) of (a) TiO_2 nanocolloid particles; (b) Degussa P25 TiO_2 powder; (c) Degussa P25 TiO_2 powder nitrided with triethylamine; and (d) nitrided $TiO_2–xNx$ nanocolloid particles. Reproduced with permission from Ref. 47.

applications [11]. A typical dye-sensitized nanocrystalline solar cell consists of MO, TiO_2 nanoparticles deposited on a highly porous wide-bandgap semiconductor electron acceptor layer. Dye molecules are adsorbed to the surface of TiO_2. Under irradiation, the photoexcited dye molecules inject electrons to the TiO_2 layer that are transported through the porous TiO_2 layer and collected by a conductive SnO_2 layer on the glass surface. The oxidized dye is regenerated by a liquid electrolyte. With the highest reported efficiency of about 10%, the dye-sensitized TiO_2 nanocrystalline solar cell has emerged as a potentially cost effective alternative to silicon solar cell technology [51, 52]. Organic conjugated polymers have also been explored as potential solid electrolytes and sensitizers for TiO_2 solar cells [53–55]. However, most conjugated polymers are generally charac-terized by poor hole transport, low thermal stability and high oxygen sensitivity [56].

One limitation with organic dye sensitization is long term thermal or photochemical instability. Inorganic materials such as QDs are attractive

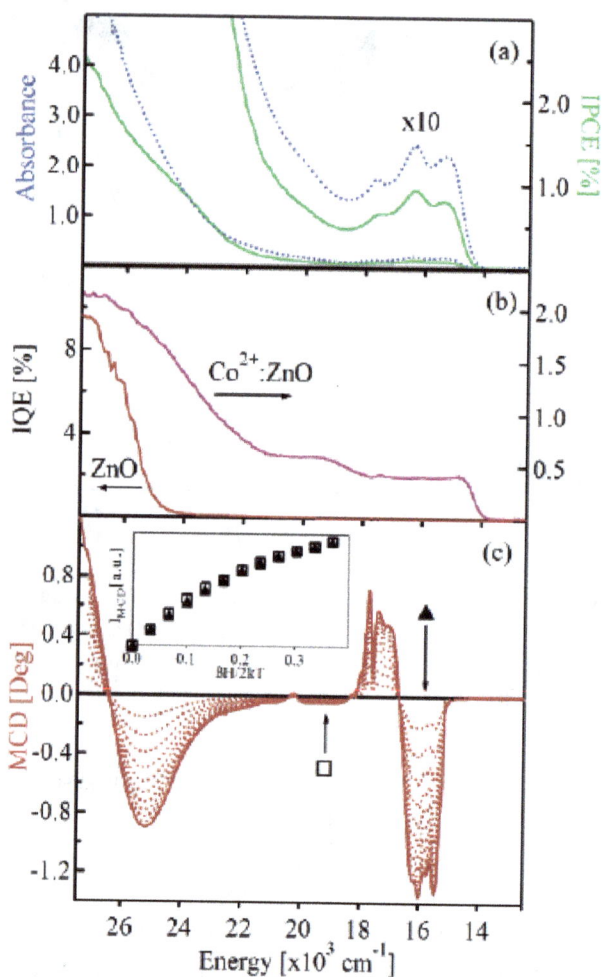

Fig. 6.11. (a) Absorption (dotted) and photocurrent action (solid) spectra of the 4.1 μm thick 3% Co^{2+}:ZnO(amine) nanocrystalline photoanode. The absorption spectrum was taken prior to solar cell assembly. (b) Photocurrent internal quantum efficiency (IQE) for the Co^{2+}:ZnO photovoltaic cell from (a), compared with the analogous undoped ZnO photovoltaic cell. (c) Variable-field (5 K, 0–6.5 T) MCD spectra of a frozen solution of 5 nm diameter 1.7% Co^{2+}:ZnO nanocrystals. Inset: 5 K MCD saturation magnetization data collected for the bands at 19,300 (□) and 15,800 cm^{-1} (▲). Reproduced with permission from Ref. 49.

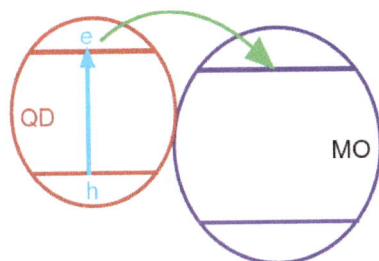

Fig. 6.12. Illustration of sensitization of a MO nanoparticle using a semiconductor QD. The QD is designed to have visible absorption and has the conduction bandedge above that of the MO to ensure efficient electron injection from QD to MO.

alternatives for dye molecules in sensitization of MOs [57, 58]. The idea is illustrated in Fig. 6.12. The key requirements are that the sensitizing QDs have visible absorption and the bottom of the CB is higher in energy than the bottom of the CB of the MOs to be sensitized.

As an example, CdTe QDs have been used successfully to sensitize TiO_2 in solar cell applications [57]. Interestingly, it has been found that the combined use of QD sensitization *and* N-doping substantially improved the photoresponse of TiO_2 as compared to using each of the two strategies separately. This has been attributed to enhanced hole transport by the occupied N energy level, as shown in Fig. 6.13. This preliminary study suggests the importance and power of engineering or manipulating the bandgap energy level for specific applications.

6.4. Nonlinear optical properties: luminescence up-conversion (LUC)

As discussed in Chapter 5, one interesting optical phenomenon involving doped materials is luminescence up-conversion where the photons emitted have a higher energy than the incident photon energy. This involves multiple photon excitation and single photon emission. The PL spectrum therefore appears in the blue of the absorption spectrum, rather than in the red as normally for the case of conventional PL spectrum. Potential

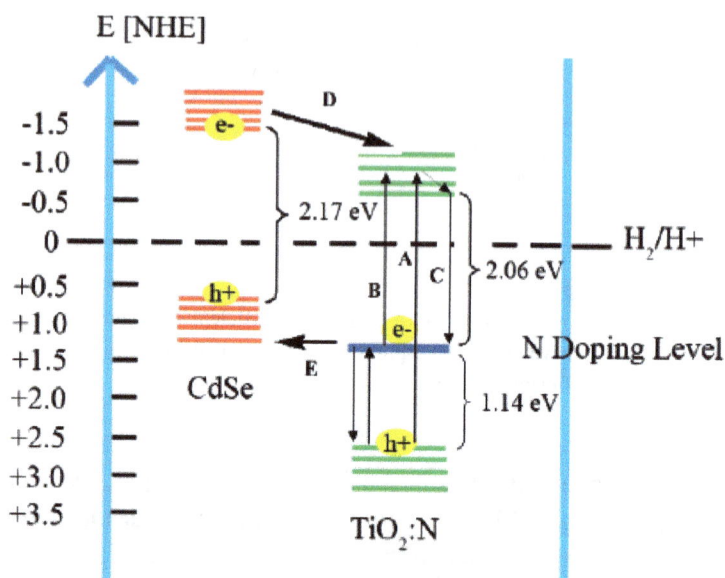

Fig. 6.13. Proposed energy diagram of TiO_2 nanoparticles sensitized with CdTe QDs and simultaneously doped with N. Electronic band structure of 3.5 nm CdSe with an effective bandgap of 2.17 eV and nanocrystalline TiO_2:N with a 3.2 eV bandgap, associated with normal TiO_2 and a N dopant state approximately 1.14 eV above the valence band. Different electron and hole creation, relaxation and recombination pathways are illustrated, including (A) photoexcitation of an electron from the valence band (VB) to the conduction band (CB) of TiO_2, (B) transition or photoexcitation of an electron from the N energy level to the CB of TiO_2, (C) recombination of an electron in the CB of TiO_2 with a hole in the N energy level, (D) electron transfer or injection from the CB of a CdSe QD to the CB of TiO_2, and (E) hole transfer from the VB of a CdSe QD to the N energy level. Note that not all of these processes can happen simultaneously, and many of these are competing processes. Reproduced with permission from Ref. 57.

applications of LUC include energy conversion and biomedical detection and imaging.

In nanomaterials, LUC can be enhanced by energy levels that happen to be resonant to the energy of the incident photons. Such processes have been discussed for doped and pristine semiconductors in Chapter 5. In this section, we focus on discussion of LUC in insulators or metal oxides that usually involve doping.

For example, efficient infrared-to-visible conversion is reported in thin film nano-composites, with composition 90% SiO_2–10% TiO_2, fabricated by a spin-coating sol–gel route and co-doped with Er^{3+}:Yb^{3+} and with Nd^{3+}:Yb^{3+} ions [59]. The conversion process is observed under 808 nm laser diode excitation and results in the generation of green (526 and 550 nm) and red (650 nm) emissions: from the former, and blue (478 nm) and green (513 and 580 nm) emissions from the latter. The main mechanism that allows for up-conversion is ascribed to energy transfer among Er^{3+}:Yb^{3+} or Nd^{3+}:Yb^{3+} ions in their excited states.

Other examples of doped metal oxide nanoparticles that exhibit LUC include Er^{3+}-doped $BaTiO_3$ [60], silica thin film containing $La_{0.45}$-$Yb_{0.50}Er_{0.05}F_3$ nanoparticles [61], neodymium oxalate nanoparticles/TiO_2/organically modified silane composite thin films [62], and Tm^{3+}/Yb^{3+}-co-doped heavy metal oxide-halide glasses [63]. In the case of Tm^{3+}/Yb^{3+}-co-doped glass [63], intense blue and weak red emissions centered at 477 and 650 nm, corresponding to the transitions $^1G_4 \rightarrow {}^3H_6$ and $^1G_4 \rightarrow {}^3H_4$, respectively, were observed at room temperature, which are due to three-photon absorption processes with 975 nm excitation from a laser diode. Related energy levels are shown in Fig. 6.14. It was found that the LUC was enhanced by the replacement of oxygen ion by halide ion and by the decrease of phonon density and maximum phonon energy of host glasses.

6.5. Summary

In general, the optical properties of MO semiconductor nanomaterials are similar to those of other semiconductor nanomaterials with the main difference in their bandgap energy and thereby spectral range of absorption and emission. Most MOs do not have strong visible absorption, with a few exceptions. The PL yield is usually lower for MOs compared to semiconductors QDs, due possibly to a higher density of defects or trap states. Similar to semiconductors, MOs exhibit nonlinear optical properties such as luminescence up-conversion, often requiring doping to enhance the phenomenon.

In this chapter, we have focused mainly on MOs that are considered as semiconductors in terms of electrical or optical characteristics. Insulator

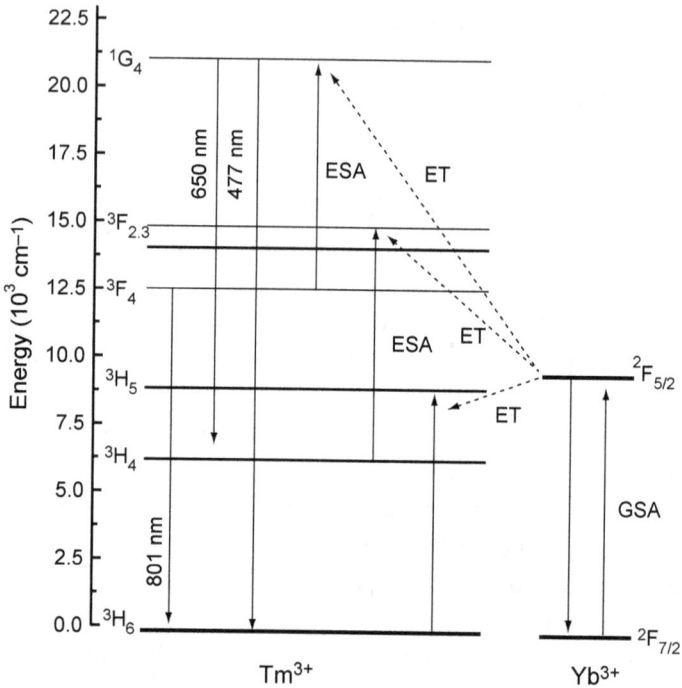

Fig. 6.14. Simplified energy level diagram of Tm^{3+} and Yb^{3+} ions and possible transition pathways in glasses. GSA, ESA and ET stand for ground state absorption, excited state absorption and energy transfer, respectively. Reproduced with permission from Ref. 63.

metal oxides are not the focus of this book due to their lack of visible absorption due to very large bandgap. With above bandgap excitation using VUV light, insulators behave like semiconductors in terms of their optical properties, absorption and emission, with the difference that the spectral range shifts toward higher energy or shorter wavelength.

References

1. M. Gratzel, *Heterogeneous Photochemical Electron Transfer*, Boca Raton: CRC Press. 176 (1989).
2. Y.S. Kang, S. Risbud, J.F. Rabolt and P. Stroeve, *Chem. Mater.* **8**, 2209+ (1996).

3. B.C. Faust, M.R. Hoffmann and D.W. Bahnemann, *J. Phys. Chem.* **93**, 6371 (1989).

4. J.K. Leland and A.J. Bard, *J. Phys. Chem.* **91**, 5076 (1987).

5. R.F. Ziolo, E.P. Giannelis, B.A. Weinstein, M.P. Ohoro, B.N. Ganguly, V. Mehrotra, M.W. Russell and D.R. Huffman, *Science* **257**, 219 (1992).

6. R.M. Cornell and U. Schwertmann, *The Iron Oxides: Structure, Properties, Reactions, Occurrence and Uses*, New York: VCH (1996).

7. T. Hyeon, S.S. Lee, J. Park, Y. Chung and H. Bin Na, *J. Am. Chem. Soc.* **123**, 12798 (2001).

8. S.Y. Zeng, K.B. Tang and T.W. Li, *J. Colloid Interf. Sci.* **312**, 513 (2007).

9. N.J. Cherepy, D.B. Liston, J.A. Lovejoy, H.M. Deng and J.Z. Zhang, *J. Phys. Chem. B* **102**, 770 (1998).

10. U. Schwertmann and R.M. Cornell, *Iron Oxides in the Laboratory: Preparation and Characterization.* 2nd ed, New York: Wiley-VCH (2000).

11. B. Oregan and M. Gratzel, *Nature* **353**, 737 (1991).

12. Y.H. Lin, D.J. Wang, Q.D. Zhao, M. Yang and Q.L. Zhang, *J. Phys. Chem. B* **108**, 3202 (2004).

13. I. Martini, J.H. Hodak and G.V. Hartland, *J. Phys. Chem. B* **102**, 607 (1998).

14. H.L. Wang, T. Lindgren, J.J. He, A. Hagfeldt and S.E. Lindquist, *J. Phys. Chem. B* **104**, 5686 (2000).

15. J.K. Leland and A.J. Bard, *J. Phys. Chem.* **91**, 5083 (1987).

16. T. Nutz and M. Haase, *J. Phys. Chem. B* **104**, 8430 (2000).

17. B.P. Zhang, N.T. Binh, Y. Segawa, Y. Kashiwaba and K. Haga, *Appl. Phys. Lett.* **84**, 586 (2004).

18. C.M. Mo, Y.H. Li, Y.S. Liu, Y. Zhang and L.D. Zhang, *J. Appl. Phys.* **83**, 4389 (1998).

19. X. Li, F.F. Zhai, Y. Liu, M.S. Cao, F.C. Wang and X.X. Zhang, *Chinese Phys.* **16**, 2769 (2007).

20. N. Serpone, D. Lawless and R. Khairutdinov, *J. Phys. Chem.* **99**, 16646 (1995).

21. N. Daude, C. Gout and C. Jouanin, *Phys. Rev. B* **15**, 3229 (1977).

22. W.Y. Choi, A. Termin and M.R. Hoffmann, *J. Phys. Chem.* **98**, 13669 (1994).

23. J. Zhang, M.J. Li, Z.C. Feng, J. Chen and C. Li, *J. Phys. Chem. B* **110**, 927 (2006).

24. W.G. Su, J. Zhang, Z.C. Feng, T. Chen, P.L. Ying and C. Li, *J. Phys. Chem. C* **112**, 7710 (2008).

25. J.F. Xu, W. Ji, X.B. Wang, H. Shu, Z.X. Shen and S.H. Tang, *J. Raman Spectroscopy* **29**, 613 (1998).

26. X.D. Guo, R.X. Li, Y. Hang, Z.Z. Xu, B.K. Yu, H.L. Ma and X.W. Sun, *Mater. Lett.* **61**, 4583 (2007).

27. A. Dieguez, A. RomanoRodriguez, J.R. Morante, N. Barsan, U. Weimar and W. Gopel, *Appl. Phys. Lett.* **71**, 1957 (1997).
28. P.V. Radovanovic, N.S. Norberg, K.E. McNally and D.R. Gamelin, *J. Am. Chem. Soc.* **124**, 15192 (2002).
29. W. Chen, J.Z. Zhang and A. Joly, *J. Nanosci. Nanotechnol.* **4**, 919 (2004).
30. J.D. Bryan and D.R. Gamelin, *Progr. Inorganic Chem.* **54**, 47 (2005).
31. D. Magana, S.C. Perera, A.G. Harter, N.S. Dalal and G.F. Strouse, *J. Am. Chem. Soc.* **128**, 2931 (2006).
32. J.L. Gole, J.D. Stout, C. Burda, Y.B. Lou and X.B. Chen, *J. Phys. Chem. B* **108**, 1230 (2004).
33. J.H. Park, S. Kim and A.J. Bard, *Nano Lett.* **6**, 24 (2006).
34. Q. Shen, K. Katayama, T. Sawada, M. Yamaguchi and T. Toyoda, *Jpn J. Appl. Phys.* **1**, 45, 5569 (2006).
35. Q.H. Yao, J.F. Liu, Q. Peng, X. Wang and Y.D. Li, *Chemistry-an Asian J.* **1**, 737 (2006).
36. Y. Murakami, B. Kasahara and Y. Nosaka, *Chem. Lett.* **36**, 330 (2007).
37. C. Burda, Y.B. Lou, X.B. Chen, A.C.S. Samia, J. Stout and J.L. Gole, *Nano Lett.* **3**, 1049 (2003).
38. X.B. Chen and C. Burda, *J. Phys. Chem. B* **108**, 15446 (2004).
39. K. Kobayakawa, Y. Murakami and Y. Sato, *J. Photoch. Photobio A* **170**, 177 (2005).
40. M. Sathish, B. Viswanathan, R.P. Viswanath and C.S. Gopinath, *Chem. Mater.* **17**, 6349 (2005).
41. Y. Nakano, T. Morikawa, T. Ohwaki and Y. Taga, *Appl. Phys. Lett.* **86**, 132104 (2005).
42. S.W. Yang and L. Gao, *J. Inorg. Mater.* **20**, 785 (2005).
43. A.R. Gandhe, S.P. Naik and J.B. Fernandes, *Micropor. Mesopor. Mat.* **87**, 103 (2005).
44. K. Yamada, H. Nakamura, S. Matsushima, H. Yamane, T. Haishi, K. Ohira and K. Kumada, *Cr. Chim.* **9**, 788 (2006).
45. P. Xu, L. Mi and P.N. Wang, *J. Cryst. Growth* **289**, 433 (2006).
46. H.Y. Chen, A. Nambu, W. Wen, J. Graciani, Z. Zhong, J.C. Hanson, E. Fujita and J.A. Rodriguez, *J. Phys. Chem. C* **111**, 1366 (2007).
47. X.B. Chen, Y.B. Lou, A.C.S. Samia, C. Burda and J.L. Gole, *Adv. Funct. Mater.* **15**, 41 (2005).
48. H. Matsui, H. Saeki, T. Kawai, H. Tabata and B. Mizobuchi, *J. Appl. Phys.* **95**, 5882 (2004).
49. W.K. Liu, G.M. Salley and D.R. Gamelin, *J. Phys. Chem. B* **109**, 14486 (2005).

50. J.D. Bryan, S.M. Heald, S.A. Chambers and D.R. Gamelin, *J. Am. Chem. Soc.* **126**, 11640 (2004).
51. A. Hagfeldt and M. Gratzel, *Acc. Chem. Res.* **33**, 269 (2000).
52. B.A. Gregg, *J. Phys. Chem. B* **107**, 4688 (2003).
53. C.D. Grant, A.M. Schwartzberg, G.P. Smestad, J. Kowalik, L.M. Tolbert and J.Z. Zhang, *J. Electroanal. Chem.* **522**, 40 (2002).
54. G.P. Smestad, S. Spiekermann, J. Kowalik, C.D. Grant, A.M. Schwartzberg, J. Zhang, L.M. Tolbert and E. Moons, *Sol. Energ. Mat. Sol. C* **76**, 85 (2003).
55. C.D. Grant, A.M. Schwartzberg, G.P. Smestad, J. Kowalik, L.M. Tolbert and J.Z. Zhang, *Synthet. Metal* **132**, 197 (2003).
56. J. Kuendig, M. Goetz, A. Shah, L. Gerlach and E. Fernandez, *Solar Energy Mater. Solar Cells* **79**, 425 (2003).
57. T. Lopez-Luke, A. Wolcott, L.-P. Xu, S. Chen, Z. Wen, J.H. Li, E. De La Rosa and J.Z. Zhang, *J. Phys. Chem. C* **112**, 1282 (2008).
58. A. Kongkanand, K. Tvrdy, K. Takechi, M. Kuno and P.V. Kamat, *J. Am. Chem. Soc.* **130**, 4007 (2008).
59. I.K. Battisha, *J. Non-Crystalline Solids* **353**, 1748 (2007).
60. H.X. Zhang, C.H. Kam, Y. Zhou, X.Q. Han, S. Buddhudu and Y.L. Lam, *Opt. Mater.* **15**, 47 (2000).
61. S. Sivakumar, F. van Veggel and P.S. May, *J. Am. Chem. Soc.* **129**, 620 (2007).
62. W.X. Que and C.H. Kam, *Opt. Mater.* **19**, 307 (2002).
63. H.T. Sun, Z.C. Duan, G. Zhou, C.L. Yu, M.S. Liao, L.L. Hu, J.J. Zhang and Z.H. Jiang, *Spectrochim. Acta. A* **63**, 149 (2006).

Chapter 7

Optical Properties of Metal Nanomaterials

Metal nanomaterials are an important class of nanomaterials with fascinating optical, electronic, magnetic and other properties. Due to the fundamentally different electronic band structures of the metal from that of a semiconductor and insulator, metal nanomaterials possess a number of unique features in their electronic and optical properties as compared to semiconductor or insulator nanomaterials. For example, due to the fact that the bandedge is more affected by quantum confinement and the Fermi level of metal is at the center of the band, as shown in Fig. 7.1, metals are less affected by spatial confinement than semiconductors for similar size range. In other words, quantum confinement effect becomes important for metal nanostructures only at very small sizes (1 nm or less). For the size regime of tens of nm or larger, the behavior of metal nanostructures is essentially classical, which is consistent with experimental studies.

Another unique feature is that the optical properties of metal nanostructures are, relatively speaking, more sensitive to shape and less sensitive to size, in contrast to semiconductor or insulator nanomaterials with optical properties more sensitive to size and less sensitive to shape. This feature is related to that, as mentioned before, metal nanomaterials behave more or less classically until they attain a very small size and their optical properties are predominantly determined by the collective properties of the conduction band electrons. Therefore, classical electrodynamics

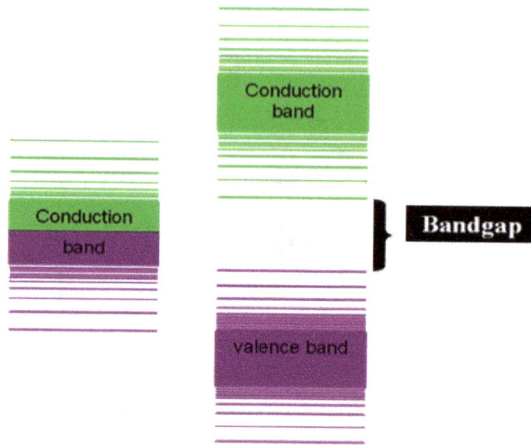

Fig. 7.1. Comparison of electronic band structure of metal (left) versus semiconductor (right) nanomaterials. The Fermi level is in the middle of the bandgap for the semiconductor while it is at the middle of the conduction band for the metal. The purple color represents the electron occupied states and the green represents empty states.

or mechanics can be used to successfully describe most of the behaviors of metal nanostructures, including electronic as well as phonon properties. Very small metal nanoparticles or clusters behave more like molecules and will not be the focus in this book. Compared to bulk metal, the very large surface-to-volume ratio is a prominent feature that affects the properties of metal nanostructures such as chemical reactivity, thermodynamic and kinetic stability, interaction with the embedding environment, and electron and phonon relaxation dynamics. Some of the key interesting optical and associated properties of metal nanostructures will be discussed in more detail next, with emphasis on surface plasmon resonance (SPR) absorption and surface enhanced Raman scattering (SERS).

7.1. Strong absorption and lack of photoemission

Metal nanostructures generally have strong optical absorption but weak photoemission or luminescence. This is true typically when the photoexcitation is intraband or within the same electronic energy band. If excitation is interband, i.e. between two bands, especially when high energy or short wavelength light is used, strong photoemission is possible, and is likely, because this situation then resembles that of bandgap excitation in a semiconductor or insulator. However, due to the usually high density of states, photoluminescence (PL) in metal even with interband transitions is often weaker than in semiconductors or insulators.

There are a number of reports of relatively strong PL in metal nanostructures [1–4]; and the origin of the observed PL is not always very clear. First of all, strong is relative and is used sometimes with reference to the very weak or lack of PL expected from metal nanostructures. Second, in some cases, the origin of the photoemission is not clearly identified. For example, surface species associated with the metal nanostructures could be photoemissive when excited. This is particularly true when short wavelength light is used, where species that are usually non-emissive with visible excitation become emissive due to higher energy photons used. This can happen with solvent or small amount of impurities as well. Therefore, care must be taken to conduct proper control experiments to understand the origin of photoemission whenever the sample contains multiple components of chemical species.

7.2. Surface plasmon resonance (SPR)

The optical properties of metal nanostructures in the visible region are dominated by the surface plasmon absorption caused by collective conduction band electron oscillation in response to the electrical field of the light radiation. In a typical experimental setup, the light absorbance, A,

$$A(\lambda) = \log_{10}(e^{\sigma_{ext}(\lambda)D}), \tag{7.1}$$

is measured in a sample composed of nanoparticles dispersed in a homogeneous matrix. Here, D is the optical path length, and $\sigma_{ext}(\lambda)$ is the extinction coefficient, which is given by the sum of scattering and absorption of the electromagnetic (EM) wave going through the sample. In the dilute limit where interaction between nanoparticles is weak and can be ignored, the extinction coefficient of the whole sample can be modeled as the optical response of one immersed nanoparticle times the particles concentration [5]:

$$\sigma_{ext}(\lambda) = f(C_{abs}(\lambda) + C_{sca}(\lambda)) \qquad (7.2)$$

where f is the number of particles per unit volume or volume-filling fraction, $C_{abs}(\lambda)$ and $C_{sca}(\lambda)$ are the absorption and scattering cross-sections of a single particle, respectively. The optical response of a nanoparticle, or $C_{abs}(\lambda)$ and $C_{sca}(\lambda)$, can, in principle, be obtained by solving the Maxwell equations. Nanoparticles are large enough for classical EM theory to be employed, but they are still small enough to exhibit dependence of their optical properties on size and shape. This means that each point of the nanoparticle can be described in terms of a macroscopic dielectric function, $\varepsilon(\lambda,R)$, which depends on the light wavelength or frequency and, sometimes, on nanoparticle radius, R, as will be discussed further later.

In 1908, Gustav Mie found the exact solution to the Maxwell equations for the optical response of a sphere of arbitrary size immersed in a homogeneous medium, and subject to a plane monochromatic wave [6]. In the Mie theory, the electric and magnetic field vectors are given in a series of the ratio (R/λ) in terms of Legendre Polynomials and Bessel spherical functions. From this series, one can interpret the absorbed and scattered EM fields in terms of the different multipolar contributions [7]. When the size of the particle is much smaller than the wavelength of the incident light, the nanoparticle experiences a field that is spatially constant, but with a time dependent phase, which is known as the quasi-static limit. In this limit, the displacement of the charges in a sphere is homogeneous, yielding a dipolar charge distribution on the surface, while for larger spherical particles higher multipolar distributions are excited. In this case, the particle only absorbs energy by the excitation of the *surface plasmon resonance* (*SPR*), and the *absorption cross-section* can be calculated using the first term of the Mie theory [8]:

$$C_{abs}(\lambda) = \frac{18\pi f\, \varepsilon_m^{3/2}}{\lambda} \frac{\varepsilon_2}{(2\varepsilon_m + \varepsilon_1)^2 + \varepsilon_2^2}, \qquad (7.3)$$

where f is the volume-filling fraction of the metal, ε_m is the dielectric constant of the medium in which the particles are embedded, ε_1 and ε_2 are the real and imaginary parts of the complex dielectric constant of the metal particle with a radius R. This SPR of dipolar character dominates the absorption of small spherical particles, and its position is independent of the particle size and direction of the incident light such that only one proper mode is found. The Mie theory has been subsequently modified by others to incorporate other effects including quantum effect [8, 9].

From Eq. (7.3), one can reasonably predict the position and line shape of the plasmon absorption for spherical metal particles [10]. According to this equation, when ε_2 is small or does not change so much around the band, the absorption coefficient has a maximum value at a resonance frequency that satisfies:

$$\varepsilon_1 = -2\varepsilon_m. \tag{7.4}$$

The location of this plasmon resonance is therefore determined by the wavelength dependence of $\varepsilon_1(\omega)$, and its width by $\varepsilon_2(\omega)$ and ε_m. This resonance is generally around 400 nm for Ag and 520 nm for Au spherical particles in water.

If the particle size becomes comparable to or smaller than the mean free path of the conduction band electrons or "free" electrons, the collisions of the electrons with the particle surface becomes important and the effective mean free path is less than that in bulk materials. This usually results in broadening and blue-shift of the plasmon band for particles smaller than about 10 nm [7, 11–13].

On the other hand, rigorous solutions of the Maxwell equations for nonspherical particles are not straightforward. Only a few exact solutions are known: the case of spheroids [14] and infinite cylinders [15]. Thus, the optical properties of nanoparticles with other arbitrary shapes can be found only in approximate ways. Because of the many possible complex shapes of nanoparticles, efficient computational methods capable of treating these structures are essential. In the last few years, several numerical methods have been developed for calculating the optical properties of nonspherical nanoparticles, such as the discrete dipole approximation (DDA) [16–19] and the finite-difference time-domain method (FDTD) [20–22], which have been extensively used for calculations.

Because spherical particles are in principle completely symmetrical, there is only one dipolar plasmon resonance, or, in other words, all the

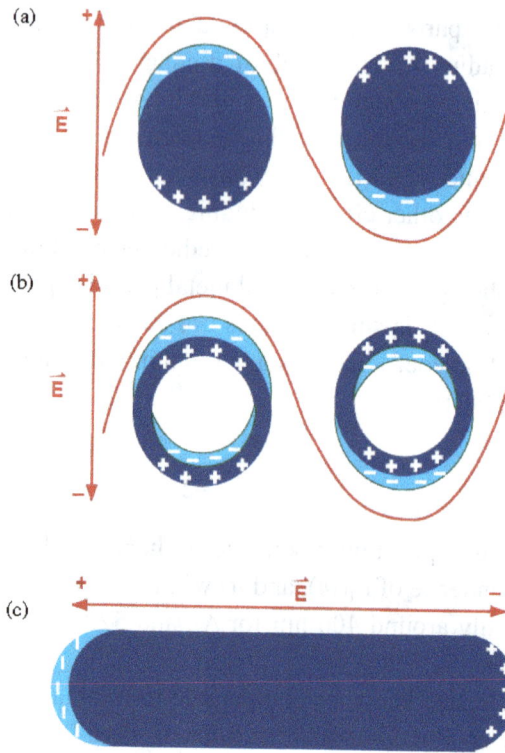

Fig. 7.2. Schematic illustration of dipolar SPR in spherical nanoparticles (a), core/shell structures (b), and nanorods (c) in a polarized light field. For the nanorod, only the longitudinal mode is shown. The transverse mode is essentially the same as in spherical nanoparticles. The lower the symmetry of the structure, the more nondegenerate modes there are. Both the solid spherical nanoparticles and hollow nanospheres have one resonance due to high symmetry. The nanorods with one symmetry axis have two nondegenerate modes (only one, the longitudinal mode, is shown). Complex structures such as strongly aggregated nanoparticles typically have multiple nondegenerate dipole modes, resulting in broad plasmon absorption spectra. Reproduced with permission from Ref. 25.

possible dipolar modes are degenerate [Fig. 7.2(a)]. Similar argument indicates a single dipolar mode for hollow nanospheres, [Fig. 7.2(b)]. By elongating a spherical particle in one dimension, however, it is possible to form a second, lower energy dipolar resonance mode in the longitudinal direction while the original transverse mode persists, as illustrated schematically in Fig. 7.2(c). As the nanostructures become more complex, many more nondegenerate dipolar modes arise and their optical absorption becomes more complex with many bands over a broad spectral region. For example, small ellipsoids with three different axes have three different dipole modes. Furthermore, as the particle becomes less symmetric, the induced charge distribution on the surface can result in not only dipolar modes with different resonant frequencies but also higher multipolar charge distributions, even in the quasi-static limit. For instance, the induced electronic cloud on nonspherical nanoparticles is not distributed homogeneously on the surface such that higher multipolar charge distributions are clearly induced [16–19, 23, 24].

The higher multipolar SPRs are always located at shorter wavelengths with respect to the dipolar ones, which, additionally, are always red-shifted by the presence of the electric field generated by higher multipolar charge distributions. Examples include nanorods, nanoprims, nanocages, aggregates and hollow nanospheres [26–40]. When nanoparticles are dispersed in a matrix, their random orientation leads to an average absorption spectrum containing all SPRs associated to the corresponding geometry. On the other hand, when all the nanoparticles in a system are oriented in the same direction, it is possible to distinguish between the different resonances by using polarized light [41]. The orientation of the nanoparticles means a macroscopic anisotropy that can be related to optical birefringence [42, 43] and, under some circumstances, to optical noncentrosymmetry, giving rise to nonlinear optical properties [44].

Figure 7.3 shows the electronic absorption spectra of typical gold and silver nanoparticles. Their characteristic SPR bands are around 520 nm for gold and 400 nm for silver as have been well-documented.

As an example to illustrate strong shape dependence of SPR, Fig. 7.4 shows the electronic absorption spectra of gold nanorods with different aspect ratios [45]. It is clear that the SPR bandwidth and peak position are both sensitive to the nanorod aspect ratio. The blue band is due to the transverse plasmon resonance while the redder band is due to longitudinal plasmon resonance, which red-shifts with increasing nanorod length.

Aggregates are frequently encountered and can be useful for applications such as optical sensing. For aggregates with weak or moderate interparticle interaction, a typical signature is red-shift and some broadening

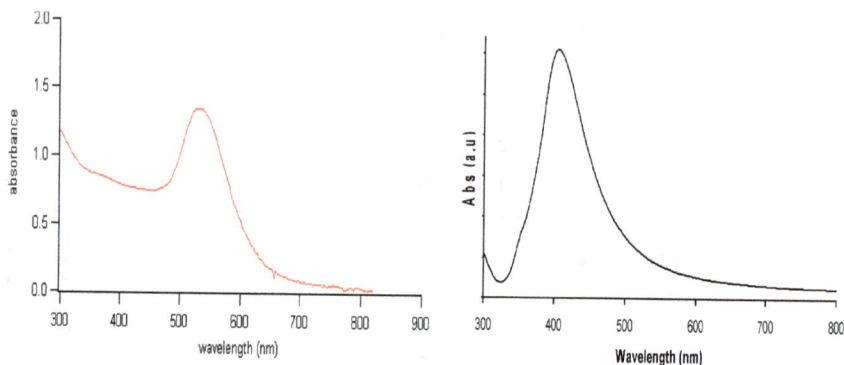

Fig. 7.3. Representative electronic absorption spectra of gold (left) and silver (right) nanoparticles synthesized in aqueous solution, with featured SPR at 525 nm for gold and 400 nm for silver. The exact SPR peak position and width are somewhat dependent on the solvent and particle size and distribution and therefore the specific synthetic methods used.

of the transverse SPR band. For aggregates with strong interparticle interaction, new bands appear due to electronic coupling between particles. Figure 7.5 shows an example of absorption spectra of gold nanoparticle aggregates produced by the reaction of $HAuCl_4$ and Na_2S [46]. While the blue band is due to transverse plasmon of isolated gold particles in the solution or aggregates, the red band is attributed to extended surface plasmon band of aggregates of gold nanoparticles with strong interaction. This reaction has been the subject of debate since the initial and several follow-up reports suggested an Au_2S/Au core/shell structure as the explanation for the observed near-IR absorption band [47]. However, later studies have suggested that aggregates are the more likely explanation [46]. Aggregation of nanoparticles can be induced by molecules on the particle surface and this, in turn, can be used as a means to detect molecules of interest based on change in SPR, usually quantitatively without molecular specificity capability [48, 49].

As discussed in Chapter 1, hollow gold nanospheres (HGNS) represent an excellent example of metal nanostructures that exhibit narrow and tunable SPR controllable by simply varying the shell thickness and sphere diameter (Fig. 1.2) [50]. The results are qualitatively in very good agreement with earlier theoretical calculations [20], demonstrating the power of theoretical work in understanding optical properties of metal nanostructures.

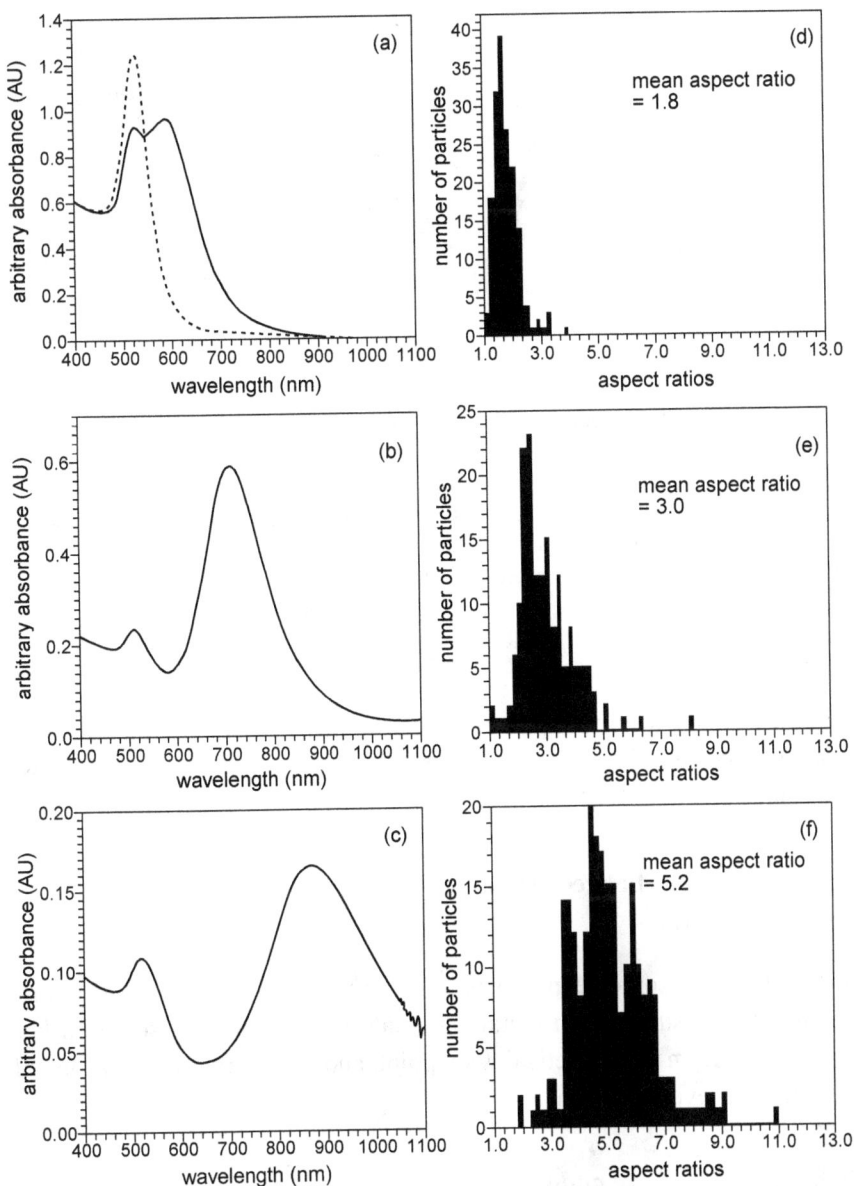

Fig. 7.4. Aspect ratio dependent absorption spectra of gold nanorods synthesized electrochemically. (a–c) Absorption spectra of suspended gold nanorods solutions with increasing mean aspect ratios. (d–f) Distributions of aspect ratios analyzed from the corresponding TEM micrographs; mean/FWHM = 1.8/0.9 (d), 3.0/1.9 (e), 5.2/3.0 (f). Reproduced with permission from Ref. 45.

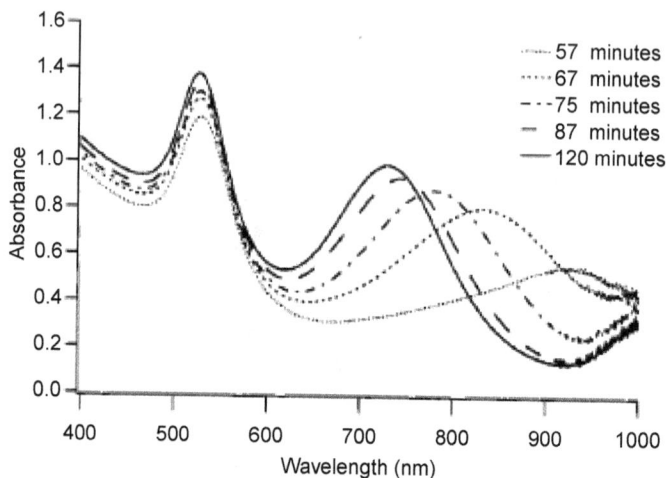

Fig. 7.5. UV-vis electronic absorption spectra of gold nanoparticle aggregates produced by the reaction of $HAuCl_4$ and Na_2S. The time indicated in the inset is the time of reaction. It is clear that the structure of the aggregates change during the first two hours of the reaction. The spectrum and aggregate size/shape stopped changing after about two hours. Reproduced with permission from Ref. 46.

Figure 7.6 shows some representative UV-visible spectra and a HRTEM image of the HGNS. The color is controlled by varying the wall thickness and outer diameter [50].

7.3. Correlation between structure and SPR: a theoretical perspective

In order to gain some deeper insight into the plasmonic absorption and its relation to the structure of metal nanomaterials, we provide some further discussion from a theoretical viewpoint and compare to experimental results whenever appropriate.

7.3.1. *Effects of size and surface on SPR of metal nanoparticles*

As mentioned before, the extinction coefficient responsible for the meas- ured UV-vis spectrum of metal nanoparticles has contributions from electronic absorption and scattering. The scattering effect becomes increas-

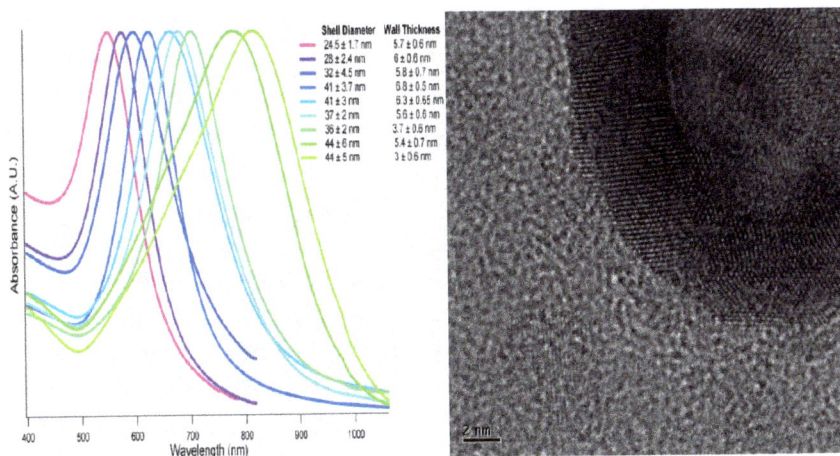

Fig. 7.6. UV-visible spectra (left) and HRTEM image of HGNS. For the HRTEM, only part of the sphere is shown. Adapted with permission from Ref. 50.

ingly more important for larger particles. For metallic nanoparticles of less than about 30 nm of diameter, the scattering processes are usually negligible, and the particle mainly absorbs light energy through either SPR or interband electronic transitions. SPR is dependent on the particle shape, size, embedding environment and chemical nature of the material. Due to the small size and thereby large surface-to-volume (S/V) ratio, surface dispersion or scattering of the "free" electrons, especially when their mean free path is comparable to the dimension of the nanoparticles, is an important process to consider, e.g. in electronic energy relaxation or dissipation.

For metal NPs smaller than ~10 nm, collision of electrons with the particle surface becomes important, which usually results in broadening and blue-shift of the SPR absorption band [7, 11, 12]. From a quantum viewpoint, for particles containing about <100 atoms the electronic energy bands are quantized and energy level spacing may become comparable to thermal energy kT. This affects intraband transitions of the conduction electrons and leads to a damping of electron motion, which, in turn, influences the dielectric constant [51].

One way to account for the size or surface effect is to divide the contribution to the dielectric constant into two parts: one from interband transitions and the second from intraband transitions including the surface effect [11, 18, 52, 53]:

$$\varepsilon_1(\omega) = \varepsilon_1^{\text{intra}}(\omega) + \varepsilon_1^{\text{inter}}(\omega)$$
$$\varepsilon_2(\omega) = \varepsilon_2^{\text{intra}}(\omega) + \varepsilon_2^{\text{inter}}(\omega)$$

(7.5)

The intraband contribution can be calculated using the Drude model for the nearly free electrons [53]:

$$\varepsilon_1^{\text{intra}}(\omega) = 1 - \frac{\omega_p^2}{\omega^2 + \Gamma^2}$$

$$\varepsilon_2^{\text{intra}}(\omega) = 1 - \frac{\omega_p^2 \Gamma}{\omega(\omega^2 + \Gamma^2)}$$

(7.6)

where ω_p is the plasmon frequency, which is related to the free electron density N, and Γ is a damping constant. For metals such as Al and Ag, where the onset of interband transitions is well separated from the plasmon band, the plasmon bandwidth is controlled by Γ [7]. For metals such as Au and Cu, where the interband transitions occur in the same spectral region as the plasmon band, the bandwidth is determined by both Γ and $\varepsilon^{\text{intra}}(\omega)$, which makes the analysis more difficult. In either case, however, the dependence of the bandwidth on particle size can be accounted for by expressing [7, 12, 52, 54]:

$$\Gamma = \Gamma_0 + A \frac{v_F}{R}$$

(7.7)

where R is the particle radius, v_F is the Fermi velocity of the electrons and is 1.4×10^8 cm/s for Au and Ag [11], A is a constant dependent on the electron-surface interaction and is usually on the order of unity [11, 12], and Γ_0 stands for the frequency of inelastic collisions of free electrons within the bulk metal, e.g. electron–phonon coupling or defects, and is on the order of hundredths of an eV [11]. For small-size particles, the second term in Eq. (7.7), which accounts for dephasing due to electron-surface scattering, can greatly exceed the bulk scattering frequency Γ_0, which accounts for bulk contribution to the electronic dephasing. For example, it has been determined experimentally for Au nanoparticles that $A = 0.43$ and $\Gamma_0 = 4.4 \times 10^{14}$ Hz, which implies an intrinsic, bulk electronic dephasing time of 2.3 fs [12]. It is clear from Eq. (7.7) that the smaller the particle size, the higher the surface collision frequency, which usually leads to a blue-shift of the plasmon peak. A more detailed discussion of dephasing and other dynamic processes will be given in Chapter 9.

It should be noted that the S/V ratio for spherical particles is inversely proportional to the particle radius R. Thus, as R decreases, the S/V ratio increases. In other words, change in S/V ratio is always associated with change in R and these two factors cannot be independently varied. This is important to keep in mind when interpreting data. In general, the SPR depends weakly on the particle radius R. As R increases, one generally expects a red-shift and broadening of the SPR and the absorption cross-section per particle will increase mainly due to an increase in particle volume. In the meantime, surface effect becomes less dominant and scattering increases with increasing R. Scattering effects become particularly important for nanoparticles of more than 30 nm in diameter, where electrons, accelerated due to the incident electromagnetic field, can radiate energy in all directions. Due to this secondary radiation, the electrons lose energy and experience a damping effect on their motion. It was found that SPRs are less intense, wider and red-shifted when the particle size increases because of these scattering effects [13]. Scattering dominates the optical response of nanoparticles of 100 nm in diameter and larger. In general, for larger particles, the SPR shifts to the red, decreases in magnitude, and is broadened.

7.3.2. The effect of shape on SPR

It is well established that the shape of metal NPs strongly affects their SPR. For many metal NPs, there are indications of the presence of polyhedral shapes with well-defined facets and vertices, including icosahedral (ih), decahedral (dh), face-centered cubic (fcc), and truncated cubes (tc) [24, 55–61]. The SPRs for polyhedral nanoparticles have been recently studied and a general relationship between the SPR and the morphology or the shape of each nanoparticle can be established [19]. For instance, the optical properties have been investigated computationally for cubes and decahedra as well as for their different truncations [19]. Figure 7.7 shows the extinction coefficients for cubic and decahedral Ag NPs whose volume is equal to that of a sphere of 4.5 nm in diameter. The extinction coefficients were calculated using DDA [16, 19]. The results clearly show a strong dependence of SPR on the shape of Ag NPs.

As another example, for elongated nanoparticles, it has been found that the positions of longitudinal and transversal modes depend not only on the aspect ratio but also on the shape of the ends of the nanoparticles [43]. For a given aspect ratio, up to 100 nm of difference in the location of the SPR can be found by just changing the cap-ends shape.

7.3.3. *The effect of substrate on SPR*

Besides the influence of nanoparticle morphology such as size and shape, the physical environment also modifies SPRs [62]. It has been shown that SPRs are shifted if the dielectric properties of the surrounding media are changed. In particular, the SPRs in a medium with $n > 1$ are red-shifted with respect to those in vacuum [18]. The SPR is also dependent on the geometry of the system including distance between the metal NP and substrate [63, 64]. The effects of both the dielectric properties and the physical environment of the nanoparticles are often of interest, since they have implications in processes such as SPR detection [48, 49, 65–67] and SERS [68–70], catalytic processes [71, 72], and plasmonic devices [73, 74]. Some of these will be further discussed in Chapter 10.

7.3.4. *Effect of particle–particle interaction on SPR*

There is intense and growing interest in controlled patterning and alignment of nanostructures from the perspective of their synthesis and unique properties, especially with respect to their optical response. In these systems, one crucial issue is the interaction among particles. The width and frequency of the SPRs depend on the distance between particles, their size and polarization of the external field. In this case, one should consider the

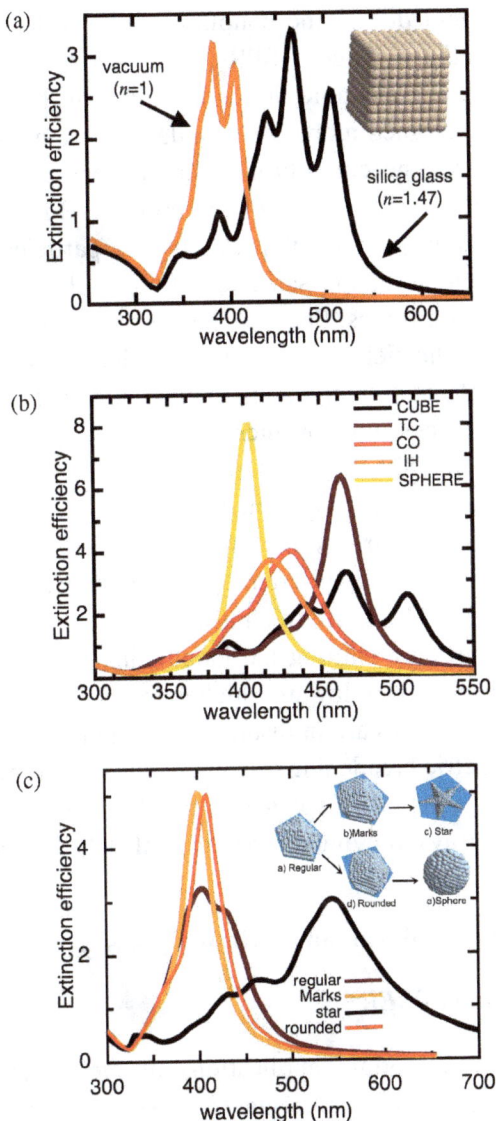

Fig. 7.7. Extinction efficiency of (a) silver cubic nanoparticle of 4.5 nm in diameter immersed in vacuum and in silica glass, adapted in part from Fig. 4 of Ref. 18 with permission; (b) silver cube, truncated cube (TC), cuboctahedron (CO), icosahedron (IH) and a sphere, adapted in part from Fig. 5 of Ref. 18 with permision; and (c) regular decahedron and its truncated morphologies: Marks, rounded, and star decahedra, adapted in part from Fig. 9(a) of Ref. 19 with permission.

so-called local field due to the coupling of the SPRs on a particular nanoparticle to all the induced SPR in the rest of the particles upon polarization [75]. For example, it has been shown that when spherical nanoparticles are aligned along a particular direction forming a chain, the symmetry of the system is broken, and different proper modes can be found for light polarized parallel and transversal to the chain. When the external field is parallel, it polarizes the particles such that the induced local field is in the same direction as the applied field and against the restoring forces, thus decreasing the frequency of the SPRs (red-shift). When the field is transversal to the chain, it polarizes the induced local field against the applied field but in the same direction of the restoring forces, thereby increasing the frequency of the SPRs (blue-shift) [18, 76–78].

Lazarides and Schatz [79] demonstrated that red-shifts occur for three-dimensional cubic arrays of 13 nm gold nanoparticles with separations of 5–15 nm. However, different types of disorder in 3D will lead to different results [80–82]. In 2D systems, the interaction splits the SPRs of individual nanoparticles, shifting the longitudinal modes towards lower and the transversal modes towards higher frequencies. The splitting is larger when nanoparticles are in disordered positions, where fluctuations induce an asymmetric broadening of both modes [83], as observed for 1D systems. These predictions are consistent with experimental data in 1D and 2D ordered arrays of cylinders, and crystal slabs [31, 84, 85].

7.4. Surface enhanced Raman scattering (SERS)

7.4.1. *Background of SERS*

One of the most important applications of metal nanostructures is surface enhanced Raman scattering (SERS). SERS is based on the enhancement of Raman scattering of an analyte molecule near or on a roughened metal substrate surface. It is an extremely useful method for chemical and biochemical analysis and detection. The origin of the Raman enhancement is largely due to an enhanced EM field at the metal surface due to increased absorption of the incident light. This is related to several other surface enhancement phenomena, including surface plasmon resonance, discussed in the last section and surface-enhanced fluorescence [33, 66, 86–95].

Normal Raman scattering, first discovered in 1928, is a widely used spectroscopic technique for chemical analysis, detection and imaging. Its key advantage is molecular specificity or selectivity that allows for unique identification of samples. The major limitation of normal Raman is the very small signal or low quantum yield of scattering (on the order of 10^{-7}) [96]. SERS, discovered in the middle 1970s, has made Raman or SERS more popular for practical applications due to the combined advantages of molecular specificity *and* large signal [97–102]. Since the initial discovery, extensive experimental and theoretical work has been conducted on both understanding the fundamentals of SERS and its practical applications for chemical and biochemical analysis. With enhancement of many orders of magnitude compared to normal Raman scattering, SERS allows measurements with very low analyte concentration, low laser intensity, and short data collection time, with detection of single molecules suggested as a possibility [103–109]. However, due to dependence of the SERS activity on the metal substrate structure (e.g. distance from the surface, orientation and conformation of the molecules, and strength of interaction between the molecule and substrate), SERS spectrum can differ substantially from normal Raman spectrum for the same molecule in terms of peak intensity distribution and particular vibrational modes detected. Therefore, care needs to be taken to interpret SERS spectra [110, 111].

7.4.2. *Mechanism of SERS*

There are several different theories developed over the years to explain the origin of SERS. It is generally agreed today that the main source of enhancement is due to amplified electromagnetic (EM) field at the surface of the metal substrate, usually in the form of nanostructures [69, 112–115]. The second, and less prominent, mechanism is the so-called "chemical enhancement", which was suggested to involve metal-molecule charge transfer that improves resonance with the Raman excitation laser [116]. The chemical enhancement is sensitive to the surface properties of the SERS substrate and the nature of the analyte molecules. It is thought to be responsible for at most two orders of magnitude of the total enhancement that can be greater than 10^{13} [117].

We will use a simple model, the quasistatic treatment of an isolated sphere, to gain some insight into the EM enhancement mechanism. This model assumes that the metal particle is composed of positive charges with a cloud of electrons that can move freely. When the system is polarized, e.g. by an EM field, the electrons will move in response to the EM field but the positively charged nuclei will remain essentially static. Assuming that light is incident on a spherical metal particle imbedded in some medium of dielectric constant ε_0 with the EM field vector E_0 pointing along the z-axis and independent of coordinates for distances on the order of the size of the sphere, one can determine the EM field inside and outside of the sphere by the quasi-static approximation of Maxwell's equations [7, 17, 114]. The field outside the sphere is written as:

$$E_{out} = E_0 z - \alpha E_0 \left[\frac{z}{r^3} - \frac{3z}{r^5}(zz + xx + yy) \right].$$

(7.8)

α is the metal polarizability x, y and z are the Cartesian coordinates, r is the radial distance, and x, y and z are the unit vectors. E_0 is the magnitude of E_0. The first term in Eq. (7.8) is the applied field and the second is the induced dipole of the polarized sphere. If the metal sphere has a dielectric constant of $\varepsilon_i = (\varepsilon_1 + i\,\varepsilon_2)$ and a radius of R, one can write the polarizability as:

$$\alpha = gR^3$$

(7.9)

with

$$g = \frac{\varepsilon_i - \varepsilon_0}{\varepsilon_i + 2\varepsilon_0}$$

(7.10)

when the real part of ε_i, ε_1, is equal to $-2\varepsilon_0$ and the imaginary part of ε_i, ε_2, is small, g and α becomes large, making the induced field large. This implies that the dielectric of the particle must be large and negative for SERS to take place. This is the case for metals with relatively "free" electrons, e.g. gold and silver at long wavelengths of light. When the resonance condition is met, i.e. the light frequency matches the intrinsic collective electron oscillation frequency, absorption of light will take place, which corresponds to the surface plasmon absorption. This resonance is essential for effective SERS enhancement. It is worth noting that the plasmon resonance is not only determined by ε_i but also by ε_o. If the dielectric function of the imbedding medium increases by $\Delta\varepsilon_o$, the resonance condition will shift accordingly to $\varepsilon_1 = -2(\varepsilon_o + \Delta\varepsilon_o)$. Since the dielectric of the metal generally becomes more negative with increasing wavelength, an increase in imbedding dielectric will shift the plasmon resonance to the red. This has been observed experimentally and can be useful for various applications [33, 62, 87, 90, 118–120].

The SERS intensity depends on the absolute square of E_{out}:

$$E_{out}^2 = E_o^2\left[|1 - g|^2 + 3\cos^2\theta(2Re(g) + |g|^2)\right] \tag{7.11}$$

where θ is the angle between the excitation field vector and the dipole moment of the analyte molecule in question on the surface of the particle. When g is large, e.g. when the resonance condition is met, this equation reduces to:

$$E_{out}^2 = E_o^2|g|^2(1 + 3\cos^2\theta) \tag{7.12}$$

It is clear that the SERS intensity will be greatest on the axis with the excitation light, at $\theta = 0°$ and $180°$. Besides enhancement due to the original excitation light, additional enhancement can result from Raman shifted photons being reabsorbed by the particle, enhanced and reemitted. The overall enhancement from both the incident and scattered fields can be expressed as:

$$E_r = \frac{E_{out}^2 E_{out}'^2}{E_o^4} = 16|g|^2|g'|^2 \tag{7.13}$$

Here, the primed variables represent the reradiated fields. This is based on the assumption that the Raman photons are close in frequency to the excitation light, i.e. Raman photons with small Raman shifts. With larger shifts, the reradiation effect becomes less pronounced. More detailed theoretical treatment of SERS can be found in several review articles [17, 33, 69, 112, 114].

7.4.3. Distance dependence of SERS

In practical SERS measurements, several critical factors need to be taken into consideration. The first and most important factor is the distance between the molecule and the SERS substrate, usually silver or gold, and, to a lesser degree, other metals [112]. The closer the molecule is to the surface, the stronger the SERS intensity, given all other factors being the same. To date, there is not a simple analytical mathematical equation to relate SERS intensity to distance. As a matter of fact, different distance dependences of SERS have been reported experimentally and rationalized theoretically using different models. This is in part because it is actually challenging to unambiguously determine the distance dependence of SERS because of the complications caused by changes in other factors when the distance is varied on the atomic scale [110, 111]. One theoretical model based on the mechanism of EM enhancement has suggested a SERS distance dependence approximately as [121]:

$$I = \left(1 + \frac{r}{a}\right)^{-10}$$

where I is the intensity of the Raman mode, a is the average size of the field enhancing features on the surface, and r is the distance from the surface to the adsorbate. This theoretical prediction has been corroborated by recent experimental SERS study of pyridine adsorbed on Ag film over nanospheres (AgFON) using Al_2O_3 films as a spacer to vary the distance between the molecules and the AgFON substrate surface [70]. Figure 7.8(a) shows the SERS spectra for pyridine adsorbed on AgFON surfaces coated with four different thicknesses of deposited Al_2O_3. A plot of the relative intensity of the 1594 cm^{-1} band as a function of Al_2O_3 thickness is shown in Fig. 7.8(b). Fitting the experimental data in Fig. 7.8(b) to Eq. (7.14) implied the average size of the enhancing particle $a = 12.0$ nm.

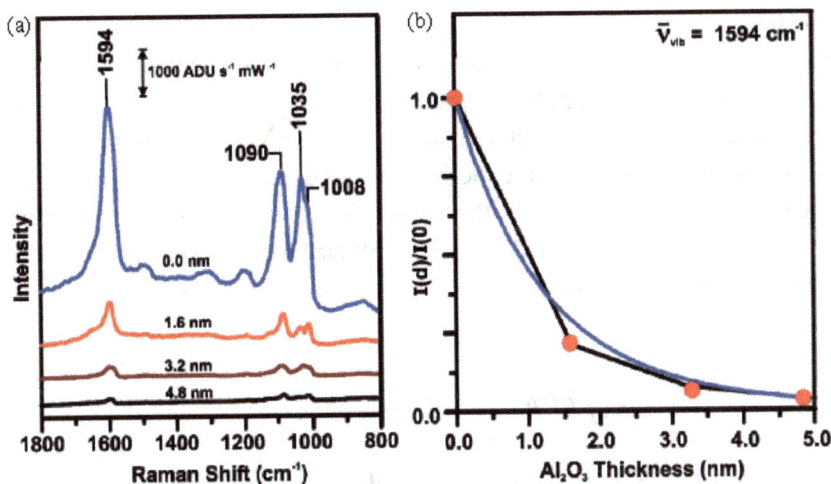

Fig. 7.8. (a) SERS spectra of pyridine adsorbed to silver film over nanosphere (AgFON) samples treated with various thicknesses of alumina (0.0 nm, 1.6 nm, 3.2 nm, 4.8 nm). $\lambda_{ex} = 532$ nm, $P = 1.0$ mW, acquisition time = 300 s. (b) Plot of SERS intensity as a function of alumina thickness for the 1594 cm^{-1} band (filled circles and straight line segments). The solid curved line is a fit of this data to Eq. (7.14). Reproduced with permission from Ref. 70.

7.4.4. *Location and orientation dependence of SERS*

A second important factor is the location of the analyte molecule on the substrate surface. As an example, for spherical particles with incident laser light polarized in a particular direction, the EM field enhancement is anisotropic. Therefore, the location of the molecule will significantly affect how strong the SERS will be for a given molecule. The third factor is the orientation and conformation of the molecule with respect to the surface normal of the metal substrate. Given the anisotropic EM field distribution on the substrate surface, the SERS activity depends strongly on the orientation and conformation of the molecule. The strongest enhancement will occur when the polarizability or, to a good approximation, the dipole moment of the molecule is parallel to the electrical field vector of the enhanced light field. Finally, the strength of interaction between the molecule and the substrate is also important to SERS. Stronger interaction usually results in shorter distance between the molecule and the substrate, and thereby stronger SERS. When the interaction is very strong, charge transfer between the molecule and metal substrate can take place, which is often used to explain the "chemical enhancement". Quantification of this effect is usually challenging. Besides the factors discussed above, a practical experimental problem to consider is the presence of impurities. Due to the large signal of SERS, trace amount of impurities that are not detectable ordinarily can be easily observed, especially if the impurities are polarizable molecules, e.g. aromatic compounds.

7.4.5. *Dependence of SERS on substrate*

SERS spectra are clearly sensitive to the metal substrate. The size and shape as well as their distribution will affect the sensitivity as well as the consistency of SERS measurement. Much effort has been made to improve SERS spectra consistency in practical measurements by developing metal substrates that are uniform or monodisperse in size and shape.

For example, recently a small SERS probe using hollow gold nanospheres (HGNs or HAuNSs) has been developed and found to be more SERS active than solid gold nanoparticles [50]. The unique optical and structural properties of HGNs afford several advantages for SERS applications, including small size, spherical shape, tunable and narrow absorption from the visible to near IR [40]. Furthermore, compared to aggregates or nanoparticles with inhomogeneous distribution in size and shape [103, 104, 122], the HGNs afford SERS spectra that are highly consistent when measured by the peak ratio of SERS spectra of single HGNs. The high consistency is a direct result of the narrow plasmon absorption band and high structural uniformity. Figure 7.9 shows a comparison of SERS spectra of mercapto benzoic acid (MBA) obtained using hollow gold nanospheres versus normal silver nanoparticle aggregates. The inset shows a histogram detailing the distribution of the peak ratio of two SERS peaks for the two different substrates. The HGNs show much narrower and thus better distribution (red), indicating highly uniform structure and narrow SPR, and thereby consistent SERS for single nanostructure SERS [40].

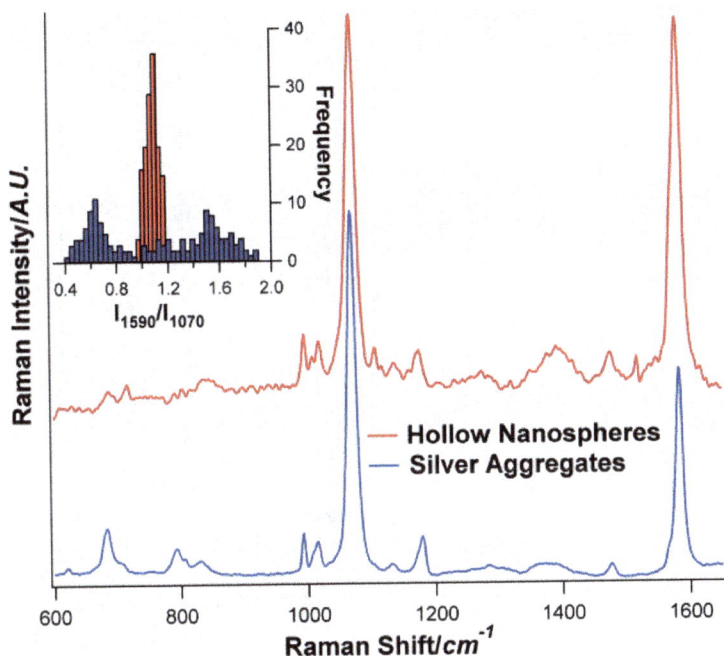

Fig. 7.9. Single-particle SERS spectrum of MBA on HGSs (red trace, top) and silver aggregates (blue trace, bottom). The inset is a histogram of the relative intensity of the two most intense peaks of MBA at 1070 and 1590 cm^{-1} of 150 HGNs (red bars) and 150 silver aggregates (blue bars). Reproduced with permission from Ref. 40.

While the HGNs are indeed very uniform and show highly consistent SERS, they provide relatively weak SERS signal compared to Ag nanoparticles [40]. Ideally, one would produce hollow silver, instead of gold, nanospheres to improve their SERS sensitivity. To achieve this, HGNs have been used as seed templates and coated with a thin layer of Ag on the outer surface [123]. This approach has resulted in successful generation of unique hollow gold-silver double nanoshells that show significantly shifted and broadened plasmon band. The plasmon absorption changes from predominantly that of gold to that of silver when increasing the silver coating on the outer shell. The double nanoshells have shown greatly increased Raman enhancement over HGNs attributed to a combination of the increased volume of metal and the improved enhancement factor of silver [123].

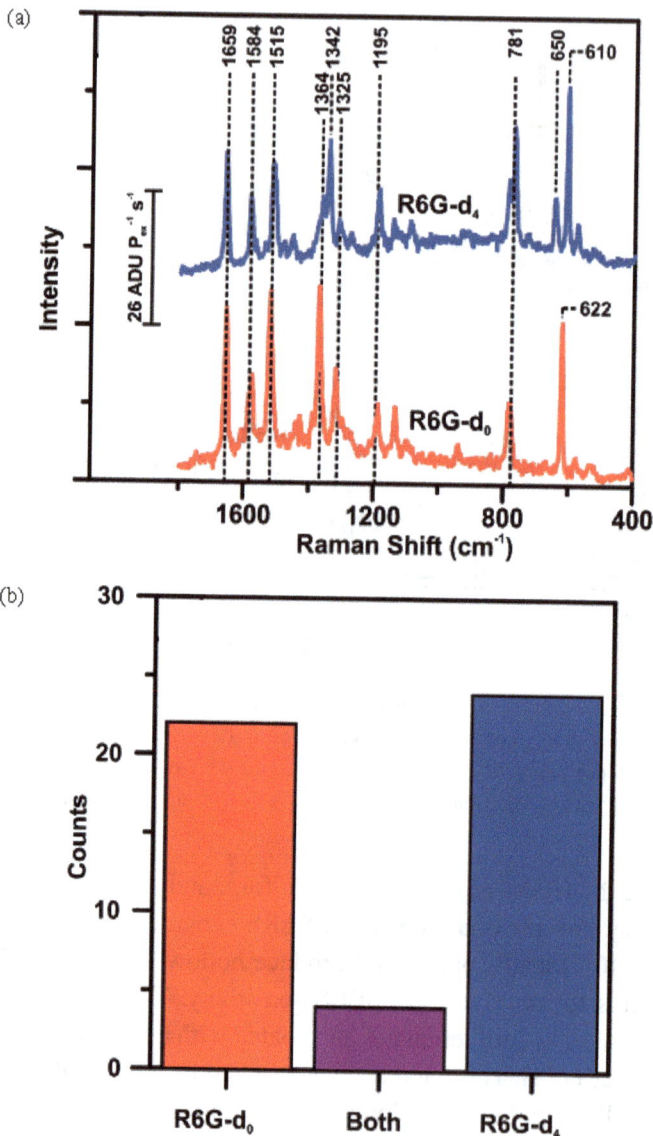

Fig. 7.10. (a) Two representative spectra from the single-molecule results where one contains uniquely R6G-d_0 (red line) and the other uniquely R6G-d_4 (blue line) vibrational character. (λ_{ex} = 532 nm, t_{aq} = 10 s, P_{ex} = 2.4 W/cm^2, grazing incidence) (b) Histogram detailing the frequency with which only R6G-d_0, only R6G-d_4 and both R6G-d_0 and R6G-d_4 vibrational modes were observed with low adsorbate concentration under dry N$_2$ environment [127]. R6G-d_0 represents the R6G with no deuterium substitution while R6G-d_4 represents R6G with four hydrogen atoms on the isolated benzene ring replaced with four deuterium atoms. Reproduced with permission from Ref. 127.

7.4.6. *Single nanoparticle and single molecule SERS*

One of the most intriguing issues in SERS studies is whether one can observe SERS from a single molecule. While it is reasonably easy to establish single nanoparticle SERS, it is often challenging to prove beyond doubt of single molecule SERS [107, 124–126]. A recent experiment based on isotope substitution is perhaps the most elegant to show that it is possible to observe SERS from a single molecule. This is demonstrated using two isotopologues of Rhodamine 6G (R6G) that offer unique vibrational signatures. When an average of one molecule was adsorbed per silver nanoparticle, only one isotopologue was typically observed [127]. Figure 7.10 shows the SERS spectra of R6G with and without isotope substitution. When a mixture of unsubstituted and substituted R6G are used and only one, unsubstituted or substituted, was observed, this is essentially a proof that only one R6G molecule was detected by SERS.

7.5. Summary

Metal nanomaterials have interesting optical properties mainly due to strong surface plasmon absorption and field enhancement effects. The SPR is useful for chemical and biological detection applications. SERS is both a fundamentally intriguing phenomenon and an important tool for analytical purposes. Given the over 30 years of history of SERS, there are still basic scientific issues that need to be further investigated.

References

1. A. Mooradian, *Phys. Rev. Lett.* **22**, 185 (1969).
2. M.B. Mohamed, V. Volkov, S. Link and M.A. El-Sayed, *Chem. Phys. Lett.* **317**, 517 (2000).
3. O. Varnavski, R.G. Ispasoiu, L. Balogh, D. Tomalia and T. Goodson, *J. Chem. Phys.* **114**, 1962 (2001).
4. L.A. Peyser, A.E. Vinson, A.P. Bartko and R.M. Dickson, *Science* **291**, 103 (2001).
5. C.F. Bohren and D.R. Huffman, *Absorption and Scattering of Light by Small Particles*, New York: John Wiley & Sons. 546 (1983).

6. G. Mie, *Annalen der Physik* **25**, 377 (1908).
7. U. Kreibig and M. Vollmer, *Optical Properties of Metal Clusters*, Springer Series in Materials Science 25, Berlin: Springer. 532 (1995).
8. W.P. Halperin, *Rev. Mod. Phys.* **58**, 533 (1986).
9. L. Genzel, T.P. Martin and U. Kreibig, *Z. Physik B* **21**, 339 (1975).
10. J.A. Creighton and D.G. Eadon, *J. Chem. Soc. Faraday T* **87**, 3881 (1991).
11. M.M. Alvarez, J.T. Khoury, T.G. Schaaff, M.N. Shafigullin, I. Vezmar and R.L. Whetten, *J. Phys. Chem. B* **101**, 3706 (1997).
12. J.H. Hodak, A. Henglein and G.V. Hartland, *J. Phys. Chem. B* **104**, 9954 (2000).
13. C. Noguez, *Opt. Mater.* **27**, 1204 (2005).
14. S. Asano and G. Yamamoto, *App. Opt.* **14**, 29 (1975).
15. A.C. Lind and J.M. Greenber, *J. Appl. Phys.* **37**, 3195 (1966).
16. I.O. Sosa, C. Noguez and R.G. Barrera, *J. Phys. Chem. B* **107**, 6269 (2003).
17. K.L. Kelly, E. Coronado, L.L. Zhao and G.C. Schatz, *J. Phys. Chem. B* **107**, 668 (2003).
18. C. Noguez, *J. Phys. Chem. C* **111**, 3806 (2007).
19. A.L. Gonzalez and C. Noguez, *J. Comput. Theoret. Nanosci.* **4**, 231 (2007).
20. E. Hao, S.Y. Li, R.C. Bailey, S.L. Zou, G.C. Schatz and J.T. Hupp, *J. Phys. Chem. B* **108**, 1224 (2004).
21. E.K. Payne, K.L. Shuford, S. Park, G.C. Schatz and C.A. Mirkin, *J. Phys. Chem. B* **110**, 2150 (2006).
22. W.H. Yang, G.C. Schatz and R.P. Vanduyne, *J. Chem. Phys.* **103**, 869 (1995).
23. R. Fuchs, *Phys. Rev. B* **11**, 1732 (1975).
24. A.S. Kumbhar, M.K. Kinnan and G. Chumanov, *J. Am. Chem. Soc.* **127**, 12444 (2005).
25. J.Z. Zhang and C. Noguez, *Plasmonics* **3**, 127 (2008).
26. B. Nikoobakht and M.A. El-Sayed, *J. Phys. Chem.* **107**, 3372 (2003).
27. N.R. Jana, L. Gearheart and C.J. Murphy, *J. Phys. Chem. B* **105**, 4065 (2001).
28. C.J. Johnson, E. Dujardin, S.A. Davis, C.J. Murphy and S. Mann, *J. Mater. Chem.* **12**, 1765 (2002).
29. C.J. Orendorff, A. Gole, T. Sau and C.J. Murphy, *Anal. Chem.* **77**, 3261 (2005).
30. C.J. Orendorff, L. Gearheart, N.R. Jana and C.J. Murphy, *Phys. Chem. Chem. Phys.* **8**, 165 (2006).
31. C.L. Haynes, A.D. McFarland, L.L. Zhao, R.P. Van Duyne, G.C. Schatz, L. Gunnarsson, J. Prikulis, B. Kasemo and M. Kall, *J. Phys. Chem. B* **107**, 7337 (2003).

32. C.L. Haynes, C.R. Yonzon, X.Y. Zhang and R.P. Van Duyne, *J. Raman Spectroscopy* **36**, 471 (2005).

33. L.J. Sherry, R.C. Jin, C.A. Mirkin, G.C. Schatz and R.P. Van Duyne, *Nano Lett.* **6**, 2060 (2006).

34. B. Wiley, Y.G. Sun, B. Mayers and Y.N. Xia, *Chem.-Eur. J.* **11**, 454 (2005).

35. M. Hu, H. Petrova, A.R. Sekkinen, J.Y. Chen, J.M. McLellan, Z.Y. Li, M. Marquez, X.D. Li, Y.N. Xia and G.V. Hartland, *J. Phys. Chem. B* **110**, 19923 (2006).

36. M. Hu, J.Y. Chen, Z.Y. Li, L. Au, G.V. Hartland, X.D. Li, M. Marquez and Y.N. Xia, *Chemical Soc. Rev.* **35**, 1084 (2006).

37. S.J. Oldenburg, S.L. Westcott, R.D. Averitt and N.J. Halas, *J. Chem. Phys.* **111**, 4729 (1999).

38. C.L. Nehl, N.K. Grady, G.P. Goodrich, F. Tam, N.J. Halas and J.H. Hafner, *Nano Lett.* **4**, 2355 (2004).

39. C. Radloff and N.J. Halas, *Nano Lett.* **4**, 1323 (2004).

40. A.M. Schwartzberg, T.Y. Oshiro, J.Z. Zhang, T. Huser and C.E. Talley, *Anal. Chem.* **78**, 4732 (2006).

41. A. Oliver, J.A. Reyes-Esqueda, J.C. Cheang-Wong, C.E. Roman-Velazquez, A. Crespo-Sosa, L. Rodriguez-Fernandez, J.A. Seman and C. Noguez, *Phys. Rev. B* **74** (2006).

42. J.A. Reyes-Esqueda, C. Torres-Torres, J.C. Cheang-Wong, A. Crespo-Sosa, L. Rodriguez-Fernandez, C. Noguez and A. Oliver, *Opt. Exp.* **16**, 710 (2008).

43. A.L. Gonzalez, A. Reyes-Esqueda and C. Noguez, *J. Phys. Chem. C*, in press, (2008).

44. P. Figliozzi, L. Sun, Y. Jiang, N. Matlis, B. Mattern, M.C. Downer, S.P. Withrow, C.W. White, W.L. Mochan and B.S. Mendoza, *Phys. Rev. Lett.* **94** (2005).

45. Y.Y. Yu, S.S. Chang, C.L. Lee and C.R.C. Wang, *J. Phys. Chem. B* **101**, 6661 (1997).

46. T.J. Norman, C.D. Grant, D. Magana, J.Z. Zhang, J. Liu, D.L. Cao, F. Bridges and A. Van Buuren, *J. Phys. Chem. B* **106**, 7005 (2002).

47. H.S. Zhou, I. Honma, H. Komiyama and J.W. Haus, *Phys. Rev. B-Condensed Matter* **50**, 12052 (1994).

48. R.J. Heaton, P.I. Haris, J.C. Russell and D. Chapman, *Biochem. Soc. Trans.* **23**, S502 (1995).

49. J.S. Yuk, J.W. Jung, Y.M. Mm and K.S. Ha, *Sensor Actuat. B-Chem.* **129**, 113 (2008).

50. A.M. Schwartzberg, T.Y. Olson, C.E. Talley and J.Z. Zhang, *J. Phys. Chem. B* **110**, 19935 (2006).

51. A. Taleb, C. Petit and M.P. Pileni, *J. Phys. Chem. B* **102**, 2214 (1998).

52. P.B. Johnson and R.W. Christy, *Phys. Rev. B* (*Solid State*) **6**, 4370 (1972).

53. N.W. Ashcroft and N.D. Mermin, *Solid State Physics*, Philadelphia: Saunders College. 826 (1976).

54. D.M. Wood and N.W. Ashcroft, *Phys. Rev. B* (*Condensed Matter*) **25**, 6255 (1982).

55. A. Tao, P. Sinsermsuksakul and P.D. Yang, *Angew Chem. Int. Edit* **45**, 4597 (2006).

56. F. Baletto and R. Ferrando, *Rev. Mod. Phys.* **77**, 371 (2005).

57. Y.G. Sun and Y.N. Xia, *Science* **298**, 2176 (2002).

58. M. Zhou, S. Chen and S. Zhao, *J. Phys. Chem. B* **110**, 4510 (2006).

59. Y. Chen, X. Gu, C.-G. Nie, Z.-Y. Jiang, Z.-X. Xie and C.-J. Lin, *Chem. Commun.* **33**, 4181 (2005).

60. C. Salzemann, A. Brioude and M.P. Pileni, *J. Phys. Chem. B* **110**, 7208 (2006).

61. I.A. Banerjee, L.T. Yu and H. Matsui, *Proc. Nat. Acad. Sci. USA* **100**, 14678 (2003).

62. C. Novo, A.M. Funston, I. Pastoriza-Santos, L.M. Liz-Marzan and P. Mulvaney, *J. Phys. Chem. C* **112**, 3 (2008).

63. C.E. Roman-Velazquez, C. Noguez and R.G. Barrera, *Phys. Rev. B* **61**, 10427 (2000).

64. C.E. Roman-Velazquez, C. Noguez and R.G. Barrera, in *Materials Research Society Symposium on Nanophase and Nanocomposite Materials III*. 2000: MRS.

65. R. Rella, J. Spadavecchia, M.G. Manera, P. Siciliano, A. Santino and G. Mita, *Biosens Bioelectron* **20**, 1140 (2004).

66. I.C. Stancu, A. Fernandez-Gonzalez and R. Salzer, *J. Optoelectr. Adv. Mat.* **9**, 1883 (2007).

67. J.B. Beusink, A.M.C. Lokate, G.A.J. Besselink, G.J.M. Pruijn and R.B.M. Schasfoort, *Biosens Bioelectron* **23**, 839 (2008).

68. A.J. Haes, C.L. Haynes, A.D. McFarland, G.C. Schatz, R.R. Van Duyne and S.L. Zou, *MRS Bulletin* **30**, 368 (2005).

69. G.C. Schatz, M.A. Young and R.P. Van Duyne, in *Surface-Enhanced Raman Scattering: Physics and Applications*. 19 (2006).

70. J.A. Dieringer, A.D. McFarland, N.C. Shah, D.A. Stuart, A.V. Whitney, C.R. Yonzon, M.A. Young, X.Y. Zhang and R.P. Van Duyne, *Faraday Discuss* **132**, 9 (2006).

71. M. Haruta, H. Kageyama, N. Kamijo, T. Kobayashi and F. Delannay, in *Successful Design of Catalysts: Future Requirements and Development*, ed., T. Inui, Elsevier Science Ltd 33 (1989).

72. M. Haruta, T. Kobayashi, H. Sano and N. Yamada, *Chem. Lett.* **405** (1987).

73. K.R. Li, M.I. Stockman and D.J. Bergman, *Phys. Rev. B* **72**, 153401 (2005).

74. A. Cvitkovic, N. Ocelic, J. Aizpurua, R. Guckenberger and R. Hillenbrand, *Phys. Rev. Lett.* **97** (2006).

75. J.C.M. Garnett, Philosophical Transactions of the Royal Society of London Series a-Containing Papers of a Mathematical or Physical Character, **203**, 385 (1904).

76. C. Noguez and R.G. Barrera, *Phys. Rev. B* **57**, 302 (1998).

77. J.J. Penninkhof, A. Polman, L.A. Sweatlock, S.A. Maier, H.A. Atwater, A.M. Vredenberg and B.J. Kooi, *Appl. Phys. Lett.* **83**, 4137 (2003).

78. L.A. Sweatlock, S.A. Maier, H.A. Atwater, J.J. Penninkhof and A. Polman, *Phys. Rev. B* **71** (2005).

79. A.A. Lazarides and G.C. Schatz, *J. Phys. Chem. B* **104**, 460 (2000).

80. R.G. Barrera, C. Noguez and E.V. Anda, *J. Chem. Phys.* **96**, 1574 (1992).

81. S. Kumar and R.I. Cukier, *J. Phys. Chem.* **93**, 4334 (1989).

82. B. Cichocki and B.U. Felderhof, *J. Chem. Phys.* **90**, 4960 (1989).

83. R.G. Barrera, M. Delcastillomussot, G. Monsivais, P. Villasenor and W.L. Mochan, *Phys. Rev. B* **43**, 13819 (1991).

84. E.M. Hicks, S.L. Zou, G.C. Schatz, K.G. Spears, R.P. Van Duyne, L. Gunnarsson, T. Rindzevicius, B. Kasemo and M. Kall, *Nano Lett.* **5**, 1065 (2005).

85. D. Nau, A. Schonhardt, C. Bauer, A. Christ, T. Zentgraf, J. Kuhl, M.W. Klein and H. Giessen, *Phys. Rev. Lett.* **98** (2007).

86. E.G. Matveeva, Z. Gryczynski, J. Malicka, J. Lukomska, S. Makowiec, K.W. Berndt, J.R. Lakowicz and I. Gryczynski, *Anal. Biochem.* **344**, 161 (2005).

87. A.J. Haes, S.L. Zou, J. Zhao, G.C. Schatz and R.P. Van Duyne, *J. Am. Chem. Soc.* **128**, 10905 (2006).

88. M.V. Meli and R.B. Lennox, *J. Phys. Chem. C* **111**, 3658 (2007).

89. F. Toderas, M. Baia, L. Baia and S. Astilean, *Nanotechnology* **18** (2007).

90. K.A. Willets and R.P. Van Duyne, *Ann. Rev. Phys. Chem.* **58**, 267 (2007).

91. E. Fu, T. Chinowsky, K. Nelson, K. Johnston, T. Edwards, K. Helton, M. Grow, J.W. Miller and P. Yager, in *Oral-Based Diagnostics*. 335 (2007).

92. E.G. Matveeva, I. Gryczynski, A. Barnett, Z. Leonenko, J.R. Lakowicz and Z. Gryczynski, *Anal. Biochem.* **363**, 239 (2007).

93. K. Ray, H. Szmacinski, J. Enderlein and J.R. Lakowicz, *Appl. Phys. Lett.* **90**, 251116 (2007).
94. J. Zhang, Y. Fu, M.H. Chowdhury and J.R. Lakowicz, *Nano Lett.* **7**, 2101 (2007).
95. J. Zhang, Y. Fu, M.H. Chowdhury and J.R. Lakowicz, *J. Phys. Chem. C* **111**, 11784 (2007).
96. J.R. Ferraro and K. Nakamoto, *Introductory Raman Spectroscopy*, Boston: Academic Press. xi (1994).
97. M. Fleischmann, P.J. Hendra and A.J. McQuillan, *Chem. Phys. Lett.* **26**, 163 (1974).
98. M.G. Albrecht and J.A. Creighton, *J. Am. Chem. Soc.* **99**, 5215 (1977).
99. D.L. Jeanmaire and R.P. Vanduyne, *J. Electroanal. Chem.* **84**, 1 (1977).
100. M. Moskovits, *J. Chem. Phys.* **69**, 4159 (1978).
101. Z.Q. Tian, B. Ren, J.F. Li and Z.L. Yang, *Chem. Commun.* **3514** (2007).
102. T. Wang, X.G. Hu and S.J. Dong, *Small* **4**, 781 (2008).
103. A.M. Michaels, J. Jiang and L. Brus, *J. Phys. Chem. B* **104**, 11965 (2000).
104. J. Jiang, K. Bosnick, M. Maillard and L. Brus, *J. Phys. Chem. B* **107**, 9964 (2003).
105. S.M. Nie and S.R. Emery, *Science* **275**, 1102 (1997).
106. K. Kneipp, G.R. Harrison, S.R. Emory and S.M. Nie, *Chimia* **53**, 35 (1999).
107. K. Kneipp, Y. Wang, H. Kneipp, L.T. Perelman, I. Itzkan, R. Dasari and M.S. Feld, *Phys. Rev. Lett.* **78**, 1667 (1997).
108. K. Kneipp, H. Kneipp, R. Manoharan, I. Itzkan, R.R. Dasari and M.S. Feld, *J. Raman Spectroscopy* **29**, 743 (1998).
109. K. Kneipp, H. Kneipp, I. Itzkan, R.R. Dasari and M.S. Feld, *Chem. Phys.* **247**, 155 (1999).
110. L. Seballos, T.Y. Olson and J.Z. Zhang, *J. Chem. Phys.* **125**, 234706 (2006).
111. L. Seballos, N. Richards, D.J. Stevens, M. Patel, L. Schuresko, S. Lokey, G. Millhauser and J.Z. Zhang, *Chem. Phys. Lett.* **447**, 335 (2007).
112. M. Moskovits, *Rev. Mod. Phys.* **57**, 783 (1985).
113. M. Moskovits, *J. Raman Spectroscopy* **36**, 485 (2005).
114. G.C. Schatz and R.P. Van Duyne, *Electromagnetic Mechanism of Surface-enhanced Spectroscopy*, Chichester: John Wiley and Sons Ltd. (2002).
115. A. Otto, I. Mrozek, H. Grabhorn and W. Akemann, *J. Phys-Condens Matter* **4**, 1143 (1992).
116. D.Y. Wu, X.M. Liu, S. Duan, X. Xu, B. Ren, S.H. Lin and Z.Q. Tian, *J. Phys. Chem. C* **112**, 4195 (2008).

117. A. Otto, *J. Raman Spectroscopy* **36**, 497 (2005).

118. A.V. Whitney, J.W. Elam, S.L. Zou, A.V. Zinovev, P.C. Stair, G.C. Schatz and R.P. Van Duyne, *J. Phys. Chem. B* **109**, 20522 (2005).

119. J. Rodriguez-Fernandez, I. Pastoriza-Santos, J. Perez-Juste, F.J.G. de Abajo and L.M. Liz-Marzan, *J. Phys. Chem. C* **111**, 13361 (2007).

120. I. Pastoriza-Santos, A. Sanchez-Iglesias, F.J.G. de Abajo and L.M. Liz-Marzan, *Adv. Funct. Mater.* **17**, 1443 (2007).

121. B.J. Kennedy, S. Spaeth, M. Dickey and K.T. Carron, *J. Phys. Chem. B* **103**, 3640 (1999).

122. C.E. Talley, L. Jusinski, C.W. Hollars, S.M. Lane and T. Huser, *Anal. Chem.* **76**, 7064 (2004).

123. T.Y. Olson, A.M. Schwartzberg, C.A. Orme, C.E. Talley, B. O'Connell and J.Z. Zhang, *J. Phys. Chem. C* **112**, 6319 (2008).

124. A. Otto, *J. Raman Spectroscopy* **33**, 593 (2002).

125. W.E. Doering and S.M. Nie, *J. Phys. Chem. B* **106**, 311 (2002).

126. K. Kneipp and H. Kneipp, *Biophys. J.* **88**, 365A (2005).

127. J.A. Dieringer, R.B. Lettan, K.A. Scheidt and R.P. Van Duyne, *J. Am. Chem. Soc.* **129**, 16249 (2007).

Chapter 8

Optical Properties of Composite Nanostructures

Composite nanostructures refer to nanomaterials that consist of more than one type of the conventional inorganic, organic or biological materials, or other mixtures of materials with apparently different chemical and physical properties. The composites could have two or more components. There are many examples of composite materials in nature as well as in synthetic materials. For instance, crab shells and bones are composite materials of inorganic, organic and biological materials in nature. Teeth are composites of inorganic and organic materials.

If we follow the conventional classification of materials as metal, semiconductor and insulator, then we can have composites of metal–semiconductor, metal–insulator, semiconductor–insulator, and so on. They could also be composed of two different metals or two different semiconductors, etc. Composite materials often have properties and functionalities quite different from those of the individual components. Therefore, they can be very useful in meeting application requirements that single component materials cannot. Also, by changing the ratio and nature of the individual components, a practically unlimited number of composite materials can be created. This is particularly true for nanomaterials since, in many composite materials, the features that are critical to their properties are on the nanometer scale. Nanocomposites have many interesting and useful properties including mechanical, electronic, magnetic, thermal and optical.

In this book, we will specifically focus on their optical properties and related electronic properties.

The difference in optical properties between composite materials and their constituent components depends largely on the chemical nature of each component as well as how the two or more components interact with each other. The interaction between the components depends strongly on characteristics such as interface, size, shape and structure. In the extreme case of no or little interaction between the components, the optical properties of the composite should be equivalent to a simple sum of the optical properties of the individual components. In cases where the interaction between the components is strong, the optical properties of the composite system can differ substantially from the simple sum of the properties of the individual components. The characteristics of the individual components are lost and new features arise as a result of the strong interaction. For example, new absorption and emission bands may appear in the absorption and emission spectra of composites as compared to the individual components. This can be rationalized by considering the interaction as a perturbation and the starting individual components as the zeroth order system.

For example, in a simple two-component system consisting of two different materials A and B, their interaction can be considered as a perturbation energy V. If the zeroth order wavefunctions are labeled as ψ_A and ψ_B and the new composite system wavefunctions are labeled as ψ_1 and ψ_2 for the lower and higher energy states resulting from the two initial zeroth order states, applying perturbation theory gives us the following:

$$\psi_1 = c_1 \psi_A + c_2 \psi_B \qquad (8.1)$$

$$\psi_2 = c_1' \psi_A + c_2' \psi_B \qquad (8.2)$$

The coefficients in the above two equations depend on the strength of perturbation V. This is similar to the treatment of molecular orbitals for a heterodiatomic molecule from atomic orbitals of two different atoms. In general, the stronger V is, the more different are the new states from the zeroth-order states. Figure 8.1 shows a schematic of the energy level involved in such a simple two-component system.

Most cases fall in between the two extremes of no interaction and very strong interaction, with moderate interaction. In this case, the optical

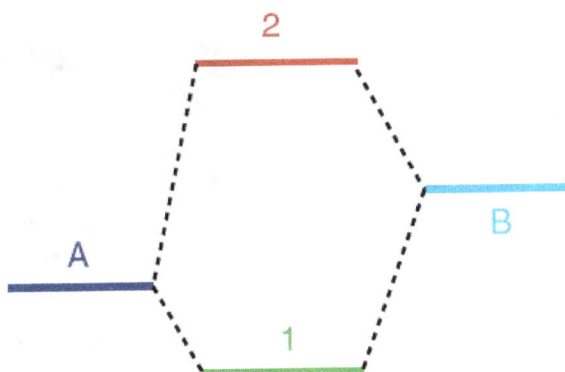

Fig. 8.1. Illustration of the energy levels of the zeroth order systems, A and B, versus those of the resulting composite materials for a two-component system. The energy difference between 1 and 2 depends on how strong the interaction is between A and B; stronger interaction between A and B results in large energy difference between 1 and 2 and the new states for the composite are more dissimilar to the zeroth order states A and B.

properties of composites may still resemble those of the individual components while some changes may occur or new features may arise, but usually not to the degree as in the strongly interacting case. For example, absorption and/or emission bands of the composite system will likely exhibit modest changes compared to those of the original components, e.g. spectral shift or line width broadening.

Since this book focuses on inorganic nanomaterials, the composite systems to be discussed will have at least one component that is inorganic. We will also exclude *alloys* that are atomic level composite materials. The composites to be discussed are made up of at least two components and each composite retains its chemical identity with physical dimension on the few to a few hundred nm scale. We further restrict ourselves to situations where at least one component is nanocrystalline. By doing so, systems that are entirely amorphous on the atomic scale will be excluded. Figure 8.2 shows an illustration of a two-component system in which both components are crystalline.

8.1. Inorganic semiconductor–insulator and semiconductor–semiconductor

Inorganic semiconductors and insulators have often been used to produce nanocomposite materials for various applications including electronics

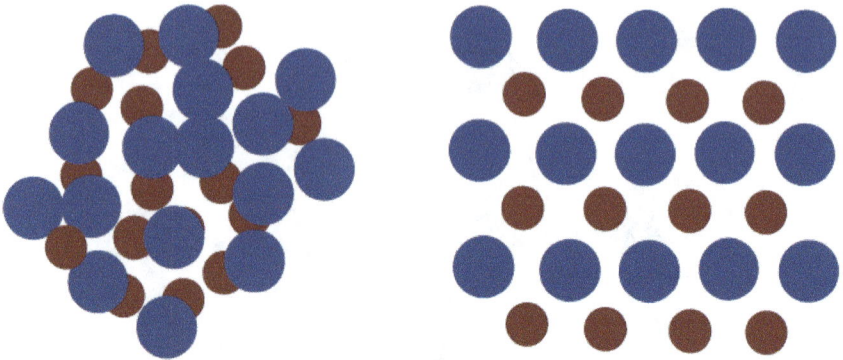

Fig. 8.2. Schematic of a two-component composite system with no structural order, e.g. aggregates (left), and with order, e.g. a superlattice (right), between the individual nanoparticles of the two components that are assumed to be spherical and crystalline. The ordered structure is basically a binary superlattice structure.

and sensors as well as in the exploration of potential new properties. For example, a film of SiO_2 nanoparticles with CdSe nanoparticles with good dispersion on the nanoscale would be considered as an inorganic semiconductor–insulator system with the classification used in this chapter. For such systems, the optical properties are usually dominated by the semiconductor, at least in the visible region, since the small bandgap of the semiconductor, compared to that of insulators, results in visible absorption and the insulator does not alter the semiconductor properties in a significant manner. However, if one considers the UV or VUV region, the insulator properties can be predominant. Therefore, the dominance of the component in the composite material properties depends on the energy or spectral region considered. This is simply because different electronic transitions are involved in different energy regions.

The interaction between the semiconductor and insulator depends on how their lowest energy levels, e.g. valence and conduction band edges, are aligned. In a typical situation, the semiconductor has a smaller bandgap where bandedges fall within the bandgap of the insulator, as shown in Fig. 8.3. In such situations, the effect of the insulator on the electronic energy levels and thereby optical properties is minimal. However, if the insulator helps to passivate the surface of the

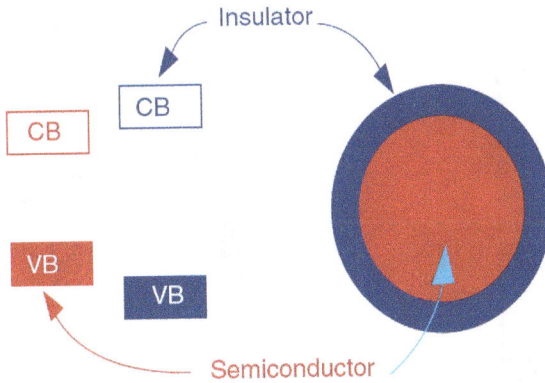

Fig. 8.3. Illustration of the energy levels (left) of a typical semiconductor–insulator composite system, exemplified by a core/shell nanostructure (right). VB and CB are for valence and conduction bands, respectively.

semiconductor as in a core/shell structures, the PL intensity of the semiconductor may be substantially enhanced by the presence of the insulator [Fig. 8.3 (right)].

On the other hand, if the bandedge of the insulator happens to fall within the bandgap of the semiconductor, energy or charge transfer could occur when the composite is subject to photoexcitation. In this case, the optical and photochemical properties of the composite could differ substantially from the semiconductor. Such situations are not as commonly encountered as the situation illustrated in Fig. 8.3.

Examples of semiconductor–insulator composites are numerous. We will provide two examples to highlight their optical properties. The first example is silica-coated CdSe/ZnS core/shell QDs useful for biolabeling and other applications [1]. Figure 8.4 shows some UV-vis and PL spectra of these silanized QDs. It was found that the silica coating does not significantly modify the optical properties of the nanocrystals. The silanized nanocrystals exhibit enhanced photochemical stability over organic dye molecules and display high stability in buffers at physiological conditions (>150 mM NaCl), which are desired for conjugation to biological molecules.

The second example is CdSe nanocrystal QDs (NQDs) embedded into TiO_2 matrix with TiO_2 essentially functioning as an insulator even though

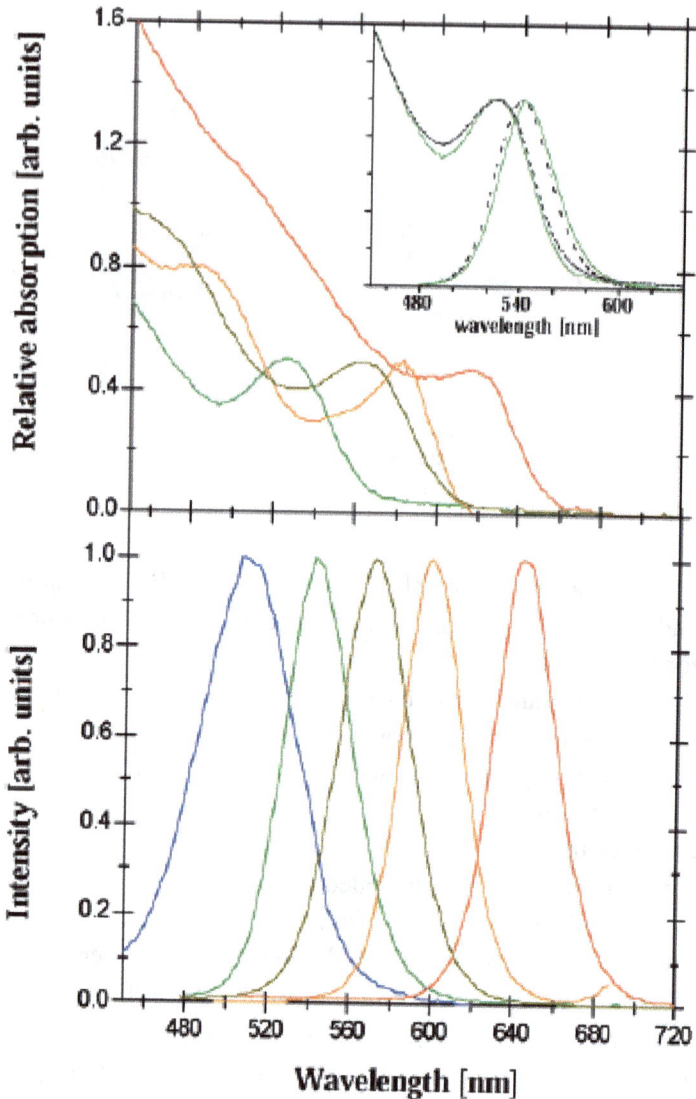

Fig. 8.4. Absorption (upper panel) and emission (lower panel) spectra of a series of silanized CdSe/ZnS in 10 mM PBS buffer, pH ~ 7. The data are normalized for the convenience of the display. From right to left, red, orange, yellow, green and blue emitting nanocrystals are shown. For blue emitting particles, the absorption spectrum does not show features above 450 nm and is therefore omitted. Inset: Absorption and emission of silanized green nanocrystals in 10 mM PB (solid lines), and of the same green CdSe/ZnS particles in toluene (dashed lines). Reproduced with permission from Ref. 1.

A

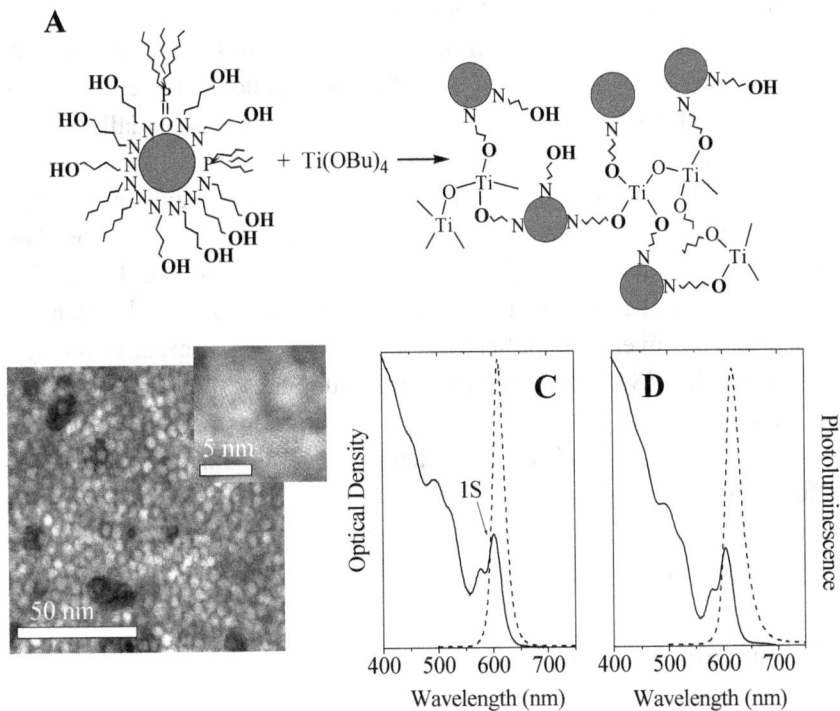

C

D

Optical Density

1S

Photoluminescence

400 500 600 700   400 500 600 700
Wavelength (nm)   Wavelength (nm)

Fig. 8.5. (a) Schematic illustration of the pathway via which CdSe NQDs are incorpo-rated into a TiO$_2$ matrix. (b) Low and high (inset) resolution STEM of a thin sol-gel film containing CdSe NQDs (white area). (c) UV-vis (solid line) and PL (dashed line) spectra of as-prepared CdSe NQDs (R=2.5 nm) in a hexane solution (the arrow indicates the low-est energy 1S exciton absorption resonance). (d) UV-vis and PL spectra of the same NQDs incorporated into TiO$_2$. Reproduced with permission from Ref. 2.

in some settings TiO$_2$ is considered as a semiconductor. This is because the bandgap of CdSe is much smaller than that of TiO$_2$ and there is weak electronic interaction between CdSe and TiO$_2$. Figure 8.5 shows a schematic for the nanocompsoite synthesis (a), STEM (b), and UV-vis and PL (c and d) of the NQDs before (c) and after (d) they are incorporated into TiO$_2$ [2]. A comparison of Figs. 8.5(c) and 8.5(d) clearly shows that TiO$_2$ has little effect on the absorption and PL of the CdSe NQDs. This composite material has been found to exhibit good photostability and easy processibility for nonlinear optical applications.

The situation for two semiconductors composited together is very similar to that of semiconductor–insulator composites with the main difference being in the bandgap of one of the semiconductors versus that of the insulator. Core/shell structures such as CdSe/ZnS is essentially a semiconductor–semiconductor composite nanostructure. The properties of such structures strongly depend on their relative bandgap energies and their interaction. Their interaction could depend sensitively on their structural and interfacial characteristics. As an example, CdSe-TiO$_2$ nanocomposites have been studied and explored as a potential system for solar energy conversion applications [3]. This is very similar to the system in which CdSe, in conjunction with N-doping, was used to sensitize TiO$_2$ nanoparticles to enhance visible absorption for solar energy conversion application [4]. Such semiconductor–semiconductor nanocomposites offer the opportunity to manipulate the band structure of the overall system for various applications of interest. Some application examples will be given in Chapter 10.

8.2. Inorganic metal–insulator

Similar to insulator–semiconductor composites, the optical properties of insulator–metal composites are often dominated by the metal in the visible spectral region since metal nanoparticles, such as Ag and Au, tend to have strong visible absorption due to the surface plasmon absorption, as discussed in Chapter 7, and the insulators have little or no absorption in the visible region. In most cases, the insulator will have some but limited effect on the absorption of the metal nanostructure and this can be reasonably explained by the dielectric properties of the insulator, since the plasmon absorption (line width and peak position) depend somewhat on the dielectric constant of the embedding media. Significant change to the optical properties of the metal nanostructure could occur if there is strong electronic interaction between the metal and insulator as a result of their electronic structure or energy levels. However, this is expected to be rare.

The most common example of inorganic metal–insulator nanocomposites is perhaps gold, silver or cupper nanoparticles coated or embedded in an insulator such as silica, alumina or other metal oxides

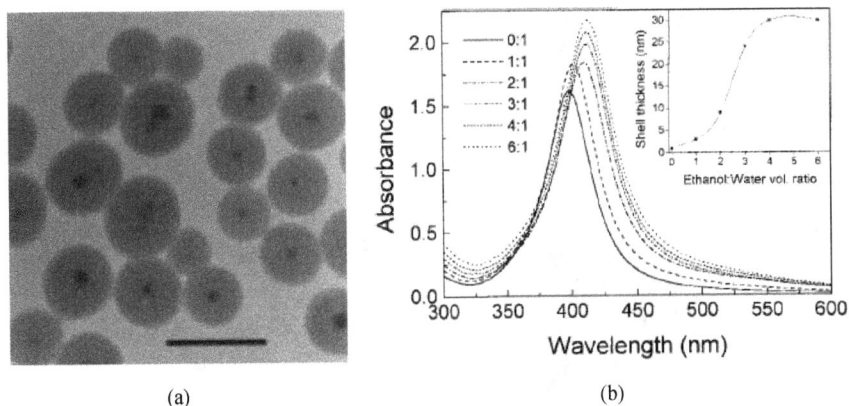

(a)   (b)

Fig. 8.6. (a) An electron micrograph of silica-coated silver nanoparticles with silicate ion concentration of 0.02% and at pH 12: (top) 0.02%. Scale bar: 100 nm. (b) Effect of ethanol/H$_2$O volume ratio on the UV-vis absorption spectra of silica-coated silver colloidal nanoparticles. There is an increase in band intensity and a red-shift of the peak position with increasing ethanol content. The inset shows the thickness of the silica shell as a function of the added ethanol content. Reproduced with permission from Ref. 8.

with large bandgaps and little or no absorption in the visible [5–12]. For instance, Ag nanoparticles have been successfully coated with a thick layer of silica (SiO$_2$) [8]. Some representative EM images and UV-vis spectra are shown in Fig. 8.6. The outcome of the synthesis depends on reaction parameters such as pH, silicate ion concentration and ethanol/water ratio. The UV-vis absorption spectra shown in Fig. 8.6(b) clears shows that the SPR peak position and width are affected by the ethanol/water ratio used in the silica coating process. Overall, silica coating does not significantly change the SPR of the Ag nanoparticles.

Recently, similar methods have been used to coat silica onto gold nanorods [12]. It has been observed that whereas the transverse plasmon band remains almost unaltered, the longitudinal surface plasmon band red-shifts upon the first silica deposition, which is attributed to an increase in the local refractive index around the rods produced by the silica shell. Figure 8.7 shows TEM images and SPR absorption spectra of the silica-coasted gold nanorods [12].

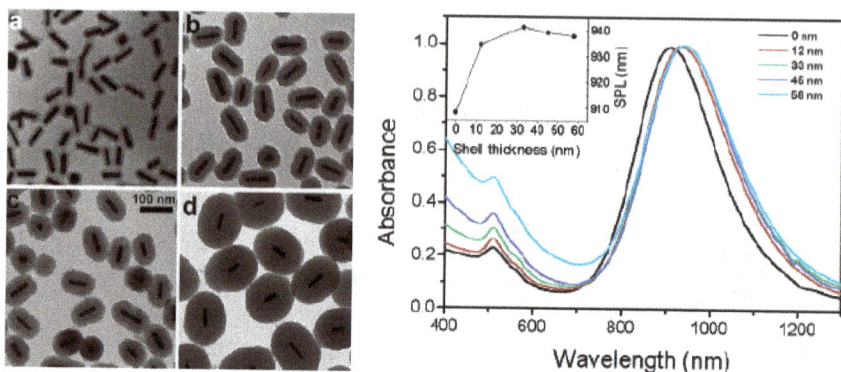

Fig. 8.7. (left) Transmission electron micrographs (TEM) of silica-coated gold nanorods, with silica shell thickness increasing from *a* to *d*. The scale is the same for all images. (right) Normalized experimental (left) UV-visible-NIR spectra of Au@SiO$_2$ gold nanorods in 2-propanol with increasing silica shell thickness. The insets summarize the shift of the longitudinal surface plasmon with increasing shell thickness. Reproduced with permission from Ref. 12.

8.3. Inorganic semiconductor-metal

Both semiconductor and metal are usually considered as active both optically and electronically since they have visible absorption due to electronic transitions in the relatively low energy region (1–3 eV). Their composites can have interesting and varied optical properties depending on the nature of the two components, their relative energy levels, their interactions, as well as the ratio between the two components. Their interactions are expected to depend on the details of their structure and the interface between them. Semiconductor–metal nanocomposites are generally more complex than semiconductor–insulator or metal–insulator composites.

While the absorption spectrum of semiconductor–metal composites tend to be close to the simple sum of the spectra of the two individual components, at least when their electronic interaction is not strong, their PL properties tend to change significantly, especially in terms of intensity, either quenched or enhanced depending on the interaction and distance in between the two components, compared to that of the semiconductor (metal has no or very weak PL). When one component is

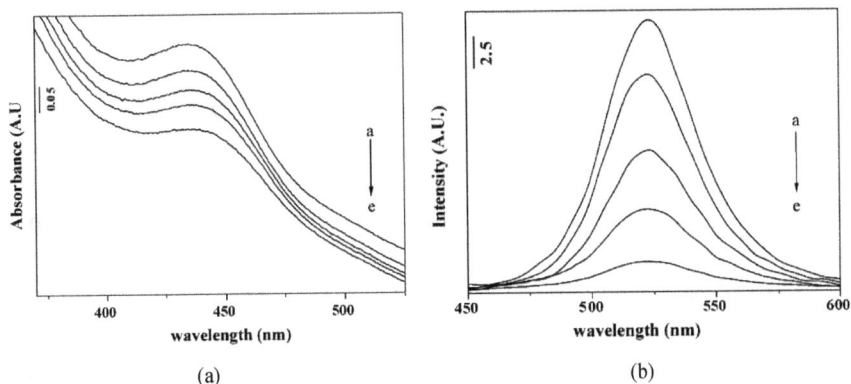

Fig. 8.8. UV-vis (a) and PL (b) spectra of Ag-CdTe nanocomposites in water with increasing Ag ratio compared to the original CdTe nanocrystals template: (a) CdTe template, (b) 1:50, (c) 1:20, (d) 1:10 and (e) 1:5. Reproduced with permission from Ref. 13.

substantially dominant in size or weight over the other, the optical properties of the composite tend to be primarily determined by the predominant component.

For example, Fig. 8.8 shows the UV-vis absorption and PL spectra of Ag-CdTe nanocomposites formed mainly through electrostatic interaction based on opposite charges on the Ag (negatively charged) and CdTe (positively charged) nanoparticles [13]. As can be seen, the absorption spectra exhibit an excitonic absorption band peaked around 433 nm typical for spherical CdTe nanocrystals. Upon the addition of Ag nanoparticles, the CdTe excitonic absorption band red-shifted by 2.5 nm and the full width at half-maximum (FWHM) became wider. The surface plasmon band around 400 nm characteristic of Ag nanoparticles is not noticeable due to the strong absorption of CdTe QDs and the fact that Ag nanoparticles are the minor component. However, the PL of CdTe QDs is strongly influenced by the presence of Ag nanoparticles. The CdTe PL is significantly quenched when increasing Ag nanoparticle content. The PL quenching is attributed to nonradiative photoinduced electron transfer to Ag nanoparticles. Similar observations have been reported for analogous systems such as Au-CdS [14] and Au-CdSe nanocomposites [15]. It should be noted that while the QD PL intensity is quenched significantly by the metal nanoparticle, the PL spectrum does not change much. This suggests that the bandedges of the QDs are not affected substantially by the presence of the metal nanoparticles.

In a different scenario, PL from the semiconductor can be enhanced by metal nanoparticles at appropriate distance or with appropriate interaction. For example, the enhancement of luminescence of CdSe/ZnS core/shell QDs by gold nanoparticles has been studied as a function of distance between the QDs and gold nanoparticles [16]. This distance is controlled using a layer-by-layer polyelectrolyte deposition technique to insert well-defined spacer layers between gold nanoparticles and QDs. The maximum enhancement by a factor of 5 is achieved for a distance around 11 nm. The PL enhancement has been attributed to enhanced QD excitation within the locally enhanced electromagnetic field produced by the gold nanoparticles. This is somewhat similar to the EM enhancement mechanism responsible for SERS, discussed in Chapter 7. Quenching or enhancement of PL of the semiconductor by the metal nanoparticles depends sensitively on the distance or interaction. In the same study, PL quenching was observed at distances of 3 nm and 19 nm. This nonmonotonic distance dependence has been suggested to arise from a competition between EM field enhancement and resonant energy transfer (RET) as well as possibly electron transfer to the gold nanoparticles. Resonant energy transfer is strongly dependent on distance and expected to be particularly effective at short distance. A related technique, FRET (fluorescence or Foster resonance energy transfer), will be discussed in Chapter 10.

A relevant theoretical study of exciton–plasmon interaction in hybrid semiconductor QD and metal nanoparticle complex, e.g. CdTe QDs and Au nanorods (NRs), has shown that both the radiative rate of exciton in the QD and the nonradiative energy transfer rate from the QD to the Au NRs vary significantly with distance between them and the orientation of the NRs [17]. Generally, both rates increase quickly as distance between the QD and NR decreases. Quantitative experimental verification of the theoretical results is yet to be conducted.

Other common semiconductor–metal nanocomposites include Pt-TiO_2 and Au-TiO_2 that are important for photocatalytic and photoelectrochemical applications [18–20]. In such systems, the optical properties, e.g. UV-vis absorption, in the visible is dominated by the metal nanoparticles due to weak absorption of TiO_2. There is no indication of new absorption bands, indicating weak or moderate interaction between the metal and TiO_2. However, there is evidence of improved photoinduced charge separation due to the presence of the metal nanoparticles, which is useful for photocatalytic reactions [19].

Another example of metal–semiconductor nanocomposites is Co-CdSe core/shell structures that exhibit magnetic, due to Co, and optical, due to CdSe properties [21]. In this system, the optical properties, absorption and PL, in the visible are dominated by CdSe due to weak absorption of Co. Such bifunctional nanostructures, magnetic and optical, are potentially useful for sensing and other magneto-optical applications. Nanocomposite structures are particularly suitable for introducing multiple functionalities.

8.4. Inorganic–organic (polymer)

8.4.1. *Nonconjugated polymers*

Both small and large organic molecules, such as polymers, have been used to modify the surface of inorganic NCs of semiconductor and metal or to create essentially inorganic–organic composite materials. There are two different classes of organic molecules: conjugated (aromatic with extended π-bonding) and nonconjugated, that tends to have quite different optical and electronic properties. Generally, for nonconjugated organic molecules composited with inorganic, their influence on the properties of the inorganic is small, especially for metal and semiconductor materials. This is mainly because such organic molecules typically have no or weak visible absorption and their energy levels are not near those of the semiconductor or metal nanostructures, and thereby there is little energy level mixing or interaction. Their effect on optical properties of the metal or semiconductor is thus weak. In this situation, the effect of the organic molecules is similar to that of inorganic insulators, as illustrated in Fig. 8.3.

A variety of polymers, mostly non-conjugated, have been used as matrix for creating composite nanostructures or for passivating the particle surface to stabilize their structure and optical properties, including metal [22–26] and semiconductor [27–37] nanomaterials. Interaction between the polymer and nanoparticle is often through functional groups such as -SH, -OH, -NH$_2$, -COOH and so on. Polymers have advantages such as low cost, flexible and large possible variations in structure, and ease to use. However, the nature of interaction between nanoparticles and the polymers in composites tends to be complex and the surface passivation is usually not complete, resulting in a high density of surface trap states on the nanoparticle surface, as indicated by low PL yield and trap state PL in semiconductor nanoparticles.

For inorganic metals with organic nonconjugated polymers, the optical properties are usually dominated by the metal nanostructure [22–26]. Similarly, optical properties of inorganic semiconductors with organic nonconjugated polymers are dominated by the semiconductor nanostructures [27, 29–36]. One difference is that semiconductor nanostructures tend to be photoluminescent and the PL properties are sensitive to surface characteristics that can be strongly influenced by the polymer. This does not occur in most metal nanoparticles unless when the size is extremely small and the metal clusters can be luminescent. In this case, they behave like semiconductors or even molecules [38, 39].

As an example, optical absorption of Au nanorods incorporated into poly(vinyl alcohol) (PVA) has been dominated by that of Au nanorods [24]. The surface plasmon absorption spectra strongly depends on the polarization of the incident light due to the anisotropy introduced by the Au nanorods that are aligned in the polyvinyl alcohol (PVA) film by stretching, as shown in Fig. 8.9. This can be explained using a dipole model.

8.4.2. *Conjugated polymers*

For conjugated organic molecules, including conjugated polymers with extended π-bonding, the energy level spacing between HOMO and LUMO is small, making them absorb visible light. Their energy levels are often close to those of inorganic metal or semiconductors [37]. The conjugated polymers themselves behave like organic semiconductors. This makes it more likely to mix their energy levels and their interaction is strong and can result in significant change in optical properties of the

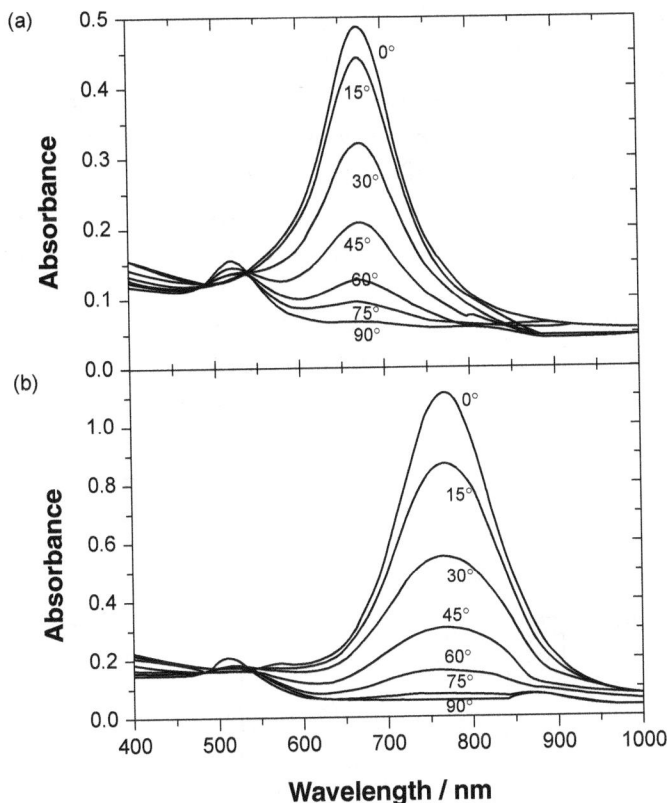

Fig. 8.9. Experimental UV-vis-NIR spectra of stretched PVA films containing aligned Au nanorods with two different aspect ratios: 2.23 (a) and 2.94 (b) for varying polarization angle of the electric field of polarized incoming light, as indicated in the figure. Reproduced with permission from Ref. 24.

composites compared to their isolated components. The changes depend on the details of their energy levels with respect to each other.

For inorganic semiconductor–organic conjugated polymer composites, the situation resembles that of a composite of two different inorganic semiconductor nanomaterials since the conjugated polymer is practically a semiconductor. Their absorption spectra are expected to be the sum of the two materials, and the PL of the composite may differ substantially from those of the two isolated components because of the strong interaction expected between the components. A number of studies have been done on semiconductor QD-conjugated polymer composites to gain a better understanding of their fundamental properties as well as to exploit their potential new properties for applications including sensors, light emitting diodes, and solar cells [40–48]. The change of optical properties of the composite as to the isolated QD or conjugated polymer depends sensitively on the relative electronic energy levels of the two components as well as how strongly they interact. Their electronic interaction also depends on their structural and interfacial characteristics. For example, in the composite of CdS QDs and polyfluorene copolymers, CdS QDs have been found to enhance the PL as well as electroluminescence (EL) of the polyfluorene copolymer [46]. In the case of InP QDs with polythiolphene, the PL of the QDs has been quenched by the conjugated polymer, as shown in Fig. 8.10 [44]. The PL quenching is attributed to photoinduced

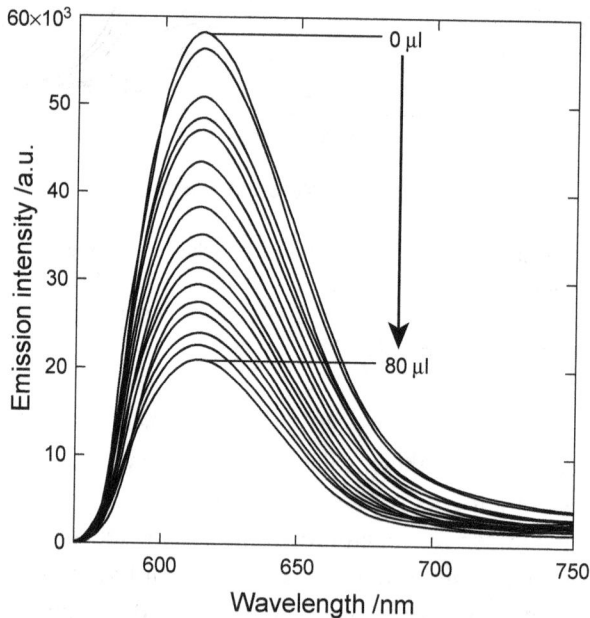

Fig. 8.10. Steady-state PL measurement of an InP QD solution (1.52×10^{-6} M) as aliquots of poly(3-hexylthiolphene), P3HT, (0.8 g/L) are titrated in. The PL of the InP QDs is quenched as P3HT is added. Reproduced with permission from Ref. 44.

hole transfer from the QD to the conjugated polymer due to the strong interaction between them. This finding has interesting implications in solar cell applications of this type of nanocomposite materials.

For inorganic metal–organic conjugated polymer nanocomposites, the situation is similar to that of inorganic metal–inorganic semiconductor. PL of the conjugated polymer can be quenched by the metal NPs, similar to PL quenching of inorganic QDs by metal NPs. While the absorption spectrum tends to be a mix or sum of that of the polymer and that of the metal (depending on relative weight or volume content), the PL is dominated by the conjugated polymer since there is little PL from the metal. If the metal content is high, complete PL quenching is expected for the polymer by the metal. For instance, when poly(p-phenylenevinylene) (PPV) film was doped with gold-coated silica (SiO_2@Au) nanoparticles, with optical resonance of SiO_2@Au nanoparticles specifically designed to interact with the triplet excitons that play a primary role in the photooxidation of PPV, the rate of photooxidation in PPV was drastically reduced, even though the absorption and PL spectra only change slightly between pristine PPV and the PPV nanocomposite film (due to the low SiO_2@Au content, ~0.05–0.1% by volume) [49]. This observation has important implications in enhancing the lifetime of conjugated polymers for optoelectronics applications. In the above example, the metal NP is the minor component. An opposite example is one with the metal NP as the major component and the conjugated polymer as the minor component. In this case, the optical property, e.g. absorption, tends to be dominated by the metal, as in the system of metal (Au, Pd, and Pt) nanoparticles capped with poly(dithiafulvene) (PDF) [50].

In principle, one could have inorganic insulator–organic composites such as SiO_2/polymer. Since these are less common and usually do not have strong visible absorption unless the organic molecules, including polymers, are optically active in the visible. Due to space limitation, we will not discuss this class of nanocomposites.

8.5. Inorganic–biological materials

As mentioned in the Introduction of this chapter, there are many examples of inorganic–biological composite materials in nature. Many synthetic nanocomposite materials have been developed and explored over the years for various applications including biomedical such as tissue engineering, bone replacement, detection and treatment of diseases. We will illustrate some of these exciting developments using inorganic metal or semiconductor nanomaterials conjugated with biological molecules, e.g. proteins or DNA, as examples. It is noted that the size of many biological molecules is also on the order of a few to a few hundred nm. Thus they are also nanomaterials and naturally compatible with other inorganic or organic nanomaterials in size and dimension.

We will first discuss semiconductor–biological nanocomposite materials that involve at least an inorganic, nanocrystalline semiconductor nanostructure and a biological molecule. Such composites could form based on electrostatic interaction or mechanical force. However, in general, since most inorganic and biological systems are not naturally compatible, it is often necessary to provide a bridge or linker between the two components, as illustrated in Fig. 8.11. A specific example is II-VI semiconductor QD coated with bifunctional linker molecules for conjugating to protein or DNA molecules. The PL from QDs can be used for detection or imaging applications. For instance, if the conjugated protein is an antibody, the system can be used for detecting the corresponding antigen of interest. Likewise, a single strand DNA conjugated to the QD can be used to detect the complementary DNA. Some of the chemistry involved in bioconjugation has been discussed in Chapter 4. Specific application of bioconjugation will be discussed in Chapter 10.

In the context of discussing nanocomposites, semiconductor–biomolecule conjugates can be considered as composite materials. Similar to QD-nonconjugated polymers, as long as the biological molecules do not contain chromophores that cause absorption in the visible region of the spectrum, the QDs with strong visible absorption will dominate the optical properties of the composite system in the visible region. Indeed, in many respects, many biomolecules such as DNA, RNA and proteins are nonconjugated polymers. The change to the optical property of the QDs due to the biological molecules is usually small, at least in terms of spectral features. The same is true for other nonconjugated or nonaromatic linker molecules, if present in the system.

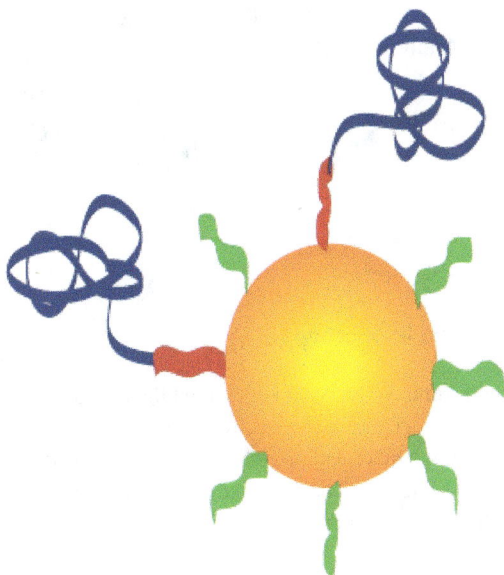

Fig. 8.11. Illustration of conjugation of biomolecules (blue) to semiconductor QD or metal nanoparticle (MN) (orange) surface via bifunctional linker molecules (red). The green represents surface passivating or surfactant molecules. Sometimes, the passivating molecules can serve as linker molecules as well. The linker molecules are usually bifunctional with one end bound to the QD or MN and the other end attached to biomolecules through covalent bonding.

However, it has been noted in many cases that the linker molecule and/or biological molecule could have substantial influence on the PL yield, and to a lesser degree, on PL spectrum, due to changes of the surface properties as a result of their presence on the QD surface. In some other cases, PL enhancement was observed, possibly due to better surface passivation when the biological or linker molecules were present. What has been described is true for biological or linker molecules that are chemically and photochemically unreactive with the QDs, i.e. the molecules are stable on the QD surface with or without light. If the molecules are reactive, more significant changes, including degradation of the QDs, could occur.

Many QD-biomolecule conjugates have been designed and studied for bio-detection applications. It is usually necessary to have the outer

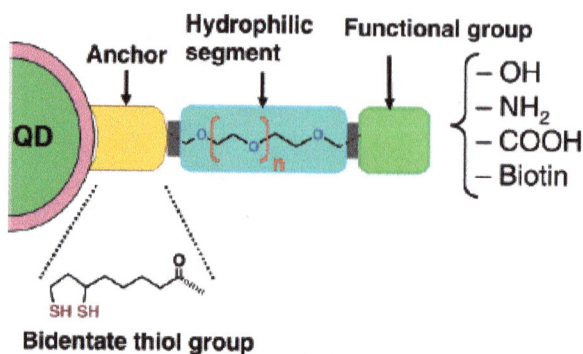

Fig. 8.12. An example of modular design of hydrophilic ligands with terminal functional groups. Dihydrolipoic acid (DHLA) is used, with one end, to cap CdSe/ZnS core/shell QDs and is linked, from the other end, to poly(ethylene glycol) (PEG). The out-pointing end of PEG is coupled with functional terminal groups to promote water-solubility and biocompatibility of the QDs. Reproduced with permission from Ref. 51.

surface rendered hydrophilic for biological applications due to the aqueous environment. Figure 8.12 shows one of the designs used to create hydrophilic surface for bio-detection based on CdSe/ZnS core/shell QDs [51]. This system is also designed to enhance the chemical and optical stability of the QDs. Because the CdSe core is well protected by the ZnS shell, the optical properties of such core/shell QDs are not sensitive to the surface molecular modification. The modular ligands based on poly(ethylene glycol) (PEG) coupled with functional terminal groups have been found to promote water-solubility and biocompatibility of QDs. The overall nanostructure is stable over a broad pH range in aqueous solution and has been demonstrated to be useful for conjugation to a variety of biological molecules.

Similar to semiconductor–biological systems, metal is usually the optically active component in metal–biological systems as long as the biological molecules are as described above, i.e. stable and no visible absorption. Metal nanostructures, especially gold and silver, have been frequently used for biological and biomedical applications in the form of composite materials. Biomolecules can bind to metal nanostructure surface through direct chemical bonding, e.g. using thiol (-SH), electrostatic force or linker molecules. For example, bioconjugates of bovine serum albumin (BSA): gold

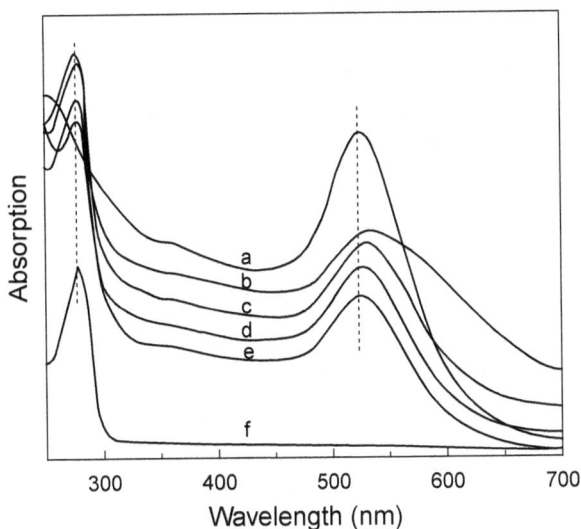

Fig. 8.13. UV-vis absorption spectra of gold nanoparticles (curve a) and the bovine serum albumin (BSA): gold nanoparticle conjugates at pH 2.7 (curve b), pH 3.8 (curve c), pH 7.0 (curve d), and pH 9.0 (curve e). Curve f is the absorption spectrum of BSA in 0.01 M PBS at pH 7.0. Reproduced with permission from Ref. 52.

nanoparticle (GNP) have been investigated to gain an understanding of possible conformational change of BSA upon adsorption onto the GNP surface [52]. Figure 8.13 shows the UV-vis absorption spectrum of the BSA-GNP conjugates at various pH values. It can be seen that the SPR of gold NPs is affected somewhat by BSA conjugation, likely due to small changes in the local medium dielectric constant. PL study shows quenching of the BSA PL by the gold NPs, attributed to energy transfer.

8.6. Summary

Several nanocomposite systems have been discussed with their optical properties highlighted. Besides the systems presented, there are other nanocomposites that we did not cover, including inorganic metal–metal, that tend to have strong interaction between their constituent components, thereby with complex optical properties. The examples covered illustrate the diversity and usefulness of nanocomposite materials. There are practically unlimited possibilities for generating composite materials due to the many parameters and material components that one can choose and vary. It is expected that the development and study of new nanocomposite materials will continue, and new applications will be exploited for emerging technologies.

References

1. D. Gerion, F. Pinaud, S.C. Williams, W.J. Parak, D. Zanchet, S. Weiss and A.P. Alivisatos, *J. Phys. Chem. B* **105**, 8861 (2001).
2. M.A. Petruska, A.V. Malko, P.M. Voyles and V.I. Klimov, *Advan. Mater.* **15**, 610 (2003).
3. A. Kongkanand, K. Tvrdy, K. Takechi, M. Kuno and P.V. Kamat, *J. Am. Chem. Soc.* **130**, 4007 (2008).
4. T. Lopez-Luke, A. Wolcott, L.-P. Xu, S. Chen, Z. Wen, J.H. Li, E. De La Rosa and J.Z. Zhang, *J. Phys. Chem. C* **112**, 1282 (2008).
5. L. Ba, L.D. Zhang and X.P. Wang, *Solid State Commun.* **104**, 553 (1997).
6. R. Serna, J.M. Ballesteros, J. Solis, C.N. Afonso, D.H. Osborne, R.F. Haglund and A.K. Petford-Long, *Thin Solid Films* **318**, 96 (1998).
7. L.M. Liz-Marzan and P. Mulvaney, *New J Chem.* **22**, 1285 (1998).
8. T. Ung, L.M. Liz-Marzan and P. Mulvaney, *Langmuir* **14**, 3740 (1998).
9. I. Pastoriza-Santos, D.S. Koktysh, A.A. Mamedov, M. Giersig, N.A. Kotov and L.M. Liz-Marzan, *Langmuir* **16**, 2731 (2000).
10. S.T. Selvan, T. Hayakawa, M. Nogami, Y. Kobayashi, L.M. Liz-Marzan, Y. Hamanaka and A. Nakamura, *J. Phys. Chem. B* **106**, 10157 (2002).
11. G.H. Fu, W.P. Cai, C.X. Kan, C.C. Li and L. Zhang, *Appl. Phys. Lett.* **83**, 36 (2003).
12. I. Pastoriza-Santos, J. Perez-Juste and L.M. Liz-Marzan, *Chem. Mater.* **18**, 2465 (2006).
13. Y.F. Wang, M.J. Li, H.Y. Jia, W. Song, X.X. Han, J.H. Zhang, B. Yang, W.Q. Xu and B. Zhao, *Spectrochim Acta A* **64**, 101 (2006).
14. P.V. Kamat and B. Shanghavi, *J. Phys. Chem. B* **101**, 7675 (1997).
15. T. Mokari, E. Rothenberg, I. Popov, R. Costi and U. Banin, *Science* **304**, 1787 (2004).
16. O. Kulakovich, N. Strekal, A. Yaroshevich, S. Maskevich, S. Gaponenko, I. Nabiev, U. Woggon and M. Artemyev, *Nano Lett.* **2**, 1449 (2002).
17. M.T. Cheng, S.D. Liu, H.J. Zhou, Z.H. Hao and Q.Q. Wang, *Opti. Lett.* **32**, 2125 (2007).

18. V. Subramanian, E.E. Wolf and P.V. Kamat, *Langmuir* **19**, 469 (2003).
19. M. Jakob, H. Levanon and P.V. Kamat, *Nano Lett.* **3**, 353 (2003).
20. T. Sasaki, N. Koshizaki, J.W. Yoon, S. Yamada, M. Koinuma, M. Noguchi and Y. Matsumoto, *Electrochemistry* **72**, 443 (2004).
21. H. Kim, M. Achermann, L.P. Balet, J.A. Hollingsworth and V.I. Klimov, *J. Am. Chem. Soc.* **127**, 544 (2005).
22. H.S. Zhou, T. Wada, H. Sasabe and H. Komiyama, *Synthet. Metal.* **81**, 129 (1996).
23. M.R. Bockstaller and E.L. Thomas, *J. Phys. Chem. B* **107**, 10017 s(2003).
24. J. Perez-Juste, B. Rodriguez-Gonzalez, P. Mulvaney and L.M. Liz-Marzan, *Adv. Funct. Mater.* **15**, 1065 (2005).
25. B. Karthikeyan, M. Anija and R. Philip, *Appl. Phys. Lett.* **88**, 053104 (2006).
26. Y. Deng, Y.Y. Sun, P. Wang, D.G. Zhang, H. Ming and Q.J. Zhang, *Physica E* **40**, 911 (2008).
27. R.E. Schwerzel, K.B. Spahr, J.P. Kurmer, V.E. Wood and J.A. Jenkins, *J. Phys. Chem. A* **102**, 5622 (1998).
28. J.H. Zhan, X.G. Yang, D.W. Wang, S.D. Li, Y. Xie, Y. Xia and Y.T. Qian, *Advan. Mater.* **12**, 1348 (2000).
29. Y. Gotoh, Y. Ohkoshi and M. Nagura, *Polymer J.* **33**, 303 (2001).
30. S.H. Yu, M. Yoshimura, J.M.C. Moreno, T. Fujiwara, T. Fujino and R. Teranishi, *Langmuir* **17**, 1700 (2001).
31. S.Y. Lu, M.L. Wu and H.L. Chen, *J. Appl. Phys.* **93**, 5789 (2003).
32. M.Z. Rong, M.Q. Zhang, H.C. Liang and H.M. Zeng, *Chem. Phys.* **286**, 267 (2003).
33. I.V. Klimenko, E.P. Krinichnaya, T.S. Zhuravleva, S.A. Zav'yalov, E.I. Grigor'ev, I.A. Misurkin, S.V. Titov and B.A. Loginov, *Russian J. Phys. Chem.* **80**, 2041 (2006).
34. J.F. Zhu, Y.J. Zhu, M.G. Ma, L.X. Yang and L. Gao, *J. Phys. Chem. C* **111**, 3920 (2007).
35. Y.B. Zhao, F. Wang, Q. Fu and W.F. Shi, *Polymer* **48**, 2853 (2007).
36. C. Inui, H. Kura, T. Sato, Y. Tsuge, S. Shiratori, H. Ohkita, A. Tagaya and Y. Koike, *J. Mater. Sci.* **42**, 8144 (2007).
37. E. Holder, N. Tessler and A.L. Rogach, *J. Mater. Chem.* **18**, 1064 (2008).
38. B.A. Smith, J.Z. Zhang, U. Giebel and G. Schmid, *Chem. Phys. Lett.* **270**, 139 (1997).
39. C.D. Grant, A.M. Schwartzberg, Y.Y. Yang, S.W. Chen and J.Z. Zhang, *Chem. Phys. Lett.* **383**, 31 (2004).
40. N.C. Greenham, X.G. Peng and A.P. Alivisatos, *Phys. Rev. B-Condensed Matter* **54**, 17628 (1996).

41. N.C. Greenham, X.G. Peng and A.P. Alivisatos, *Synthet. Metal* **84**, 545 (1997).
42. H. Skaff, K. Sill and T. Emrick, *J. Am. Chem. Soc.* **126**, 11322 (2004).
43. S.K. Hong and K.H. Yeon, *J. Korean Phys. Soc.* **45**, 1568 (2004).
44. D. Selmarten, M. Jones, G. Rumbles, P.R. Yu, J. Nedeljkovic and S. Shaheen, *J. Phys. Chem. B* **109**, 15927 (2005).
45. S.H. Choi, H.J. Song, I.K. Park, J.H. Yum, S.S. Kim, S.H. Lee and Y.E. Sung, *J. Photoch. Photobio A* **179**, 135 (2006).
46. C.H. Chou, H.S. Wang, K.H. Wei and J.Y. Huang, *Adv. Funct. Mater.* **16**, 909 (2006).
47. R. van Beek, A.P. Zoombelt, L.W. Jenneskens, C.A. van Walree, C.D. Donega, D. Veldman and R.A.J. Janssen, *Chem.-Eur. J.* **12**, 8075 (2006).
48. N.I. Hammer, T. Emrick and M.D. Barnes, *Nanoscale Res. Lett.* **2**, 282 (2007).
49. Y.T. Lim, T.W. Lee, H.C. Lee and O.O. Park, *Opt. Mater.* **21**, 585 (2003).
50. Y. Zhou, H. Itoh, T. Uemura, K. Naka and Y. Chujo, *Langmuir* **18**, 277 (2002).
51. K. Susumu, H.T. Uyeda, I.L. Medintz, T. Pons, J.B. Delehanty and H. Mattoussi, *J. Am. Chem. Soc.* **129**, 13987 (2007).
52. L. Shang, Y.Z. Wang, J.G. Jiang and S.J. Dong, *Langmuir* **23**, 2714 (2007).

Chapter 9

Charge Carrier Dynamics in Nanomaterials

Charge carrier dynamics in nanomaterials are of interest for several reasons. First, dynamic studies provide information complementary to static or equilibrium studies. Second, dynamic information is directly useful for exploiting applications, such as fast optical or photonic switches or modulators. Third, the dynamic information can be used to guide the design of nanomaterials with properties tailored towards different technological applications, e.g. sensors and solar cells. These will be illustrated using specific examples in this chapter.

9.1. Experimental techniques for dynamics studies in nanomaterials

While frequency domain measurements can provide dynamic information in an indirect manner, dynamics are usually studied directly using time-resolved techniques. As discussed in Chapter 3, one common experimental technique for probing dynamics is time-resolved laser spectroscopy that can allow resolution on the femtosecond (fs), picoseconds (ps), and nanosecond (ns) time scales. Longer time resolution can be achieved using less demanding techniques such as flash photolysis based on pulsed lamps or determined by the time resolution of photodetectors.

In typical time-resolved laser studies on the ultrafast time scale, two laser pulses are required, one for excitation or pumping and the second for interrogating or probing the sample. The time delay between the two pulses is used to gain the dynamic information of interest. In such a pump-probe setup, experimental measurements are repeated for each time delay and often many data points are averaged for a given time delay to enhance signal-to-noise ratio. The basic idea is that the pump pulse creates some excited state population that the probe pulse monitors. The time profile or dynamic information obtained reflects on the lifetime of the excited state. This is often termed transient absorption.

One can also monitor the emission from an excited state if the state is emissive. The emission can be monitored in two ways. First, the emission can be directly monitored by a photodetector such as PMT or photodiode. The time resolution is limited by the detector that is usually much longer than the excitation laser pulse width. Second, the emission can be monitored by optically gating the emission with a second short laser pulse through a process named fluorescence up-conversion (FUC). In the latter case, the emission is mixed with a gating pulse in nonlinear optical crystal to generate a new pulse of light with energy equal to the sum of the emitted light and gating light but pulse width similar to that of the gating pulse. By delaying the gating pulse with respect to the excitation pulse, dynamic information can be obtained. The time resolution for the second method is limited by the cross-correlation between the pump and gating pulses, which is much higher than that of a typical photodetector. Thus, the second method, FUC, offers higher time resolution, but is more involved technically. More details for time resolved techniques have been given in Chapter 2.

9.2. Electron and photon relaxation dynamics in metal nanomaterials

The primary dynamic processes involving excited or hot electrons in metal are electron–electron scattering (dephasing), electron–phonon coupling, and phonon–phonon scattering. While the exact time scale for these processes vary from metal to metal and depend on their detailed electronic and phonon structures, in general, electronic dephasing is the fastest, a few to a few tens of fs, electron–phonon relaxation is on the order of hundreds of fs to a few ps, and phonon scattering occurs on the tens of ps or longer time scale. Phonon scattering is particularly sensitive to the embedding environment of the materials, especially for nanomaterials due to their large S/V ratio. Each of these processes will be discussed next with specific examples as illustrations.

9.2.1. *Electronic dephasing and spectral line shape*

The homogeneous spectral line width in electronic absorption spectrum is mainly determined by the fastest event in dynamics, which is electronic dephasing. Pure dephasing, commonly referred to as the T_2 process, is considered as a dynamic process in which the energy is conserved while momentum is changed. The electronic dephasing following electronic excitation by light absorption is due to electron–electron interaction or scattering.

For metal nanoparticles, the line width or bandwidth of the observed SPR is primarily determined by dephasing. The dephasing time is on the order of a few fs to a few tens of fs [1]. As shown in Fig. 9.1, the SPR of Au nanoparticles is quite broad, even for fairly monodisperse nanoparticles. The broadness is a result of very fast dephasing, on the order of a few fs. For small metal particles, the dephasing can be described by a damping constant Γ, which is dependent on the surface as described by the following equation [2]:

$$\Gamma = \Gamma_0 + A\frac{v_F}{R} \qquad (9.1)$$

where R is the particle radius, v_F is the Fermi velocity of the electrons, and A is a constant on the order of unity. The first term on the right-hand side, Γ_0, describes the bulk contribution to the dephasing of the electrons due to electron–electron interaction, and the second term accounts for the additional dephasing due to electron-surface scattering.

Direct time-resolved study of dephasing (T_2) in metal nanoparticles is not easy due to the very short lifetime. However, various indirect measurements, e.g. based on frequency-domain methods, have provided some useful data about the dephasing time. For example, using degenerate four-wave mixing (DFWM) with temporally incoherent laser light, the electronic

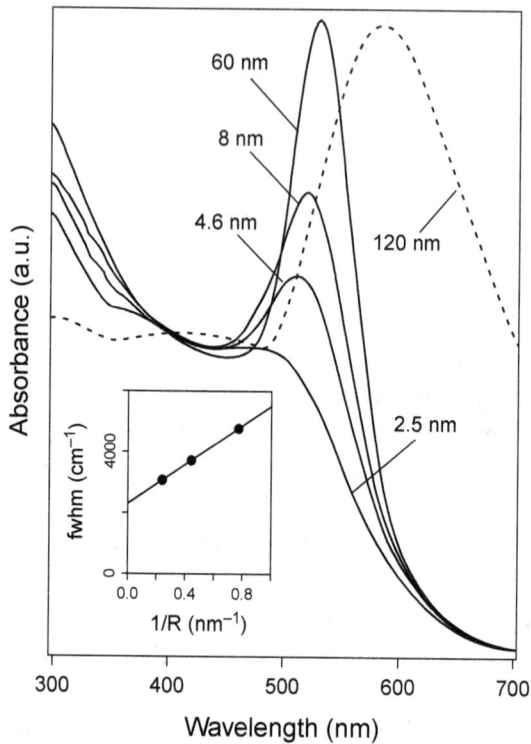

Fig. 9.1. Extinction spectra of the different-size Au particles in aqueous solutions with a Au concentration of 2×10^{-4} M. The inset shows the full width at half-maximum of the plasmon band for the 2.5, 4.6 and 8 nm diameter particles. Reproduced with permission from Ref. 3.

dephasing in gold nanoparticles has been determined to be around 20 fs [4]. Using a near-field optical antenna effect, the surface-plasmon dephasing time of 8 fs has been extracted from the near-field spectra of individual gold particles [5]. In another study, using an autocorrelation method based on second harmonic generation, the dephasing (T_2) time in both Ag and gold nanoparticles have been found to be less than 20 fs [6].

9.2.2. *Electronic relaxation due to electron–phonon interaction*

The primary dynamic process of interest in metal nanomaterials is electron energy relaxation following excitation of the SPR through electron–phonon coupling. This is a relatively simple and well-understood process that occurs on very fast time scales, a few ps. An interesting issue is the possible dependence of electronic relaxation time or electron–phonon interaction on the size and/or shape of the metal nanostructures. As size decreases, the density of states for both the electron and phonon decreases. This should lead to weaker electron–phonon coupling and thereby longer electron relaxation time. The extreme would be particles with only a few metal atoms, which are essentially molecules. Another factor is the increase in surface-to-volume ratio with decreasing size. Increased surface scattering or surface phonon contribution could shorten the electron lifetime.

The relaxation of the hot electrons can be understood by the two-temperature model that describes the energy exchange between the electrons and phonons by the two-coupled equations [7–9]:

$$C_e(T_e)\frac{\partial T_e}{\partial t} = -g\,(T_e - T_l)$$
$$C_l\frac{\partial T_l}{\partial t} = g\,(T_e - T_l) \tag{9.2}$$

where T_e and T_l are the electronic and lattice temperatures, $C_e(T_e) = gT_e$ is the temperature-dependent electronic heat capacity, $\gamma = 66\ \mathrm{Jm^{-3}\,K^{-2}}$ for Au [10], C_l is the lattice heat capacity, and g is the electron–phonon coupling constant. The equations indicate that the relaxation time should increase with higher excitation laser power due to higher initial electronic temperature, which is consistent with experimental data [3].

To date, experimental results have shown that the electron relaxation lifetime for Au and Ag nanoparticles are very similar to that of bulk metal (about 1.5 ps) and this is true at least down to a few nm in diameter. There seems to be a substantial increase in the lifetime only when the particles reach very small size with less than 50 atoms. Early electron relaxation measurements by Zhang *et al.* suggested a possible size dependence of the electron relaxation in the 1–40 nm range [11]. However, later on Harland *et al.* and El-Sayed *et al.* found no size dependence on the relaxation down to the size of 2.5 nm [12–14]. The relaxation time in Au NPs was reported to be the same as that of bulk gold (~1 ps). Excitation intensity dependence of the relaxation time has been found and could be the reason for discrepancy. It is likely that the excitation intensities used in the earlier work are much higher than those used in later measurements. Another possible explanation for the difference is surface and/or solvent environment, which was found to affect electron relaxation of Au NPs [15–17]. It seems convincing that the electron–phonon coupling constant is the same for Au NPs as for bulk, at least for particles down to ~2 nm in diameter [3].

When the particle size is smaller than 1 nm, the electron relaxation time seems to become significantly longer than that of bulk. An earlier study of Au_{13} and Au_{55} found that the electron lifetime becomes significantly longer for Au_{13} nanoclusters than for Au_{55} or larger nanoparticles, indicating bulk to molecule transitions in the size regimes of 55 and 13 Au atoms [11]. A study of Au_{28} clusters found biexponential decays with a subpicosecond and a ns component [18]. The fast component was attributed to relaxation from a higher lying excited state to lower electronic state, and the longer nanosecond component was assigned to radiative lifetime of the lower electronic state Au_{28} clusters [18]. More recently, studies of electron relaxation in Au_{11} clusters have revealed a similar long-lived component (~1 ns) [19]. The longer lifetime again suggests that for very small metal clusters, particles are becoming molecule-like in nature and the electron–phonon interaction becomes weaker, similar to that found for Au_{13} [11].

Due to difficulty in the synthesis and sample stability, the study of very small particles is not as well established as that of larger particles and further research is probably warranted. In any case, the results seem to suggest that metal nanoparticles behave more or less classically as bulk metal down to a few nm in diameter, and then start to behave differently, more molecule-like, only for very small sizes.

Different shaped metal nanostructures have also been studied in terms of their dynamic properties. Possible shape dependence of the hot relaxation time has been studied in several nonspherical metal nanostructures, mainly silver and gold, including nanorods and nanocages. It has been found that the relaxation time is similar to that in spherical nanoparticles and bulk [20–24]. This indicates that the electron–phonon coupling is not affected by the shape of the nanostructure in any significant manner.

This is in contrast to subsequent phonon relaxation that is strongly shape-dependent, as discussed in the next subsection (Sec. 9.2.3).

Crystallinity and grain boundaries of the nanoparticles are expected to affect the phonon frequency and distribution and thereby electron–phonon interaction and electron relaxation. For example, in a comparative study of polycrystalline versus single crystalline gold nanoparticles, it has been found that electron–phonon relaxation rate decreases greatly when polycrystalline prismatic gold nanoparticles are annealed and transformed into nearly single-crystalline nanospheres [25]. The results are explained by the presence of high-density grain boundaries with dense, high frequency phonons that are effective in removing the energy of the excited electrons in the polycrystalline prismatic nanoparticles.

9.2.3. *Photon relaxation dynamics*

It has been found that the electron relaxation can be coupled to and thereby modulated by phonon vibration. As the electrons relax and the phonons get excited, the electron density changes with phonon vibration because vibration causes volume change. This results in change in the surface plasmon absorption. In a typical transient absorption (TA) measurement, the shift or change of the plasmon absorption with phonon vibration is reflected on intensity undulation of the probe light. The oscillation period is strongly dependent on the size and shape of the nanostructures. Therefore, this feature can be used to probe the mechanical property of the nanostructure.

Lattice vibrational oscillations have been observed in the electron relaxation dynamics of Au and bimetallic core/shell particles [3, 26–29]. Femtosecond transient absorption data for Au particles probed at 550 nm following 400 nm pulse excitation show clear modulations with a period of about 16 ps. The frequency of the oscillation was found to increase linearly with decreasing particle size. The oscillations have been attributed to a coherent excitation of the radial breathing vibrational modes of the particles. Photoexcited electrons can transfer their energy into the lattice, heating up the particle and causing a rapid expansion. The expansion and contraction of particle volume over time cause the electron density of particle to change leading to a periodic shift of the surface plasmon

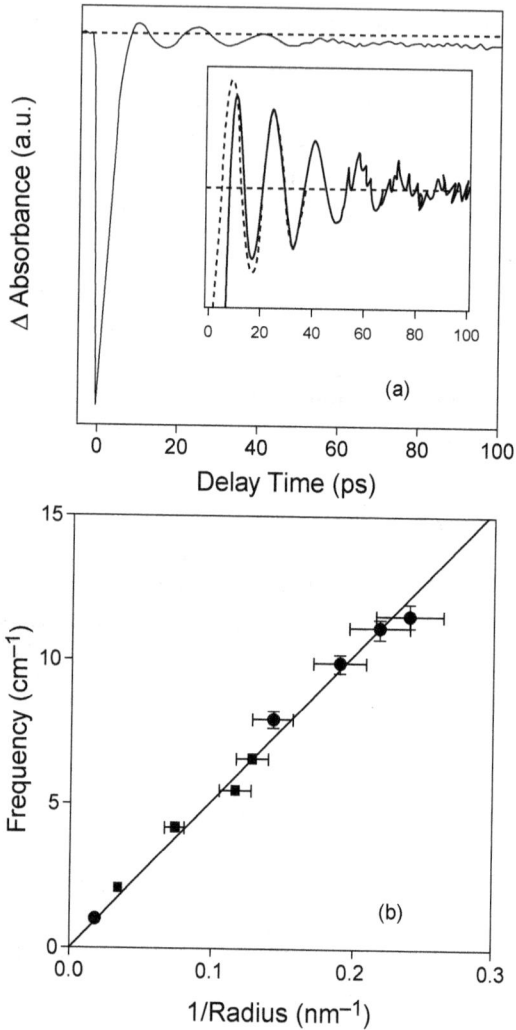

Fig. 9.2. (a) Transient absorption data for 60 nm diameter Au particles recorded with 400 nm pump and 550 nm probe pulses. Modulations due to the coherently excited breathing motion can be clearly seen. (b) Plot of the measured frequencies versus $1/R$ for different Au samples. A straight line fit to the data is also shown. Reproduced with permission from Ref. 3.

absorption band, manifesting itself as a modulation in the transient absorption signal.

As an example, Fig. 9.2 shows transient absorption data for 60 nm diameter Au particles recorded with 400 nm pump and 550 nm probe pulses [3].

Clear modulations to the TA decay due to the coherently excited breathing motion can be seen. The measured frequencies as a function of $1/R$ for different-size Au nanoparticles can be fit nicely to a straight line, indicating a simple linear relationship between particle size and the oscillation period or frequency.

Similar oscillation has been observed more recently in silver ellipsoids [30], Au nanorods [31], strongly coupled Au aggregates [32], and Au and Ag nanocages [21, 22]. Coherent excitation of the acoustic vibrational modes in Au nanorods results in the oscillation of transient absorption signals. The period of the oscillation has been found to be $2L/c_t$, where L is the length of the rod and c_t is the transverse speed of sound in bulk gold. This is different from the results of silver ellipsoids, where the period is determined to be $2d/c_l$, d is the length or width of the ellipsoid, and c_l is the longitudinal speed of sound in silver [30]. The discrepancy has been explained by the different natures of the vibrational motion and elastic properties [31].

Aggregates of metal NPs are fundamentally interesting in that they can be used to study interparticle interactions. The interaction between particles can be roughly divided into three regimes: weak, moderate and strong. In weakly or moderately interacting systems such as DNA-linked Au NPs [33, 34] and superlattice structures of Au NPs [35], the transverse plasmon band shifts noticeably to the red. In strongly interacting systems such as Au nanoparticle aggregates, a whole new absorption band, termed extended plasmon band (EPB), appears near 700–950 nm in addition to the transverse surface plasmon band (~520 nm) [36, 37]. The EPB is similar to the longitudinal plasmon absorption in nanorods and its appearance is a signature of strong interaction between nanoparticles. In a recent dynamic study of strongly interacting Au aggregates, the electron relaxation time appeared to be similar to those of isolated particles and bulk [32]. Surprisingly, periodic oscillations and probe wavelength dependence of the oscillation period were observed in the dynamic profiles of Au nanoparticle aggregates as shown in Fig. 9.3. The oscillations have been attributed to the coherent excitation of vibration of the aggregates. The dependence of oscillation period on probe wavelength indicates that the broad EPB in the static absorption spectrum is inhomogeneously broadened due to different aggregate sizes, supported by a spectral hole burning experiment [32]. Samples with such inhomogeneously broadened spectrum is undesired for SERS since only a subset, often small percentage, of the aggregates will have plasmon absorption resonant with the incident light and is thereby effective for SERS. One interesting observation is that the oscillation period is longer than that predicted based on an elastic sphere model. A possible explanation is that the vibrational motion in aggregates is "softer" than that of isolated hard spherical particles. This study shows that TA measurements can provide information on size distribution and vibrational frequencies of metal nanoparticle aggregates or similar systems.

It should be noted that the phonon relaxation dynamics is expected to be sensitive to the embedding environment of the metal nanoparticles, e.g solvent or solid matrix. In general, solvent or matrix that can best accept phonon energy from the metal will facilitate phonon relaxation in the metal nanostructure and the phonon relaxation rate will be larger or the time constant will be shorter. Crystallinity or grain boundaries of the nanoparticles also affect the phonon frequency and distribution, and thereby electron–phonon as well as phonon–phonon interaction and phonon relaxation [25, 38].

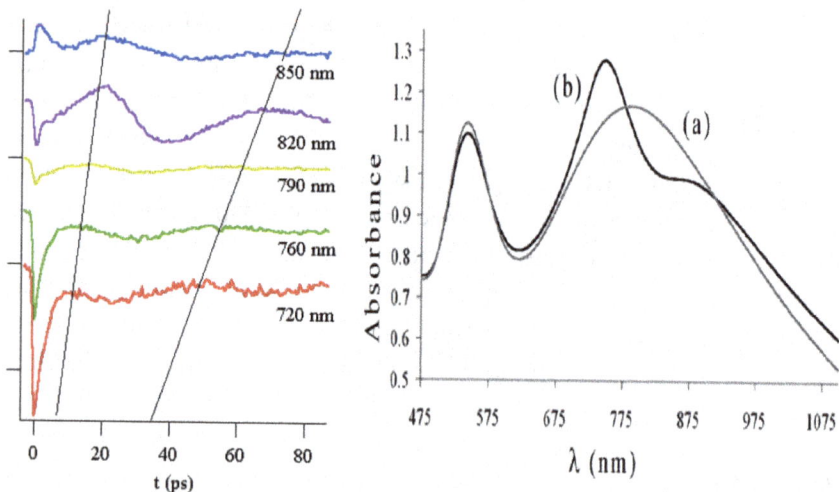

Fig. 9.3. (left) Femtosecond (fs) electronic relaxation and phonon oscillation dynamics in Au nanoparticle aggregates following excitation at 390 nm and probed at different wavelengths indicated in the near IR region where only aggregates absorb, and (right) spectral hole-burning in Au nanoparticle aggregates before (a) and after (b) excitation with an 810 nm fs laser. Adapted with permission from Ref. 32.

9.3. Charge carrier dynamics in semiconductor nanomaterials

Dynamic processes in semiconductor nanoparticles are generally different and more complex than in metal nanoparticles. This is mainly due to their fundamentally different band structures. For metal, the relevant optical absorption and dynamics studied usually only involve intraband processes. Of course, depending on the specific metal and wavelength of excitation, interband processes can take place as well. Sometimes, this is actually an area of confusion in the literature. Some interband processes have been misunderstood or are interpreted as intraband processes. For example, excitation of Ag and Au nanostructures with light wavelength near their plasmon absorption bands results in excitation of the surface plasmon or excitation of conduction band electrons collectively. The life-time determined reflect the plasmon lifetime that is short and there is usually no light emission associated with this process. However, if one uses a shorter wavelength of light, interband transition can occur. The electron can now be excited to an unoccupied band that has long lifetime and strong light emission. The behavior in this case is like a bandgap exci-tation of a semiconductor. It is very important that this is not confused with the intraband plasmon excitation. There are claims of strong emis-sion and long lifetime for electron relaxation in metal nanostructures that are described as intraband processes but really are interband processes. Therefore, care should be taken to understand and distinguish interband versus intraband processes.

Back to semiconductor nanostructures, their dynamic properties are largely determined by the fact that there is a bandgap as well as the real-ity that there are usually a substantial number of bandgap states, often referred to *as surface states or trap states*, which are absent in ideal, per-fect crystals. The bandgap states complicate matters but can also make the system interesting and useful. For example, many bandgap states are responsible for the reactivity of the material and thereby useful for appli-cations such as catalysis. Optically, the bandgap states alter the photoemission and, to a lesser degree, absorption properties. In particular, photoemission from bandgap states can serve as a probe of these states. Bandgap states can be introduced intentionally as in a doped system for applications such as phosphors.

Just as for metal nanomaterials, time-resolved laser spectroscopy is powerful for probing charge carrier dynamics in semiconductor nanoparticles, such as charge carrier dephasing, cooling, electron-hole recombination, carrier trapping and charge transfer. In general, the following dynamic processes are expected (approximately from the shortest to the longest lifetimes): intraband carrier dephasing, intraband relaxation, trapping, interband recombination or trapped carrier recombination. In case of multiple excitons generated per nanoparticle due to high excitation light intensity, exciton–exciton annihilation or Auger recombination can occur. Similar to metal, electronic dephasing occurs on the few fs to tens of fs time scales. Intraband energy relaxation takes place on the hundreds of fs time scale, similar to hot electron relaxation in the conduction band of metal. Trapping occurs on the few ps to tens of ps time scales, depending on the nature of trap states and energy barrier, if present. Usually, trapping into shallow traps close to the bandedge is faster than into deep trap states near the middle of the bandgap. Recombination from trap states or the conduction and valence bands have a typical lifetime of a few ns for dipole allowed or direct bandgap transitions, if the decay is primarily radiative, and nonradiative pathways play a minor role. The recombination lifetime becomes shorter, e.g. hundreds of ps, if nonradiative pathways become more dominant, e.g. due to a high density of trap states. For dipole forbidden or indirect bandgap transitions, the lifetime can be much longer, microseconds or even milliseconds. As discussed in Chapter 5, the higher the density of trap states, the shorter the observed lifetime for the excitonic or bandedge states and the lower the overall PL quantum yield. This is mainly because the trap states tend to cause the carriers to nonradiatively decay. The exact time scales for these processes are dependent on the nature of the materials. The numbers given above and in Fig. 9.4 are only to provide a rough order of magnitude for major dynamic processes in semiconductor nanomaterials based on typical cases studied. Exceptions to these numbers are expected. The following provides a more detailed discussion of the general points made above using examples as illustrations.

9.3.1. *Spectral line width and electronic dephasing*

Same as for metal, homogeneous spectral line width in electronic absorption spectrum of semiconductor nanomaterials is usually determined by the

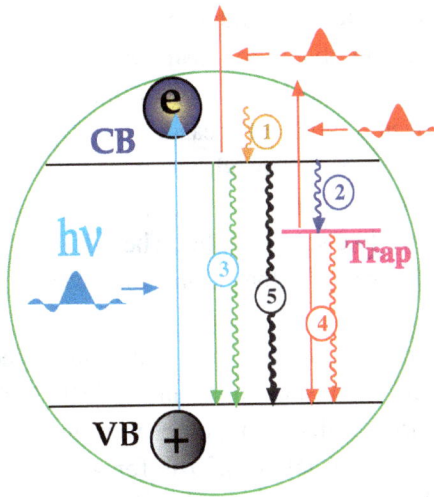

Fig. 9.4. Different dynamic processes involved in charge carrier relaxation in photoexcited semiconductor nanoparticles: (1) electronic relaxation in the conduction band (CB) through electron–phonon interaction; (2) electron trapping; (3) radiative and nonradiative bandedge electron-hole recombination; (4) radiative and nonradiative electron-hole recombination; and (5) nonlinear exciton–exciton annihilation. Electronic dephasing that occurs before process (1) is not shown. Likewise, hole trapping, that can also occur, is not illustrated. Reproduced with permission from Ref. 39.

fastest event in dynamics, which is electronic dephasing. The homogeneous line widths and time for pure dephasing can be determined using experimental techniques such as hole burning [40–42] or photon echo [43–45]. The hole burning technique has been successfully applied to measure homogeneous line widths and line shapes of semiconductor nanoparticles, including CdSe, CdS, CuCl, CuBr [40, 41, 46, 47], and CdSe/ZnS core/shell [42]. The hole burning studies, mostly carried out at low temperature, e.g. 2 K, also found that quantum confinement of carriers and resulting strong Coulomb interaction between confined carriers and trapped carriers are essential for the energy change as reflected in the observed persistent spectral hole burning phenomenon [40]. Exciton localization and photoionization of the nanocrystals have been suggested to play an important role in the observed hole burning and hole filling processes. Homogeneous line width as narrow as 32 meV were obtained at low temperature (10 K) for CdSe/ZnS core/shell structures [42]. At low temperature, the temperature dependence of the homogeneous line width was found to deviate from the usual linear dependence, which was considered as a reflection of the effects of phonon quantization. Time-resolved hole burning has been applied to measure the dephasing time in CdSe nanocrystals and the energy dependence in the gain region was found to be rather constant for nanocrystal while increasing toward the transparency point for bulk-like samples [46]. The difference was attributed to different gain mechanisms in the strong and weak quantum confined regimes.

Time-resolved photon echo experiments have been conducted to directly measure the dephasing times in CdSe [43, 44] and InP [45] nanocrystals. For CdSe the dephasing times were found to vary from 85 fs for 2 nm diameter nanocrystals to 270 fs for 4 nm nanocrystals at low temperature. The dephasing times can be affected by several dynamical processes that are dependent sensitively on nanocrystal size, including trapping of the electronic excitation to surface states, which increases with increasing size, and coupling of the excitation to low-frequency vibrational modes, which peaks at intermediate size [44]. Contributions from acoustic phonons were found to dominate the homogeneous line width at room temperature [43]. Dephasing time in CuCl nanocrystals has been measured using fs transient DFWM and for $R < 5$ nm a reservoir correlation time of 4.4–8.5 ps, which increases with increasing R, was found based on the stochastic model [47]. The results are explained based on an increase of exciton-acoustic phonon coupling strength and a change in the acoustic phonon DOS due to quantum confinement of acoustic phonons.

9.3.2. Intraband charge carrier energy relaxation

Above bandgap photoexcitation of a semiconductor creates an electron in the CB and a hole in the VB. This is a nonequilibrium situation. To go back to equilibrium, the photogenerated carriers have a number of pathways. The simplest is radiative electron-hole recombination, releasing a photon. Usually, the electron and hole lose some of their initially acquired extra kinetic energy within their bands (intraband) before they recombine, mainly through electron–phonon interaction (process 1 in Fig. 9.4).

As a result, the emitted photon has a lower energy, or red-shifted, than the incident photon. For single crystal semiconductors, radiative electron-hole recombination is the dominant mechanism of relaxation and the material is thus highly luminescent, with quantum yield >50%. The time scale for intraband relaxation for the electron and hole is similar to that of electron relaxations in metals, typically on the order of tens to hundred of fs [48–50].

If the crystal has impurities or defects, it tends to have electronic states within the bandgap that can trap the carriers. Trapping is a major nonradiative pathway for semiconductors with a high density of trap states. It can have strong effect on both intraband as well as interband relaxations. The effect of surface is usually difficult to quantify theoretically. Qualitatively, however, the surface is expected to have two effects. First, the surface phonon frequencies and distribution are expected to change relative to bulk. This will affect the overall electron–phonon interaction and thereby primarily the intraband relaxation time. Second, the surface introduces a high density of surface trap (electronic) states within the bandgap that will act to trap the charge carriers. This usually shortens the carrier lifetime significantly and is also the dominant pathway of relaxation. As a result, more interband relaxation is affected, as to be discussed later. Trapped carriers will further relax radiatively or nonradiatively (process 4 in Fig. 9.4). The following sections will provide a number of examples of semiconductor nanoparticle systems that have been studied to illustrate the different time scales for various trapping and interband relaxation mechanisms.

9.3.3. *Charge carrier trapping*

One of the most important characteristics of nanoparticles is their extremely large surface relative to volume. This creates a high density of surface states due to surface defects and dangling bonds. These surface states may fall energetically within the bandgap of a semiconductor and serve to trap charge carriers. Charge carrier trapping thus plays a critical role in the electron relaxation process of photoinduced carriers. The trapped electrons and holes significantly influence the optical properties, e.g. emission and chemical reactivities of the nanoparticles. One measure of the presence of trap states is trap state emission, which is usually substantially red-shifted relative to bandedge emission, as a result of relaxation (trapping) of the electron and/or hole. The time scale for trapping is in general very short, on the order of a few hundred fs to tens of ps, depending on the nature of the nanoparticles. A higher DOS leads to faster trapping. The trapping time also depends on the energy difference between the trap states and the location of the bandedge (bottom of the CB for the electron and top of the VB of the hole). The smaller the energy difference, the faster the trapping is expected, provided that other factors are similar or the same.

For CdS NPs, an electron trapping time constant of about 100 fs has been suggested [51, 52]. A longer trapping time (0.5–8 ps) was deduced for CdSe NPs based on time-resolved photon echo experiment [44]. An even longer trapping time of 30 ps has been reported for CdS NPs based on measurement of trap state emission [53, 54]. A similar 30 ps electron trapping time for CdS NPs has been reported based on the study of effects of adsorption of electron acceptors such as viologen derivatives on the particle surface [55]. A hole trapping time of 1 ps has been reported for CdS based on time-resolved photoluminescence measurements [54]. The difference in trapping times reported could be either due to difference in the samples used or different interpretations of the data obtained. It can be concluded, however, that the trapping time is on the order of a few hundred fs to tens of ps, depending on the nature of the NPs and quality of the sample. The trapping time may also differ substantially for shallow and deep traps.

Since the time scale for trapping can be similar to that for intraband relaxation, it is sometimes not easy to distinguish between them. Careful control experiments need to be done to determine which process is truly responsible for the observed dynamics. Fortunately, with advancement of ultrafast laser technology, fs lasers with higher power and broader tunability become more easily available nowadays, and they are helpful in sorting out different dynamic processes.

9.3.4. *Interband electron-hole recombination or single excitonic delay*

Electron-hole recombination can occur before or after the electron and hole are trapped. In nanoparticles, since trapping is very fast due to a high density of trap states, most of the recombination takes place after one or both of the charge carriers are trapped. The recombination can be radiative or nonradiative. Nonradiative electron-hole recombination of excitonic decay through electron–phonon interaction is usually not efficient [56]. Nonradiative recombination mediated by trap states is dominant for nanoparticles. A number of examples will be discussed next to illustrate the similarities and differences between different nanoparticle systems in terms of their electron relaxation dynamics.

CdS is among the most extensively studied semiconductor nanoparticle systems. Earlier dynamics studies on the picosecond time scale identified a strong transient bleach feature near the excitonic absorption region of the spectrum [57–63]. It was noticed that the peak of the bleach feature shifts with time and one explanation proposed was increased screening by charge carriers for the particles [64]. It was also observed that there is a red-shift of the transient absorption features in CdSe, which was explained by some as a result of formation of biexcitons [65, 66]. Later, femtosecond measurements were carried out and a power dependence of the bleach recovery time was found for CdS [67]. The bleach recovery follows a double exponential rise with the fast component increasing with power faster than the slower component.

Subsequent work based on transient absorption measurements found a similar power dependence of the electron relaxation dynamics featuring a double exponential decay behavior with a fast (2–3 ps) and slow (50 ps) component [51, 68, 69]. As shown in Fig. 9.5, the amplitude of the fast decay component increases with excitation intensity faster than that of the slow component. It grows nonlinearly, slightly subquadratic, with excitation intensity. This nonlinear fast decay was first attributed to nongeminate electron-hole recombination at high excitation intensities [51]. Subsequent studies using fs transient absorption in conjunction with ns time-resolved fluorescence found that the bandedge fluorescence was also power dependent [68]. These results led to the proposal that the fast decay is due to exciton–exciton annihilation upon trap state saturation, as suggested previously [70], and the slow decay is due to trapped charge carrier recombination [68]. Therefore, the transient absorption signal observed has contributions from both bandedge electrons (excitons) and trapped electrons. At early times, the bandedge electrons contribute

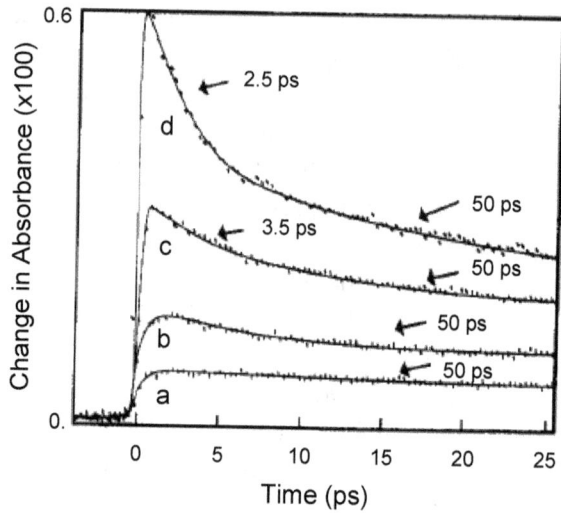

Fig. 9.5. Excitation intensity dependence of the charge carrier dynamics in CdS colloidal nanoparticles probed at 780 nm following excitation at 390 nm with photon fluence of (in photons/Å^2): (a) 0.12; (b) 0.20; (c) 0.59; (d) 1.18. The fast component increases non-linearly with excitation intensity. Reproduced with permission from Ref. 51.

significantly to the signal, especially when trap states are saturated at high excitation intensities, while as time progresses the contribution from trapped electrons becomes more dominant. On long time scales (hundreds of ps to ns), the signal is essentially all from trapped charge carriers. It was believed that this is also true for many other colloidal semiconductor nanoparticles [71–75]. A more recent study of the emission lifetime on the ns time scale of surface passivated and unpassivated CdS nanoparticles showed that the passivated sample with a lower density of surface trap states has a lower excitation threshold for observing exciton–exciton anni-hilation compared to the unpassivated sample [76]. This supports the model of exciton–exciton annihilation upon trap state saturation at high excitation intensities [68]. A more detailed discussion of nonlinear charge carrier dynamics will be discussed later in Sec. 9.3.7.

The charge carrier dynamics in CdSe nanocrystals ranging in size from 2.7 to 7 nm have recently been determined using fs fluorescence up-conversion spectroscopy [77]. It has been found that both the rise time and decay of the bandedge emission show a direct correlation to the particle size, and the rise time depends on excitation intensity. The long lifetime of the bandedge emission was suggested to originate from a triplet state. The deep trap emission that appears within 2 ps was attributed to relaxation of surface selenium dangling bond electron to the valence band where it recombines radiatively with the initial photogenerated hole. This is also believed to be responsible for the large amplitude, fast (2–6 ps) decay of the bandedge emission. This work seems to indicate that the hole trapping is slower compared to electron trapping. In a more recent study of CdSe QDs chemically adsorbed onto the inverse opal TiO_2 film as well as the common nanocrystalline TiO_2 film, both the hole and electron relaxation processes have been found to depend on the amount of CdSe QDs and the interfaces between CdSe QDs [78]. In another study, branched nanocrystal heterostructures synthesized from CdSe and CdTe have been designed with type II band structure alignment for inducing separation of charge upon photoexcitation and localizing carriers to different regions of the tetrahedral geometry [79]. Charge carrier relaxation dynamics examined with fs pump-probe spectroscopy showed that such tetrapod heterostructures have rise times and biexponential decays longer than those of nanorods with similar dimensions. This is attributed to weaker interactions with surface states and nonradiative relaxation channels allowed by the type II alignment. This experiment demonstrates the dependence of charge carrier dynamics on the detailed nanostructure and related electronic as well as surface states.

To date, no systematic studies have been reported on the temperature dependence of charge carrier dynamics in semiconductor nanoparticles on the ultrafast time scales. At low temperature, nonradiative relaxation pathways should be suppressed and radiative fluorescence quantum yield is usually enhanced. Since the nonradiative relaxation processes are typically faster than radiative processes, suppression of nonradiative pathways at low temperature is expected to result in longer lifetime of the charge carriers or slower overall relaxation. This has been clearly demonstrated in the temperature-dependent emission lifetime observed for CdSe nanocrystals measured on the ns time scale [80]. Direct study of the temperature dependence of charge carrier dynamics on the ultrafast time scale will help to gain further insight into the electron relaxation mechanisms in semiconductor nanoparticles, e.g. the rates of nonradiative trapping and trap state-mediated recombination.

Other metal sulfide nanoparticles such as PbS, Cu_xS and Ag_2S have also been studied in terms of their charge carrier dynamics [75, 81, 82]. For PbS nanoparticles, attempts have been made to study the surface and shape dependence of electron relaxation on different-shape PbS NPs. When particle shapes are changed from mostly spherical to needle and cube shape by changing the surface capping polymers, the electron relaxation dynamics remain about the same for the apparently different-shape particles [75]. For all cases studied, the electronic relaxation was found to feature a double exponential decay with time constants of 1.2 ps and 45 ps that are independent of probe wavelength and excitation intensity. The shape independence was attributed to the dominance of the surface effects on the electron relaxation. While the shapes are different, the different samples may have similar surface properties. Therefore, if the dynamics are dominated by the surface, change in shape may not affect the electron relaxation dynamics substantially. The overall short lifetime observed is indicative of high density of trap state.

Femtosecond transient absorption studies have been conducted for Ag_2S nanoparticles capped with cystine and glutathione [81]. The dynamics of photoinduced electrons feature a pulse-width limited (<150 fs) rise followed by a fast decay (750 fs) and a slower rise (4.5 ps). The signal has contributions from both transient absorption and transient bleach. An interesting excitation intensity dependence was observed for all the samples: the transient absorption contribution becomes more dominant over bleach with increasing excitation intensity. A kinetic model developed to account for the main features of the dynamics suggests that the difference in dynamics observed between the different samples is due to different absorption cross-sections of deep trap states. The observed excitation intensity dependence of the dynamics is attributed to shallow trap state saturation at high intensities [81].

Besides II–VI and other metal chalcogenide semiconductors, dynamic studies of other semiconductor NPs have also been conducted. For example, dynamics study of ion-implanted Si nanocrystals using femtosecond transient absorption identified two photoinduced absorption features, attributed to charge carriers in nanocrystal quantized states with higher energy and faster relaxation and Si/SiO_2 interface states with lower energy and slower relaxation [83]. Red emission observed in this sample was shown to be from surface trap states and not from quantized states. The faster relaxation of the blue emission relative to that of the red emission is similar to that observed for CdS NPs [68].

A few studies have been carried out on the dynamic properties of layered semiconductor nanoparticles. A picosecond transient absorption study of charge carrier relaxation in MoS_2 NPs has been reported [84] and the relaxation was found to be dominated by trap states. The relaxation from shallow traps to deep traps is fast (40 ps) at room temperature and slows down to 200 ps at 20 K. Sengupta *et al.* have conducted a fs study of charge carrier relaxation dynamics in PbI_2 nanoparticles and found that the relaxation was dominated by surface properties and independent of particle size in the size range (3–100 nm) studied [73]. The early time dynamics were found to show some signs of oscillation with a period varying with solvent but not with size (6 ps in acetonitrile and 1.6 ps in alcohol solvents). The origin of the oscillation is not completely clear and such features are rarely observed for colloidal semiconductor NPs.

Similar findings have been made for BiI_3 NPs in different solvents [85]. The electron relaxation dynamics were found to be sensitive to solvent, insensitive to particle size and independent of excitation intensity. There also appear to be oscillations at early times with a period changing with solvent but not with particle size, similar to that found for PbI_2 NPs. For BiI_3 the oscillation periods were slightly shorter and overall relaxation was somewhat faster than that in PbI_2. The decay was much faster in aqueous PVA (9 ps) and in inverse micelles (1.2 ps and 33 ps) with no oscillations observed. The results suggest that the surface plays a major role in the electron relaxation process of BiI_3 NPs, just like in PbI_2 NPs. The independence from particle size could be because the relaxation is dominated by surface characteristics that do not vary significantly with size and/or the size is much larger than the exciton Bohr radius (0.61 nm for bulk BiI_3) [86] and thereby spatial confinement is not significant in affecting the relaxation process.

Charge carrier dynamics in semiconductor quantum wires have also been investigated in a few cases, including notably GaAs [87–90]. High luminescence efficiency was found in some cases and the luminescence was found to be completely dominated by radiative electron-hole recombination [87]. A more direct, systematic comparison of electronic relaxation among 2D, 1D and 0D quantum confined systems should be interesting but is made difficult by practical issues such as sample quality, e.g. surface characteristics, and experimental conditions, e.g. excitation wavelength and intensities, that can significantly influence the measurements. Such studies should be carried out in the future, both experimentally and theoretically.

9.3.5. *Charge carrier dynamics in doped semiconductor nanomaterials*

The dynamic properties of doped semiconductor nanoparticles are complex and intriguing. The most extensively studied system is Mn-doped ZnS. In 1994, Bhargava *et al.* reported that the emission lifetime of Mn^{2+} was significantly shorter than that in bulk and the luminescence efficiency was greater in the nanocrystalline system compared with bulk [91–93]. They observed a double exponential decay with time constants of 3 ns and 20 ns, which were five orders of magnitude faster than what has been observed in bulk (1.8 ms) [94]. Since there was no significant offset or indication of slower decays on the ns or longer time scales, no attempt was made to look for or show if there is any slow decay with lifetime on the ms time scale. To explain the fast ns decay, Bhargava *et al.* had proposed that, due to quantum confinement, there is a rehybridization between the s-p conduction band of the ZnS host and the 3d states of the Mn^{2+}. Given the strong coupling between donor and acceptor, there is a rapid energy transfer and consequently fast radiative decay. There was a similar explanation for rapid decay kinetics observed in Mn^{2+} doped 2D quantum well structures [95]. This was considered as significant because sensors, display devices and lasers utilizing these nanoparticles could have much improved performance [93].

However, subsequent work has thrown some doubt as to the true time scale for radiative energy relaxation and suggested that the Mn^{2+} emission lifetime in nanoparticles is essentially the same as that in bulk (1.8 ms) [96]. First, Bol and Meijerink observed ns decay rates for the blue ZnS emission but for the orange Mn^{2+} emission a normal 1.9 ms decay time along with a small amplitude ns decay was seen [96]. Furthermore, in their system the blue 420 nm emission band was observed to have a tail that extended into the orange Mn^{2+} region and could be observed with a 2 μs gate [96]. Given this, they stated that the fast ns decay observed by Bhargava *et al.* was due to ZnS trap state emission and not the Mn^{2+} emission. Unfortunately, the apparatus used in their experiment had limited time-resolution (a 2 μs gate was used). A subsequent study by Murase *et al.* also suggested that the Mn^{2+} luminescence lifetime in ZnS:Mn nanoparticles was similar to that of bulk [97].

Later, Smith *et al.* successfully synthesized Mn^{2+}-doped ZnS nanoparticles using a new method based on reverse micelles and studied carefully the emission kinetics on the ps, ns, μs to ms time scales with adequate time-resolution for each time scale region [98]. The samples have shown Mn^{2+} emission in addition to some weak trap state emission from ZnS in the 585 nm region and that the Mn^{2+} emission lifetime in these ZnS nanoparticles is ~1.8 ms, very similar to that of bulk [98]. Faster ns and μs decays are observed for both doped and undoped particles and attributed to ZnS trap state emission, as shown in Fig. 9.6. The main difference between the doped and undoped samples is on the ms time scales. The conclusion was consistent with those of Bol and Meijerink [96]. Similar conclusions have been reached independently by Chung *et al.* [99], who have further suggested that surface bound and lattice bound Mn^{2+} have different emission lifetimes, 0.18 and 2 ms, respectively. These recent studies have consistently shown that the Mn^{2+} emission lifetime is essentially the same in nanoparticles as in bulk ZnS:Mn.

9.3.6. *Nonlinear charge carrier dynamics*

When the density of excited charge carries or excitons is very high, nonlinear dynamic processes such Auger recombination or exciton–exciton annihilation starts to occur. These can be considered and described as higher order kinetic processes. The principal consequence of such nonlinear processes is fast nonradiative decays of the charge carriers, effectively reducing the overall charge carrier lifetime and luminescence quantum yield.

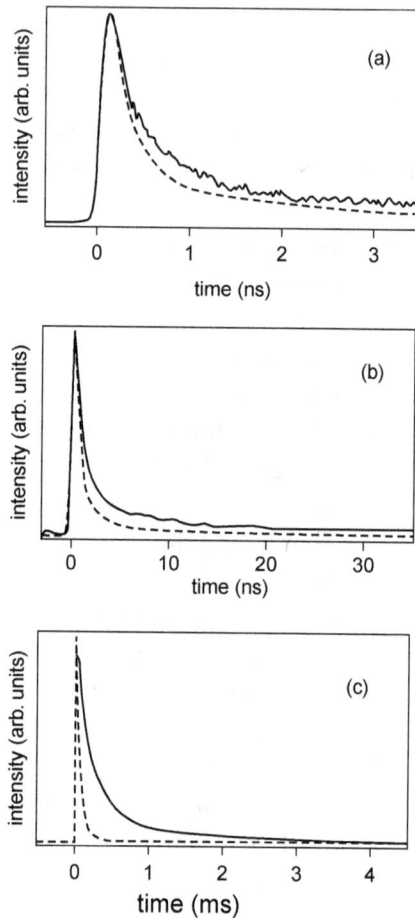

Fig. 9.6.   Time-resolved PL decay dynamics of ZnS:Mn nanoparticles. (a) 600 nm emission intensity versus time collected using the streak camera ~285 nm excitation. (b) 600 nm emission decay collected using same system as above but with a longer time sweep. (c) The data obtained from the OPO/digital oscilloscope. The data from doped clusters can be fit with a biexponential with a 200 ms and 1–2 ms time constants. Doped samples are the solid trace and undoped the dashed. Reproduced with permission from Ref. 98.

Nonlinear behavior occurs when there are multiple excitons generated in the same spatial region at the same time where there is strong interaction between the excitons. This is typically reflected as a dynamic process that depends nonlinearly (e.g. quadratically or even higher order) on the excitation light intensity [51, 52, 68, 71, 100–102].

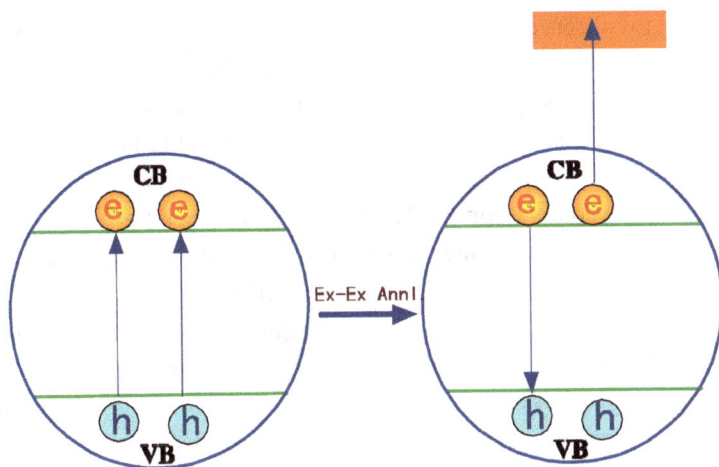

Fig. 9.7. Illustration of nonlinear and nonradiative exciton–exciton annihilation (or Auger recombination) that results in the nonradiative deexcitation or recombination of one exciton and excitation of the other exciton to higher energy. The excited exciton will eventually relax radiatively or nonradiatively. Reproduced with permission from (Fig. 4) of Ref. 39.

There have been various explanations for the observation of nonlinear dynamical behavior in nanomaterials, including higher order kinetics, Auger recombination, and exciton–exciton annihilation (illustrated in Fig. 9.7). It is challenging to assign an exact mechanism from only experimental data. All these models can explain the observations reasonably well. Some favor the exciton–exciton annihilation model since Auger recombination involves ionization and most time-resolved studies do not provide direct evidence for charge ejection or ionization. Others favor the Auger recombination model since it originates from solid state materials [102]. In the exciton–exciton annihilation model, high excitation laser intensity for the pump pulse produces multiple excitons per particle that can interact and annihilate, resulting in one exciton excited and another one deexcited. If the rate of trapping is faster than the rate of exciton–exciton annihilation, which is often the case, trapping will reduce the probability of exciton–exciton annihilation. However, when trap states are saturated, exciton–exciton annihilation will take place. Therefore, nanoparticles with a higher density of trap states have a higher threshold for observing exciton–exciton annihilation or require higher pump laser intensities to observe this nonlinear process. This behavior has been clearly demonstrated in CdS nanoparticles [68, 76].

When different-size or -shape nanoparticles are considered, comparison can be subtle and requires careful attention. This is partly because there are several factors, some competing, that need to be accounted for while making a comparison [103]. For example, when particles of different sizes but the same number of excitons per particle are compared, the smaller particles show a stronger nonlinear effect or, conversely, a lower excitation threshold for observing the nonlinear process. This is because smaller particles have stronger spatial confinement and lower density of states per particle that both facilitate exciton–exciton annihilation. On the other hand, when two samples of the same material such as CdS with the same nominal optical density or concentration but different particle sizes are compared under the same excitation intensity, the larger particles show stronger nonlinear effect. This is apparently due to a larger number of excitons per particle for the larger particles. This indicates that the volume factor dominates over the effect of trap states. In other words, larger particles have a larger molar absorptivity and thus absorb more photons to create more excitons for a given laser pulse. The observation is opposite to what is expected for a larger number of trap states per particle, which for the larger particle should raise the threshold and thereby suppress exciton–exciton annihilation [103].

As further support of the above argument, it was found that particles with similar volume but different shapes and thereby a different density/distribution of trap states show different thresholds for nonlinear effects. For nonspherical particles, the PL yield is much lower compared to that of spherical particles, indicating a higher density of trap states for nonspherical particles. Based on the model discussed above, one should expect nonspherical particles to show a higher threshold or weaker effect of exciton–exciton annihilation since it is harder to achieve trap state saturation. This is completely consistent with the experimental observation of a stronger nonlinear effect for the spherical particles [103]. Figure 9.8 shows an illustration of the size and shape dependence of exciton–exciton annihilation.

It should be pointed out that the observation on CdS NPs is qualitatively consistent with that made on CdSe nanoparticles [102]. However, the time

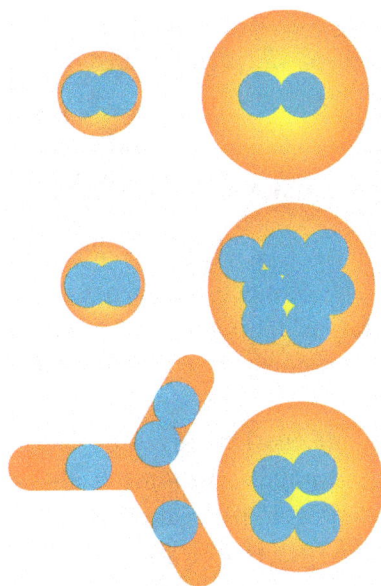

Fig. 9.8. Illustration of the size and shape dependence of exciton–exciton annihilation. Reproduced with permission from (Fig. 5) of Ref. 39.

constant for nonlinear decay is much faster for CdS (a few ps) than for CdSe NPs (as long as hundreds of ps). It is unclear if this difference is due to differences in the systems studied, experimental conditions, or even simply due to differences in data analysis and interpretation. Further studies are needed to better understand this issue.

Very recently, it has been suggested that multiple excitons can be generated using a single photon in small bandgap semiconductors such as PbSe and PbS [104–109]. This is potentially useful for solar energy conversion and other applications. Even though this is theoretically possible and there is preliminary supporting experimental evidence, it is unclear what the probability or efficiency might be in practice. Given that electronic relaxation is typically very fast (less than 100 fs), this multiexciton generation process with one photon is likely to be inefficient or has a very small cross-section unless the energy levels involved can be carefully and intelligently designed to enhance the process. Impact ionization is a scheme proposed for enhancing multiple exciton generation (MEG)

[104–108]. It is yet to be realized in practical device applications. Further research is needed to verify the feasibility of this approach and strategies to realize or enhance it.

9.4. Charge carrier dynamics in metal oxide and insulator nanomaterials

As discussed in Chapter 6, metal oxides are generally very similar to semiconductors as far as their electronic band structures are concerned. Some MOs are considered as semiconductors while others are considered as insulators because of their large bandgap energy. Therefore, the general discussions given in Sec. 9.3 about charge carrier dynamics in semiconductor nanomaterials apply qualitatively to many metal oxide nanomaterials. If there is any difference, it is usually only quantitative. This is true for processes such as electronic dephasing, intraband charger carrier cooling, charge carrier trapping, recombination, and nonlinear processes due to multiple excitons per nanoparticles. For example, similar to fast charge carrier trapping found in many non-MO semiconductor nanomaterials, trapping into deep trap states has been found to be ~200 fs for TiO_2 nanocrystalline films [110].

Different charge carrier dynamic processes have been studied for several MO nanoparticle systems, e.g. TiO_2, ZnO and Fe_2O_3. For TiO_2 NPs with excitation at 310 nm, the photoinduced electrons were found to decay following second-order kinetics with a second-order recombination rate constant of 1.8×10^{-10} cm^3/s [111]. The electron trapping was suggested to occur on the time scale of 180 fs [52]. Similar studies have been done on ZnO. For ZnO NPs with 310 nm excitation, the photoinduced electron decay was found to follow second-order kinetics [100], similar to TiO_2 [111]. It was also found that the initial electron trapping and subsequent recombination dynamics were size dependent. The trapping rate was found to increase with increasing particle size, which was explained with a trap-to-trap hopping mechanism. The electron–hole recombination is faster and occurs to a greater extent in larger particles because there are two different types of trap states. A different explanation, based on exciton–exciton annihilation upon trap state saturation, has been recently proposed for similar excitation intensity dependent and size dependent relaxation observed in CdS and CdSe NPs [68]. This explanation would also seem to be consistent with the results observed for TiO_2 [111] and ZnO [100].

Cherepy *et al.* have carried out fs dynamics studies of electron relaxation in both γ and α phased Fe_2O_3 nanoparticles with 390 nm excitation [72]. The relaxation dynamics were found to be very similar between the two types of nanoparticles, despite their difference in magnetic properties and particle shape: γ being mostly spherical and paramagnetic, while α was mostly spindle-shaped and diamagnetic [72]. The relaxation featured a multiexponential decay with time constants of 0.36 ps, 4.2 ps and 67 ps. The overall fast relaxation, in conjunction with very weak fluorescence, indicates extremely efficient nonradiative decay processes, possibly related to the intrinsic dense band structure or a high density of trap states. The fast relaxation of the photoinduced electrons is consistent with the typically low photocurrent efficiency of Fe_2O_3 electrodes, since the short lifetime due to fast electron-hole recombination does not favor charge transport that is necessary for photocurrent generation [72].

As an example of doped MO nanomaterials, TiO_2 NPs doped with transition metal ions (Cr and Fe) were studied by femtosecond transient absorption spectroscopy [112]. The transient absorption spectra were observed to shift to shorter wavelengths with time for the doped samples, while the position of the absorptions did not change on the time scale studied for the undoped sample, as shown in Fig. 9.9. The spectral shifts were the highest for the sample doped with both Cr and Fe. The transient absorption spectra right after excitation had a broad absorption band with a flat top in 550–650 nm region, whereas at 1 ns delay the maximum of the transient absorption shifted to 510–530 nm area for Cr- and Fe-doped samples. The observed new absorption bands were attributed to the trapped holes. Doping was found to reduce the photocatalytic activity of these NP films, which was attributed to doping-induced reduction in the hot and shallow trapped holes important for the photocatalytic activity. This example demonstrates the importance of studying charge carrier dynamics in understanding functionalities, e.g. photocatalysis of nanomaterials.

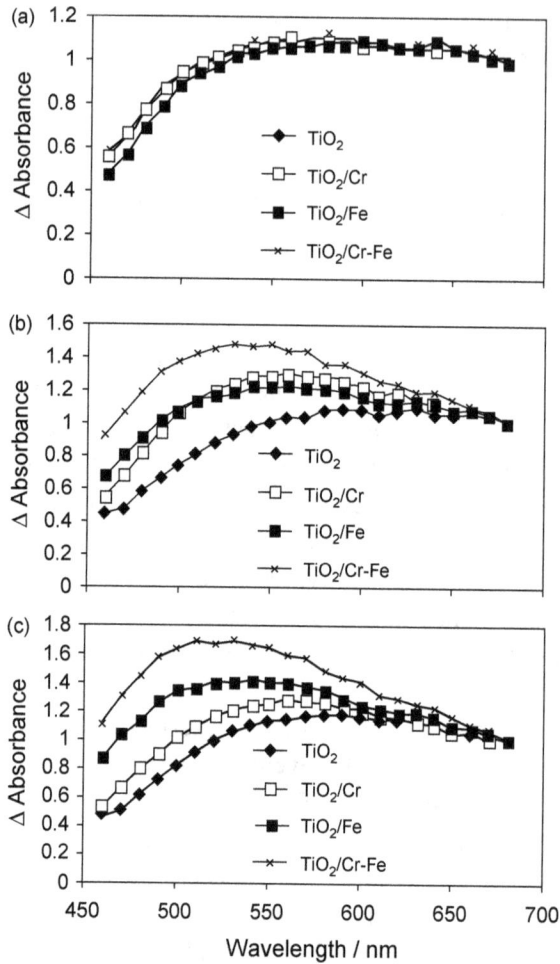

Fig. 9.9. Normalized (at 680 nm) transient absorption spectra of TiO_2 nano powder films without and with doping of metal (Cr and Fe) at (a) $t = 0$ ps, (b) $t = 5$ ps and (c) $t = 1$ ns. Reproduced with permission from Ref. 112.

9.5. Photoinduced charge transfer dynamics

Photoinduced charge transfer is one of the most important fundamental processes involved in liquid-semiconductor or liquid-metal interfaces. It plays a critical role in photocatalysis, photodegradation of wastes, photo-electrochemistry and solar energy conversion [113–118]. Charge transfer competes with relaxation processes such as trapping and recombination. In many applications, charge transfer is the desired process. For example, charge transfer has been extensively studied for dye-sensitized MO nanoparticles for solar cell applications. Dye sensitization of TiO_2 is the most notable example, primarily because of its potential use for solar energy conversion [119] as well as photocatalysis [120]. Since TiO_2 alone does not absorb visible light, dye sensitization is needed to extend the absorption into the visible region.

In dye sensitization, the electron is injected from a dye molecule adsorbed on the MO nanoparticle surface into the CB of the MO. There are several requirements for this to work effectively. First, the excited state of the dye molecule needs to lie above the bottom of the CB of the MO, e.g. TiO_2, nanoparticle. Second, strong binding of the dye onto the TiO_2 nanoparticle surface is desired for fast and efficient injection. Third, back electron transfer to the dye cation following injection should be minimal. Fourth, the dye molecule must have strong absorption in the visible region of spectrum for efficient solar energy conversion. A number of dye molecules have been studied and tested for solar energy conversion applications over the years [121]. To date, the dye molecule that has demonstrated the highest efficiency is the so-called N3 dye. Ru(4,4′-dicarboxyl-2,2′-bipyridine)$_2$ (NCS)$_2$, with a reported light-to-electricity conversion efficiency of 10% [119, 122]. This work has stimulated strong interest in understanding the mechanism of charge injection and recombination in such dye-sensitized nanocrystalline systems.

The rates of electron injection and subsequent recombination or back electron transfer in dye sensitization are expected to depend on the nature of the dye molecule and the NPs, especially the surface characteristics of the NPs. The interaction between the dye and the NP surface will determine the rates and yields of forward as well as reverse electron transfer [113, 123]. The shape (facets) and size of the particles could also be important and tend to vary from sample to sample depending on the preparation methods used.

To describe the electron transfer (ET) process from an adsorbate molecule to the nanoparticle, one can adopt an approach similar to that of Marcus and co-workers for ET between a discrete electron donor and acceptor level in solution [124–126]. The total ET rate from adsorbate with discrete states to semiconductor with a nearly continuum of product states can be expressed as [127]:

$$K_{ET} = \frac{2\pi}{\hbar} \int_0^\infty dE \, \rho(E) |\bar{H}(E)|^2 \frac{1}{\sqrt{4\pi\lambda k_B T}} \exp\left[-\frac{(\lambda + \Delta G_0 - E)^2}{4\pi\lambda k_B T}\right] \quad (9.3)$$

where $\Delta G_0 = E_{CB} - E_{ox}$ is the energy difference between conduction bandedge and the redox potential of adsorbate excited state, $\rho(E)$ is the density of states at energy E from the conduction bandedge, which can contain both the bulk states and surface states, $H(E)$ is the average electronic coupling between the adsorbate excited state and different k states in the semiconductor at the same energy E, and λ is the total reorganization energy. Furthermore, a distribution of adsorbate/semiconductor interactions exists and gives rise to a distribution of electronic coupling matrix elements, H, and thus injection rate [127]. Therefore, the ET rate depends on the detailed energetics of both the semiconductor and the adsorbate. In general, a strong electronic coupling and large energy difference $(E_{ox} - E_{CB})$ between the adsorbate excited state and the conduction band lead to faster electron transfer. It is challenging to independently examine the effects of different factors experimentally, since they are often interrelated. They can be evaluated separately under certain approximations or under special circumstances.

Experimental studies of the rate of electron injection in dye sensitization of semiconductor metal oxide nanoparticles have found that electron injection (forward electron transfer) is generally extremely fast, ~100 fs, in many systems. For instance, for the coumarin 343 dye on TiO_2, the electron injection rate was found to be around 200 fs [128]. For the N3 dye on TiO_2, the first direct fs measurement reported a hot electron injection time of <25 fs [129]. However, there has been some debate over possible degradation of the dye sample used [130, 131]. A picosecond infrared study showed an upper limit of 20 ps for the electron injection time [132]. More recent work by Ellingson et al. on N3 on TiO_2 reported an injection time of <50 fs based on transient infrared measurements [127, 133]. More recent studies of perylene derivatives on TiO_2 and ZnO NPs have also found that the injection process is very fast (<200 fs) and depends on the MO and coupling between the dye and MO NPs [134]. The experimental results are consistent with theoretical calculations using perylene on TiO_2 as a model system [135]. Cherepy *et al.* have studied an anthocyanin dye adsorbed on TiO_2 NPs and found an electron injection time of <100 fs [136]. The dye can be extracted easily from fruits such as California blackberry and is thus a potentially environmentally attractive dye for solar cell applications.

While forward electron transfer has generally been found to be very fast, back electron transfer was found to occur on a range of time scales, from about 10 ps to μs, depending on the nature of the dye and the nanoparticle [136–138]. For example, both forward and reverse electron transfers have been studied in the case of anthracenecarboxylic acids adsorbed on different types of TiO_2 NPs and were found to be dependent on the dye molecular structure and the method used to synthesize the TiO_2 particles [138].

In a series of time-resolved studies using femtosecond IR spectroscopy, Lian *et al.* systematically studied the dependence of the charge injection rate on several factors including the nature of the dye and the semiconductor as well as the coupling between them [127, 133, 139–144]. It was found out that fast electron injection (<100 fs) from two Ru dye molecules into TiO_2 can compete with vibrational energy relaxation within the electron excited state of the adsorbed dye molecule and the injection yield is thus dependent on the excited state redox potential. The injection rate from the N3 dye to different semiconductors was found to follow the trend of $TiO_2 > SnO_2 > ZnO$, indicating an interesting dependence on the nature of the semiconductor. Several factors, including the electronic coupling, density of states and driving force, that control the interfacial electron transfer rate, were examined separately. It was found that the ET rate decreased with decreasing electronic coupling strength, while the ET rate increased with the excited state redox potential of the adsorbates (increasing driving force). The results agreed qualitatively with the theoretical prediction for a nonadiabatic interfacial electron transfer process [127].

An interesting possibility in dye sensitization is hot electron injection, i.e. electron injection from the excited state of the dye molecule before vibrationally relaxing to the bottom of the potential energy surface [145].

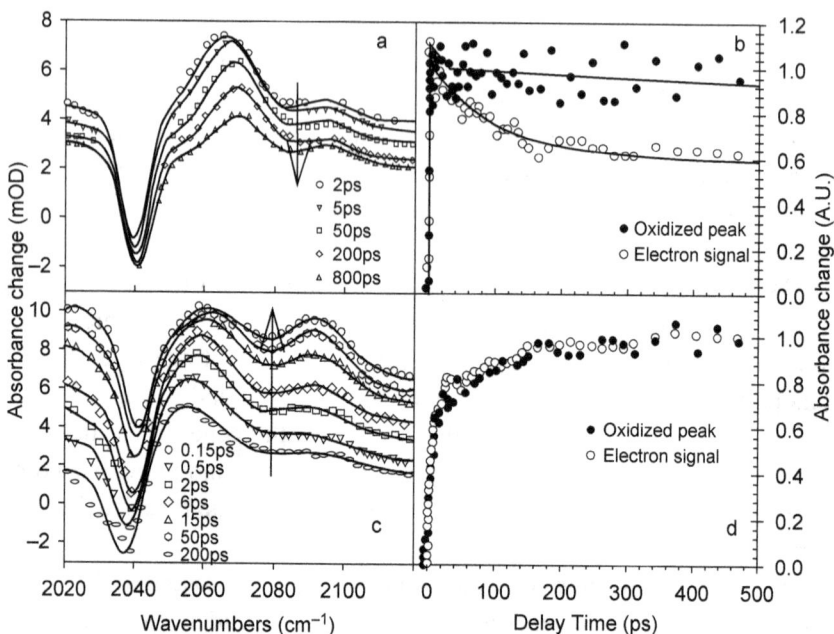

Fig. 9.10. Comparison of electron injection dynamics from ReC0 and ReC1 into TiO$_2$ nanocrystalline thin films. (a) Transient spectra of ReC0 CO stretching mode and broad infrared (IR) absorption of injected electrons as a function of delay time after 400 nm excitation. The peaks at 2040, 2060 and 2090 cm^{-1} are assigned to ground, excited and oxidized forms of ReC0, respectively. (b) Comparison of kinetics probed by IR absorption of injected electron absorption (open circles) and CO stretching band intensity of oxidized adsorbate (full circles). The oxidized peak trace shows <100 fs rise and negligible subsequent change within 1 ns. The decrease of the IR absorption of injected electrons is attributed to the relaxation of injected hot electrons in TiO$_2$. Similar transient spectral evolution and kinetics comparison for ReC1/TiO$_2$ are shown in (c) and (d). The lack of a fast injection component and electron relaxation in this case is attributed to the reduced electron coupling introduced by the methylene group resulting in injection from only the relaxed excited states, producing electrons near the bandedge. Reproduced with permission from Ref. 147.

As an example, evidence for hot injected electrons has been reported in an ultrafast IR study of ReC0 (ReCn representing Re(dcbpyCn)(CO)$_3$(Cl) with dcbpyCn = 2-2'-bipyridine-4,4'-(CH$_2$)n-COOH) on dry TiO$_2$ film [143, 146]. Figures 9.10 show transient IR spectra and kinetics for this system. The oxidized peak showed a <100 fs rise and negligible subsequent change within <1 ns, indicating fast electron injection with no recombination within the time window measured. Meanwhile, the electron absorption signal exhibited a 40% decay within 1 ns, suggesting significant electron cooling. In this system, 400 nm excitation prepared a Franck–Condon state at ca. −1.68 V versus saturated calomel electrode (SCE), high above the conduction band edge of TiO_2. Therefore, the <100 fs injection time suggests that injection occurs from the unrelaxed excited state of ReC0, producing hot electrons with relaxation indicated by the observed decay of the IR signal. The lack of injection from the relaxed excited state is attributed to its low redox potential, estimated at −0.35 V (SCE), significantly lower than the band edge of TiO_2 [147].

Other dye-sensitization studies have been conducted on SnO_2 NPs [148–150]. For instance, fs transient absorption and bleach studies have been performed on cresyl violet H-aggregate dimers adsorbed on SnO_2 colloidal particles [150]. It was found that the electron injection from the higher energy state, resulting from exciton splitting, of the dimer to SnO_2 NPs occurs in <100 fs and back electron transfer occurs with a 12 ps time constant.

Electron transfer from metal NPs such as gold to MO NPs such as TiO_2 has been demonstrated recently [151]. Using femtosecond transient absorption spectroscopy with an IR probe, plasmon induced electron transfer from 10 nm gold nanoparticles to TiO_2 nanoparticles has been directly measured and found to be faster than 240 fs with a yield of 40%. This experiment demonstrates that, even though the gold nanoparticle plasmon lifetime is very short (around 1.5 ps), it is still possible to have electron transfer from the gold nanoparticles to appropriate electron acceptors such as TiO_2 nanoparticles as long as the transfer is fast and energetically favorable. Since gold nanoparticles have an extremely high absorption coefficient compared with usual dye molecules, they may be utilized as a good electron source in nanodevices if back electron transfer can be suppressed efficiently.

Electron transfer can also take place from semiconductor NPs to an electron acceptor near or on the surface of the NPs. This is the basis of photoreduction reaction on the particle surface. This is less studied compared to electron transfer from an adsorbate to the NP. For instance, electron transfer dynamics from CdS and CdSe NPs to electron acceptors, e.g. viologen derivatives, adsorbed on the particle surface have been studied using transient absorption, transient bleach and time-resolved fluorescence [55, 152]. Electron transfer was found to take place on the time scale of 200–300 fs and competes effectively with trapping and electron-hole recombination. These results are important to understanding interfacial charge transfer involved in photocatalysis and photoelectrochemistry applications.

An interesting related system for studying ET is radicals bound to the surface of QDs. It has been found that radical species, such as persistent nitroxide free radicals 2, 2, 6, 6-tetramethylpiperidine oxide (TEMPO) or 4-amino-TEMPO, bound to QD surface can quench the PL of the QD through electron exchange interactions related to their paramagnetic properties [153, 154]. Figure 9.11 shows PL quenching of CdSe QDs with a mixture of two different sizes (2.4 and 3.2 nm) using TEMPO radicals. On the other hand, trapping of the radicals on the QD surface by other radicals, e.g. to form alkoxyamines, can restore the PL of the QD. This mechanism can therefore be used for optical sensing of radicals based on PL of the QDs. Recent experimental studies have found strong evidence

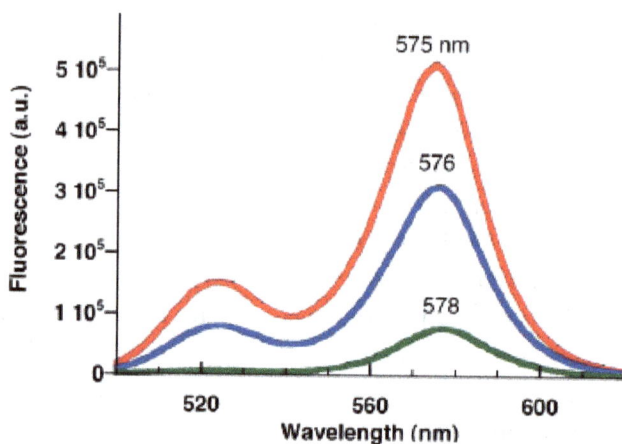

Fig. 9.11. Fluorescence spectra (λ_{ex} 380 nm) of a toluene solution of 2.4 and 3.2 nm CdSe QDs with the same dot concentrations (3.7 mM). Fluorescence intensity is decreasing as the concentration of TEMPO is increased: 0 M, 0.04 M, 0.2 M TEMPO. The shorter wavelength absorption band is due to 2.4 nm QDs while the longer wavelength band is due to 3.2 nm QDs. Reproduced with permission from Ref. 153.

for electron transfer from the QD to the radical, which is suggested to be the main mechanism for the QD PL [155]. The photoinduced ET results in consumption of the radicals as evidenced by ESR and the recovery of PL. These studies have demonstrated a simple and well-defined system for studying ET.

9.6. Summary

In summary, dynamic studies of metal, semiconductor and metal oxide nanomaterials have provided a fundamental insight into the behavior of charge carriers in confined systems and how their interact with other species at their surface through energy or charge transfer. A better understanding of the fundamental dynamic properties is useful for designing new nanomaterials with improved optical and electronic properties tailored towards emerging applications including solar energy conversion, optical sensing and imaging, nanophonics including lasers, LEDs and detectors. Some of the applications will be discussed in more detail in the next chapter.

References

1. D. Steinmuller-Nethl, R.A. Hopfel, E. Gornik, A. Leitner and F.R. Aussenegg, *Phys. Rev. Lett.* **68**, 389 (1992).
2. U. Kreibig and M. Vollmer, *Optical Properties of Metal Clusters*. Springer Series in Materials Science 25, Berlin: Springer. 532 (1995).
3. J.H. Hodak, A. Henglein and G.V. Hartland, *J. Phys. Chem. B* **104**, 9954 (2000).
4. K. Puech, F.Z. Henari, W.J. Blau, D. Duff and G. Schmid, *Chem. Phys. Lett.* **247**, 13 (1995).
5. T. Klar, M. Perner, S. Grosse, G. von Plessen, W. Spirkl and J. Feldmann, *Phys. Rev. Lett.* **80**, 4249 (1998).
6. B. Lamprecht, A. Leitner and F.R. Aussenegg, *Appl. Phys. B-Lasers Opt.* **68**, 419 (1999).
7. H.E. Elsayed-Ali, T.B. Norris, M.A. Pessot and G.A. Mourou, *Phys. Rev. Lett.* **58**, 1212 (1987).
8. R.W. Schoenlein, W.Z. Lin, J.G. Fujimoto and G.L. Easley, *Phys. Rev. Lett.* **58**, 1680 (1987).

9. C.K. Sun, F. Vallee, L.H. Acioli, E.P. Ippen and J.G. Fujimoto, *Phys. Rev. B-Condensed Matter* **50**, 15337 (1994).

10. N.W. Ashcroft and N.D. Mermin, *Solid State Physics*, Philadelphia: Saunders College. 826 (1976).

11. B.A. Smith, J.Z. Zhang, U. Giebel and G. Schmid, *Chem. Phys. Lett.* **270**, 139 (1997).

12. S. Link and M.A. El-Sayed, *J. Phys. Chem. B* **103**, 8410 (1999).

13. S.L. Logunov, T.S. Ahmadi, M.A. El-Sayed, J.T. Khoury and R.L. Whetten, *J. Phys. Chem. B* **101**, 3713 (1997).

14. J. Hodak, I. Martini and G.V. Hartland, *Chem. Phys. Lett.* **284**, 135 (1998).

15. A.E. Faulhaber, B.A. Smith, J.K. Andersen and J.Z. Zhang, *Mol. Cryst. Liq. Cryst. Sci. Tec. A* **283**, 25 (1996).

16. Q. Darugar, C. Landes, S. Link, A. Schill and M.A. El-Sayed, *Chem. Phys. Lett.* **373**, 284 (2003).

17. S. Link, D.J. Hathcock, B. Nikoobakht and M.A. El-Sayed, *Advan. Mater.* **15**, 393 (2003).

18. S. Link, M.A. El-Sayed, T.G. Schaaff and R.L. Whetten, *Chem. Phys. Lett.* **356**, 240 (2002).

19. C.D. Grant, A.M. Schwartzberg, Y.Y. Yang, S.W. Chen and J.Z. Zhang, *Chem. Phys. Lett.* **383**, 31 (2004).

20. G.V. Hartland, *Phys. Chem. Chem. Phys.* **6**, 5263 (2004).

21. M. Hu, H. Petrova, J.Y. Chen, J.M. McLellan, A.R. Siekkinen, M. Marquez, X.D. Li, Y.N. Xia and G.V. Hartland, *J. Phys. Chem. B* **110**, 1520 (2006).

22. H. Petrova, C.H. Lin, M. Hu, J.Y. Chen, A.R. Siekkinen, Y.N. Xia, J.E. Sader and G.V. Hartland, *Nano Lett.* **7**, 1059 (2007).

23. H. Petrova, C.H. Lin, S. de Liejer, M. Hu, J.M. McLellan, A.R. Siekkinen, B.J. Wiley, M. Marquez, Y.N. Xia, J.E. Sader and G.V. Hartland, *J. Chem. Phys.* **126**, 094709 (2007).

24. G.M. Sando, A.D. Berry, P.M. Campbell, A.P. Baronavski and J.C. Owrutsky, *Plasmonics* **2**, 23 (2007).

25. W.Y. Huang, W. Qian, M.A. El-Sayed, Y. Ding and Z.L. Wang, *J. Phys. Chem. C* **111**, 10751 (2007).

26. M. Nisoli, S. DeSilvestri, A. Cavalleri, A.M. Malvezzi, A. Stella, G. Lanzani, P. Cheyssac and R. Kofman, *Phys. Rev. B-Condensed Matter* **55**, 13424 (1997).

27. J.H. Hodak, I. Martini and G.V. Hartland, *J. Chem. Phys.* **108**, 9210 (1998).

28. C. Voisin, N. Del Fatti, D. Christofilos and F. Vallee, *J. Phys. Chem. B* **105**, 2264 (2001).

29. C. Voisin, D. Christofilos, P.A. Loukakos, N. Del Fatti, F. Vallee, J. Lerme, M. Gaudry, E. Cottancin, M. Pellarin and M. Broyer, *Phys. Rev. B* **69** (2004).
30. G. von Plessen, M. Perner and J. Feldmann, *Appl. Phys. B-Lasers O* **71**, 381 (2000).
31. G.V. Hartland, M. Hu, O. Wilson, P. Mulvaney and J.E. Sader, *J. Phys. Chem. B* **106**, 743 (2002).
32. C.D. Grant, A.M. Schwartzberg, T.J. Norman and J.Z. Zhang, *J. Am. Chem. Soc.* **125**, 549 (2003).
33. A.A. Lazarides and G.C. Schatz, *J. Chem. Phys.* **112**, 2987 (2000).
34. J.J. Storhoff, A.A. Lazarides, R.C. Mucic, C.A. Mirkin, R.L. Letsinger and G.C. Schatz, *J. Am. Chem. Soc.* **122**, 4640 (2000).
35. J.R. Heath, C.M. Knobler and D.V. Leff, *J. Phys. Chem. B* **101**, 189 (1997).
36. A.N. Shipway, M. Lahav, R. Gabai and I. Willner, *Langmuir* **16**, 8789 (2000).
37. T.J. Norman, C.D. Grant, D. Magana, J.Z. Zhang, J. Liu, D.L. Cao, F. Bridges and A. Van Buuren, *J. Phys. Chem. B* **106**, 7005 (2002).
38. V.A. Markel and M.R. Geller, *J. Phys. Condens Matter* **12**, 7569 (2000).
39. C.D. Grant and J.Z. Zhang, in *Annual Review of Nano Research*, G. Cao and C.J. Brinker, Editors., World Scientific Publisher: Singapore, 2, 1(2008).
40. Y. Masumoto, *J. Luminescence* **70**, 386 (1996).
41. J. Valenta, J. Dian, J. Hala, P. Gilliot and R. Levy, *J. Chem. Phys.* **111**, 9398 (1999).
42. P. Palinginis and W. Hailin, *Appl. Phys. Lett.* **78**, 1541 (2001).
43. R.W. Schoenlein, D.M. Mittleman, J.J. Shiang, A.P. Alivisatos and C.V. Shank, *Phys. Rev. Lett.* **70**, 1014 (1993).
44. D.M. Mittleman, R.W. Schoenlein, J.J. Shiang, V.L. Colvin, A.P. Alivisatos and C.V. Shank, *Phys. Rev. B-Condensed Matter* **49**, 14435 (1994).
45. U. Banin, G. Cerullo, A.A. Guzelian, A.P. Alivisatos and C.V. Shank, *Phys. Rev. B-Condensed Matter* **55**, 7059 (1997).
46. H. Giessen, B. Fluegel, G. Mohs, Y.Z. Hu, N. Peyghambarian, U. Woggon, C. Klingshirn, P. Thomas and S.W. Koch, *J. Opt. Soc. Am. B* **13**, 1039 (1996).
47. A. Nakamura and H. Ohmura, *J. Luminescence* **83–4**, 97 (1999).
48. S. Xu, A.A. Mikhailovsky, J.A. Hollingsworth and V.I. Klimov, *Phys. Rev. B* **6504**, 5319 (2002).
49. P.R. Yu, J.M. Nedeljkovic, P.A. Ahrenkiel, R.J. Ellingson and A.J. Nozik, *Nano Lett.* **4**, 1089 (2004).
50. A.W. Schill, C.S. Gaddis, W. Qian, M.A. El-Sayed, Y. Cai, V.T. Milam and K. Sandhage, *Nano Lett.* **6**, 1940 (2006).
51. J.Z. Zhang, R.H. Oneil and T.W. Roberti, *J. Phys. Chem.* **98**, 3859 (1994).

52. D.E. Skinner, D.P. Colombo, J.J. Cavaleri and R.M. Bowman, *J. Phys. Chem.* **99**, 7853 (1995).

53. M. Oneil, J. Marohn and G. McLendon, *Chem. Phys. Lett.* **168**, 208 (1990).

54. V. Klimov, P.H. Bolivar and H. Kurz, *Phys. Rev. B-Condensed Matter* **53**, 1463 (1996).

55. S. Logunov, T. Green, S. Marguet and M.A. El-Sayed, *J. Phys. Chem. A* **102**, 5652 (1998).

56. L. Jacak, A. Wójs and P. Hawrylak, *Quantum Dots*, Berlin; New York: Springer. 176 (1998).

57. N.M. Dimitrijevic and P.V. Kamat, *J. Phys. Chem.* **91**, 2096 (1987).

58. M. Haase, H. Weller and A. Henglein, *J. Phys. Chem.* **92**, 4706 (1988).

59. E.F. Hilinski, P.A. Lucas and W. Ying, *J. Chem. Phys.* **89**, 3534 (1988).

60. P.V. Kamat, N.M. Dimitrijevic and A.J. Nozik, *J. Phys. Chem.* **93**, 2873 (1989).

61. P.V. Kamat and N.M. Dimitrijevic, *J. Phys. Chem.* **93**, 4259 (1989).

62. Y. Wang, A. Suna, J. McHugh, E.F. Hilinski, P.A. Lucas and R.D. Johnson, *J. Chem. Phys.* **92**, 6927 (1990).

63. T. Vossmeyer, L. Katsikas, M. Giersig, I.G. Popovic, K. Diesner, A. Chemseddine, A. Eychmuller and H. Weller, *J. Phys. Chem.* **98**, 7665 (1994).

64. A. Henglein, A. Kumar, E. Janata and H. Weller, *Chem. Phys. Lett.* **132**, 133 (1986).

65. K.I. Kang, A.D. Kepner, S.V. Gaponenko, S.W. Koch, Y.Z. Hu and N. Peyghambarian, *Phys. Rev. B-Condensed Matter* **48**, 15449 (1993).

66. V. Klimov, S. Hunsche and H. Kurz, *Phys. Rev. B-Condensed Matter* **50**, 8110 (1994).

67. N.P. Ernsting, M. Kaschke, H. Weller and L. Katsikas, *J. Opt. Soc. Am. B* **7**, 1630 (1990).

68. T.W. Roberti, N.J. Cherepy and J.Z. Zhang, *J. Chem. Phys.* **108**, 2143 (1998).

69. J.Z. Zhang, R.H. O'Neil, T.W. Roberti, J.L. McGowen and J.E. Evans, *Chem. Phys. Lett.* **218**, 479 (1994).

70. J.P. Zheng and H.S. Kwok, *Appl. Phys. Lett.* **65**, 1151 (1994).

71. J.Z. Zhang, *Acc. Chem. Res.* **30**, 423 (1997).

72. N.J. Cherepy, D.B. Liston, J.A. Lovejoy, H.M. Deng and J.Z. Zhang, *J. Phys. Chem. B* **102**, 770 (1998).

73. A. Sengupta, B. Jiang, K.C. Mandal and J.Z. Zhang, *J. Phys. Chem. B* **103**, 3128 (1999).

74. M.C. Brelle and J.Z. Zhang, *J. Chem. Phys.* **108**, 3119 (1998).

75. A.A. Patel, F.X. Wu, J.Z. Zhang, C.L. Torres-Martinez, R.K. Mehra, Y. Yang and S.H. Risbud, *J. Phys. Chem. B* **104**, 11598 (2000).

76. F. Wu, J.Z. Zhang, R. Kho and R.K. Mehra, *Chem. Phys. Lett.* **330**, 237 (2000).

77. D.F. Underwood, T. Kippeny and S.J. Rosenthal, *J. Phys. Chem. B* **105**, 436 (2001).

78. L.J. Diguna, Q. Shen, A. Sato, K. Katayama, T. Sawada and T. Toyoda, *Materials Science & Engineering C-Biomimetic and Supramolecular Systems*, **27**, 1514 (2007).

79. P. Peng, D.J. Milliron, S.M. Hughes, J.C. Johnson, A.P. Alivisatos and R.J. Saykally, *Nano Lett.* **5**, 1809 (2005).

80. M. Nirmal, C.B. Murray and M.G. Bawendi, *Phys. Rev. B-Condensed Matter* **50**, 2293 (1994).

81. M.C. Brelle, J.Z. Zhang, L. Nguyen and R.K. Mehra, *J. Phys. Chem. A* **103**, 10194 (1999).

82. M.C. Brelle, C.L. Torres-Martinez, J.C. McNulty, R.K. Mehra and J.Z. Zhang, *Pure and Applied Chemistry* **72**, 101 (2000).

83. V.I. Klimov, C.J. Schwarz, D.W. McBranch and C.W. White, *Appl. Phys. Lett.* **73**, 2603 (1998).

84. R. Doolen, R. Laitinen, F. Parsapour and D.F. Kelley, *J. Phys. Chem. B* **102**, 3906 (1998).

85. A. Sengupta, K.C. Mandal and J.Z. Zhang, *J. Phys. Chem. B* **104**, 9396 (2000).

86. Z.K. Tang, Y. Nozue and T. Goto, *J. Phys. Soc. Japan* **61**, 2943 (1992).

87. J. Christen, E. Kapon, M. Grundmann, D.M. Hwang, M. Joschko and D. Bimberg, *Phys. Status Solidi. B* **173**, 307 (1992).

88. L. Rota, F. Rossi, P. Lugli and E. Molinari, *Semicond. Sci. Technol.* **9**, 871 (1994).

89. R. Cingolani, R. Rinaldi, M. Ferrara, G.C. Larocca, H. Lage, D. Heitmann and H. Kalt, *Semicond. Sci. Technol.* **9**, 875 (1994).

90. V. Dneprovskii, N. Gushina, O. Pavlov, V. Poborchii, I. Salamatina and E. Zhukov, *Phys. Lett. A* **204**, 59 (1995).

91. R.N. Bhargava, D. Gallagher, X. Hong and A. Nurmikko, *Phys. Rev. Lett.* **72**, 416 (1994).

92. R.N. Bhargava, D. Gallagher and T. Welker, *J. Luminescence* **60–1**, 275 (1994).

93. R.N. Bhargava, *J. Luminescence* **70**, 85 (1996).

94. H.E. Gumlich, *J. Luminescence* **23**, 73 (1981).

95. H. Ito, T. Takano, T. Kuroda, F. Minami and H. Akinaga, *J. Luminescence* **72–4**, 342 (1997).

96. A.A. Bol and A. Meijerink, *Phys. Rev. B-Condensed Matter* **58**, R15997 (1998).
97. N. Murase, R. Jagannathan, Y. Kanematsu, M. Watanabe, A. Kurita, K. Hirata, T. Yazawa and T. Kushida, *J. Phys. Chem. B* **103**, 754 (1999).
98. B.A. Smith, J.Z. Zhang, A. Joly and J. Liu, *Phys. Rev. B* **62**, 2021 (2000).
99. J.H. Chung, C.S. Ah and D.-J. Jang, *J. Phys. Chem. B* **105**, 4128 (2001).
100. J.J. Cavaleri, D.E. Skinner, D.P. Colombo and R.M. Bowman, *J. Chem. Phys.* **103**, 5378 (1995).
101. J.Z. Zhang, *J. Phys. Chem. B* **104**, 7239 (2000).
102. V.I. Klimov, A.A. Mikhailovsky, D.W. McBranch, C.A. Leatherdale and M.G. Bawendi, *Science* **287**, 1011 (2000).
103. F.X. Wu, J.H. Yu, J. Joo, T. Hyeon and J.Z. Zhang, *Opt. Mater.* **29**, 858 (2007).
104. J.M. Luther, M.C. Beard, Q. Song, M. Law, R.J. Ellingson and A.J. Nozik, *Nano Lett.* **7**, 1779 (2007).
105. A. Luque, A. Marti and A.J. Nozik, *MRS Bulletin* **32**, 236 (2007).
106. G. Allan and C. Delerue, *Phys. Rev. B* **73**, 205423 (2006).
107. J.E. Murphy, M.C. Beard, A.G. Norman, S.P. Ahrenkiel, J.C. Johnson, P.R. Yu, O.I. Micic, R.J. Ellingson and A.J. Nozik, *J. Am. Chem. Soc.* **128**, 3241 (2006).
108. R.J. Ellingson, M.C. Beard, J.C. Johnson, P.R. Yu, O.I. Micic, A.J. Nozik, A. Shabaev and A.L. Efros, *Nano Lett.* **5**, 865 (2005).
109. V.I. Klimov, *J. Phys. Chem. B* **110**, 16827 (2006).
110. Y. Tamaki, A. Furube, R. Katoh, M. Murai, K. Hara, H. Arakawa and M. Tachiya, *Cr. Chim.* **9**, 268 (2006).
111. D.P. Colombo, K.A. Roussel, J. Saeh, D.E. Skinner, J.J. Cavaleri and R.M. Bowman, *Chem. Phys. Lett.* **232**, 207 (1995).
112. M. Salmi, N. Tkachenko, R.J. Lamminmaki, S. Karvinen, V. Vehmanen and H. Lemmetyinen, *J. Photoch. Photobio. A* **175**, 8 (2005).
113. R.J.D. Miller, G.L. McLendon, A.J. Nozik, W. Schmickler and F. WIllig, eds., VCH: New York. 167 (1995).
114. A.J. Nozik and R. Memming, *J. Phys. Chem.* **100**, 13061 (1996).
115. P.V. Kamat and D. Meisel, *Semiconductor Nanoclusters — Physical, Chemical, and Catalytic Aspects*, New York: Elsevier. 474 (1997).
116. J.E. Moser, P. Bonnote and M. Gratzel, *Coordin. Chem. Rev.* **171**, 245 (1998).
117. N. Serpone, *Res. Chem. Intermediat.* **20**, 953 (1994).
118. P.V. Kamat, *Progr. Reaction Kinetics* **19**, 277 (1994).
119. B. Oregan and M. Gratzel, *Nature* **353**, 737 (1991).
120. N. Serpone and E. Pelizzetti, *Photocatalysis: Fundamentals and Applications*, New York: Wiley. 650 (1989).

121. K. Kalyanasundaram and M. Gratzel, *Coordin. Chem. Rev.* **177**, 347 (1998).
122. M.K. Nazeeruddin, A. Kay, I. Rodicio, R. Humphrybaker, E. Muller, P. Liska, N. Vlachopoulos and M. Gratzel, *J. Am. Chem. Soc.* **115**, 6382 (1993).
123. M. Gratzel, *Heterogeneous Photochemical Electron Transfer*, Boca Raton: CRC Press. 176 (1989).
124. Y.Q. Gao and R.A. Marcus, *J. Chem. Phys.* **113**, 6351 (2000).
125. S. Gosavi and R.A. Marcus, *J. Phys. Chem. B* **104**, 2067 (2000).
126. Y.Q. Gao, Y. Georgievskii and R.A. Marcus, *J. Chem. Phys.* **112**, 3358 (2000).
127. J.B. Asbury, E. Hao, Y.Q. Wang, H.N. Ghosh and T.Q. Lian, *J. Phys. Chem. B* **105**, 4545 (2001).
128. J.M. Rehm, G.L. McLendon, Y. Nagasawa, K. Yoshihara, J. Moser and M. Gratzel, *J. Phys. Chem.* **100**, 9577 (1996).
129. T. Hannappel, B. Burfeindt, W. Storck and F. Willig, *J. Phys. Chem. B* **101**, 6799 (1997).
130. J.E. Moser, D. Noukakis, U. Bach, Y. Tachibana, D.R. Klug, J.R. Durrant, R. HumphryBaker and M. Gratzel, *J. Phys. Chem. B* **102**, 3649 (1998).
131. T. Hannappel, C. Zimmermann, B. Meissner, B. Burfeindt, W. Storck and F. Willig, *J. Phys. Chem. B* **102**, 3651 (1998).
132. T.A. Heimer and E.J. Heilweil, *J. Phys. Chem. B* **101**, 10990 (1997).
133. R.J. Ellingson, J.B. Asbury, S. Ferrere, H.N. Ghosh, J.R. Sprague, T.Q. Lian and A.J. Nozik, *J. Phys. Chem. B* **102**, 6455 (1998).
134. J.M. Szarko, A. Neubauer, A. Bartelt, L. Socaciu-Siebert, F. Birkner, K. Schwarzburg, T. Hannappel and R. Eichberger, *J. Phys. Chem. C* **112**, 10542 (2008).
135. D.V. Tsivlin, F. Willig and V. May, *Phys. Rev. B* **77**, 035319 (2008).
136. N.J. Cherepy, G.P. Smestad, M. Gratzel and J.Z. Zhang, *J. Phys. Chem. B* **101**, 9342 (1997).
137. P.V. Kamat, *Progr. Inorganic Chem.* **44**, 273 (1997).
138. I. Martini, J.H. Hodak and G.V. Hartland, *J. Phys. Chem. B* **102**, 9508 (1998).
139. H.N. Ghosh, J.B. Asbury, Y.X. Weng and T.Q. Lian, *J. Phys. Chem. B* **102**, 10208 (1998).
140. H.N. Ghosh, J.B. Asbury and T.Q. Lian, *J. Phys. Chem. B* **102**, 6482 (1998).
141. J.B. Asbury, Y.Q. Wang and T.Q. Lian, *J. Phys. Chem. B* **103**, 6643 (1999).
142. J.B. Asbury, R.J. Ellingson, H.N. Ghosh, S. Ferrere, A.J. Nozik and T.Q. Lian, *J. Phys. Chem. B* **103**, 3110 (1999).
143. J.B. Asbury, E.C. Hao, Y.Q. Wang and T.Q. Lian, *J. Phys. Chem. B* **104**, 11957 (2000).
144. Y.X. Weng, Y.Q. Wang, J.B. Asbury, H.N. Ghosh and T.Q. Lian, *J. Phys. Chem. B* **104**, 93 (2000).

145. A.J. Nozik, *Ann. Rev. Phys. Chem.* **52**, 193 (2001).
146. J.B. Asbury, N.A. Anderson, E.C. Hao, X. Ai and T.Q. Lian, *J. Phys. Chem. B* **107**, 7376 (2003).
147. N.A. Anderson and T. Lian, *Ann. Rev. Phys. Chem.* **56**, 491 (2005).
148. C. Nasr, D. Liu, S. Hotchandani and P.V. Kamat, *J. Phys. Chem.* **100**, 11054 (1996).
149. D. Liu, R.W. Fessenden, G.L. Hug and P.V. Kamat, *J. Phys. Chem. B* **101**, 2583 (1997).
150. I. Martini, G.V. Hartland and P.V. Kamat, *J. Phys. Chem. B* **101**, 4826 (1997).
151. A. Furube, L. Du, K. Hara, R. Katoh and M. Tachiya, *J. Am. Chem. Soc.* **129**, 14852 (2007).
152. C. Burda, T.C. Green, S. Link and M.A. El-Sayed, *J. Phys. Chem. B* **103**, 1783 (1999).
153. M. Laferriere, R.E. Galian, V. Maurel and J.C. Scaiano, *Chem. Commun.* **257** (2006).
154. V. Maurel, M. Laferriere, P. Billone, R. Godin and J.C. Scaiano, *J. Phys. Chem. B* **110**, 16353 (2006).
155. C. Tansakul, R. Braslau, G. Millhausser, A.A. Wolcott and J.Z. Zhang, *Phys. Chem. Chem. Phys.* submitted, (2009).

Chapter 10

Applications of Optical Properties of Nanomaterials

In the previous chapters, some examples of applications have been given along with the discussion of optical properties of various nanomaterials. We believe that it is useful to devote a chapter specifically to the applications of optical properties of nanomaterials to highlight the importance and breadth of their technological use. This is important particularly for applications that require the use of different nanomaterials simultaneously.

In principle, nanomaterials can be used for any bulk material-based devices. The key question is whether nanomaterials offer any advantages or improved performance or functionality over bulk materials. The answer varies from application to application. In many cases, nanomaterials can offer improved properties and performances over bulk materials, while, in some other cases, bulk materials are more advantageous. It is thus important to recognize in what situation nanomaterials, instead of bulk materials, should be used and vice versa. For example, the large surface area offered by nanomaterials can in itself be useful or harmful, depending on the application of interest. One of the major advantages of nanomaterials over bulk crystalline materials is the processibility over large areas, especially when combined with matrix materials such as polymers, resulting in essentially nanocomposite materials. Such composite materials have many interesting new properties and potential applications. Overall, the applications of nanomaterials are vast, diverse and fast changing.

We will choose some examples to illustrate the useful aspects of their rich and fascinating optical properties.

10.1. Chemical and biomedical detection, imaging and therapy

Detection, analysis and sensing of chemicals and biochemicals are important to a number of industrial and technologic applications ranging from medicine to food, electronics, security, and chemical industry. Equipment or devices used for such analytical purposes operate based on different mechanisms. The basic idea is that the detected signal contains information about the target analyte, and can thus be used for its identification. The signal is generated when the analyzer or sensor interacts with the target analyte and the interaction usually needs to be triggered by a stimuli or source of energy. The source stimuli and the detected signal can come in many forms, e.g. light, heat, electrical current or voltage, magnetic, mechanical and sound. Among these, optical stimulation and detection are the most common and have the advantage of low cost and noninvasiveness.

We will focus our discussion on optical sensing and detection in this section. Optical sensors are analyzers that involve light as the stimuli and/or the signal for detection. For instance, light can be used to excite an analyte, an optical signal can be detected in terms of fluorescence or Raman scattering. A few specific cases will be discussed next that involve light for both stimulation and detection.

Both semiconductor and metal nanostructures have been used extensively for biomedical detection by taking advantage of their unique optical properties. The detection scheme usually makes use of their absorption, scattering (e.g. PL and Raman), or other effects such as photothermal conversion. The principles underlying these schemes have been discussed in Chapters 5 and 7. In this section, we will focus on a few specific application examples.

10.1.1. *Luminescence-based detection*

The basic mechanism behind many optical sensors is the detection of changes of optical signatures of nanomaterials when they interact with target analyte molecules. The optical signature could be absorption, luminescence or Raman scattering. Among these, luminescence is the most popular for the reason that it is highly sensitive, ubiquitous and easy to detect. Almost all semiconductors and insulators are luminescent to different degrees at the appropriate excitation wavelength. Metals are often nonluminescent or very weakly luminescent, but can be strongly luminescent when inter-band transitions are involved with appropriate excitation wavelength.

QDs have been widely used, in place of dye molecules, as fluorescence labels in biomedical detection. Compared to dye molecules, QDs offer broader absorption and PL tunability, choice of multiple colors from the same chemical composition, flexibility for surface functionalization, and, in some cases, less photobleaching or higher stability [1–7]. The optical signal detected is usually fluorescent from the QDs conjugated to appropriate biomolecules of interest, e.g. DNA and proteins.

In most cases, luminescence quenching or enhancement of a fluorophore is detected when the fluorophore interacts with a target analyte. The degree or luminescence quenching or enhancement should ideally be linearly proportional to the analyte concentration over a broad range (large dynamic range) and sensitive down to a very low concentration (sensitivity) with small background (low noise).

One of the most commonly encountered techniques based on photoluminescence is FRET (fluorescence or Förster resonance energy transfer). FRET is often used to determine the distance between two functional groups in a molecule [8]. FRET involves nonradiative transfer of energy from a donor molecule (or nanoparticle) to an acceptor molecule (or nanoparticle). Therefore, the signature of FRET is quenching of the donor fluorescence followed by lower energy or longer wavelength fluorescence of the acceptor. FRET efficiency, E_{FRET}, defined as the fraction of energy (in photons) absorbed by the donor that was subsequently transferred to the acceptor, is expressed as:

$$E_{FRET} = \frac{R_0^6}{(R_0^6 + R^6)} = \frac{1}{1 + \left(\dfrac{R}{R_0}\right)^1} \tag{10.1}$$

where R_0 is the Förster distance and R is the distance between the center of the donor and acceptable fluorophore dipole moments. From this equation, it is clear that, at the Förster distance, the FRET efficiency is 50%. The Förster distance R_0 is determined by the spectral overlap between the donor fluorescence and acceptor absorption, fluorescence quantum yield of the donor in the absence of the acceptor, the refractive index of the medium and the dipole orientation factor.

More relevant to experimental measurement is the expression:

$$E_{\text{FRET}} = 1 - \frac{F_{DA}}{F_D} = 1 - \frac{\tau_D}{\tau_{DA}} \qquad (10.2)$$

where F_{DA} and τ_{DA} are, respectively, the fluorescence intensity and lifetime of the donor in the presence of the acceptor, F_D and τ_D are the fluorescence intensity and lifetime of the donor when the acceptor is far away or absent. Equation (10.2) shows that the FRET efficiency ranges from zero when the acceptor is far away and $F_{DA} = F_D$ to 1 when the acceptor is very close to the donor and F_{DA} is near zero (complete quenching). By attaching appropriate donors and acceptors to functional groups of interest in a molecule and measuring the FRET efficiency, one can determine the distance between the functional groups. For FRET to work effectively, the donor fluorescence spectrum must overlap with the absorption spectrum of the acceptor. This is to ensure effective coupling or interaction between the donor and acceptor dipoles. In addition, the distance between the donor and acceptor cannot be too far, usually within 5 nm.

While most earlier studies of FRET are based on molecular systems, recent work has reported the success of using semiconductor nanoparticles (NPs) or quantum dots (QDs) for FRET often as donors [9–12] and, to a lesser degree, as acceptors [13–15], or both as donors and acceptors [16]. QDs are less ideal as acceptors due to their typically broad absorption that easily result in absorption of the light used to excite the donor. One limitation for QDs in FRET is their relatively large size compared to molecular systems. Their advantages include tunable absorption and emission with size as well as enhanced photostability when properly passivated, as mentioned before.

As an example, FRET has been demonstrated between two different sized QDs, with the larger, red-emitting one conjugated to an antigen (bovine serum albumin) and the smaller green-emitting one attached to the corresponding anti-BSA antibody (IgG) [16]. The formation of BSA-IgG immunocomplex resulted in FRET between the two different QDs as evidenced by the quenching of the luminescence of green-emitting QDs and simultaneous enhancement of the emission of the red-emitting QDs, as shown in Fig. 10.1. This study shows that FRET based on QDs can be potentially useful for the detection of antigens using corresponding known antibodies.

Besides FRET, QD have been used in direct detection of biological molecules such as DNA [17–19] as well as imaging [20] based on fluorescence from the QDs. They have been used for detection of metal ions [21, 22] as well as radicals [23].

10.1.2. *Surface plasmon resonance (SPR) detection*

Metal particles have been used as labels for optical detection of biomolecules such as DNA and proteins and as substrates in SERS. The optical

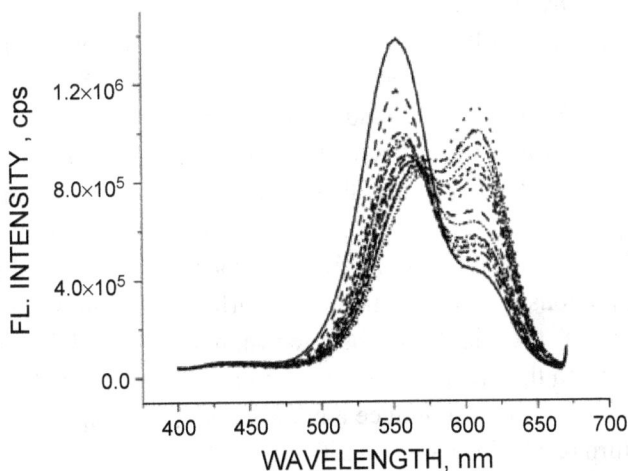

Fig. 10.1. Fluorescence emission spectra recorded at different times after mixing 10^{-7}M N-BSA with 10^{-7}M NP-IgG. The intensity of the peak at 553 nm due to green-emitting NPs decreases with time while the intensity of the peak at 611 nm due to red-emitting NPs increases. Reproduced with permission from Ref. 16.

signal is often based on either SPR or SERS. For example, spectral shift in SPR due to binding or conjugation of biomolecules onto the metal nanoparticle surface can be used to detect or image biomolecules [24–29]. As another example, SPR change has been used in a homogeneous assay for DNA detection based on reducing the interparticle distance through binding of DNA [30, 31]. In this approach, two batches of gold nanoparticles, each modified with a different oligonucleotide were mixed and single-stranded DNA was added that was complementary to both oligonucleotides immobilized on the gold nanoparticles. The gold particles were connected into a network and the interparticle distance was decreased, resulting in color change. Interestingly, this color change is reversible by heating the solution above the melting temperature of the formed DNA duplex. Different variations of such approach have been developed for detecting DNA, including single-base mismatch discrimination [32, 33]. Other spectral changes such as the appearance of new absorption peaks due to strong aggregation caused by binding of biomolecules can also be used to determine the biomolecule of interest. SPR-based techniques are usually not molecule-specific and rely on prior information for identifying the molecules detected.

Surface Plasmon Resonance (SPR) is a commercially available optical detection technique based on the measurement of changes in the surface plasmon absorption band of a thin metal layer or metal nanostructures caused by interaction with a target analyte of interest. In thin metal layers, a polarized light hits a prism covered by a thin metal, usually gold, layer and under certain conditions, including wavelength, polarization and incidence angle, free electrons at the surface absorb incident light photons and convert them into surface plasmon waves. A dip in reflectivity of the light is observed under these SPR conditions. Perturbations to the metal surface, such as an interaction with target analyte molecules of interest, induce a modification of resonance conditions, which in turn results in change in reflectivity. Measurement of this change in reflectivity is the basis for the SPR technique used widely in the form of biochips for quantification and analysis of biological molecules [34–37]. Improvements in these techniques using metal nanostructures, instead of thin films, have been explored [38]. SPR has been further developed in an imaging technique that is sensitive and label-free for visualizing the whole of the biochip via a video CCD camera [39–41]. The biochips can be prepared in an array format with each active site providing SPR information simultaneously. It captures all of the local changes at the surface of the biochip and provides detailed information about molecular binding, biomolecular interactions and kinetic processes [42].

SPR represents one of the simplest and most powerful applications of optical properties of metal nanostructures [43, 44]. It originates from the simple fact that the absorption/reflection of metal nanostructures is dependent on several factors including the dielectric constant of the imbedding material [45, 46]. Even small changes in adsorbate dielectric can have an observable effect on resonance wavelength. This can be used in a general dielectric (or refractive index) sensing mechanism. However, it is even more powerful in functionalized sensing, in which a monolayer of receptors sensitive to the desired analyte is applied to the surface of the film. When even a relatively small number of analyte molecules bind to these surfaces, the change in dielectric is enough to cause a detectable change in the plasmon resonance. This has been shown in a clinical diagnostic procedure and has the potential to be a powerful and highly sensitive technique for biomedical detection [47]. As another example, using the metal film over nanosphere lithography developed by Van Duyne *et al.*, it has been shown that the surface plasmon of these nanoparticulate films is highly sensitive to adsorbates [46, 48, 49].

10.1.3. *SERS for detection*

As discussed in Chapter 7, SERS is a powerful analytical technique for chemical and biochemical analysis and detection. In this section, we will discuss a few specific examples of applications of SERS. From an application viewpoint, SERS offers key advantages such as molecular specificity, high sensitivity, noninvasiveness, low cost and ease of measurement. The one technical limitation is that the SERS results or spectra are sensitive to the particular SERS substrate used, especially in terms of sensitivity to spectral intensity. This is often because the substrates synthesized are not completely identical or reproducible from synthesis to synthesis due to the sensitive dependence of the substrate product on the reaction conditions such as temperature, rate of reaction, impurity in reactants and cleanness of glassware. As long as such limitation can be tolerated, SERS is very useful for analytical applications.

SERS has been used to study a large number of small organic and large biological molecules including proteins and DNA. The interpretation of SERS spectra has become more complicated for large molecules and are, in general, more complicated than normal Raman spectra due to the fact that the SERS spectra are sensitive to the characteristics of the SERS substrate. For example, the distance between the analyte molecules and the surface as well as the molecular orientation and confirmation could significantly affect the SERS spectra. Even though SERS is more complex, its high sensitivity, compared to normal Raman, spectra is a major attractive feature, particularly for low concentration samples [50]. SERS has also been used to study complex system including cells, bacteria, viruses, immunoassays and tissues [51–56]. As a general rule of thumb, the larger the molecule or system, the more complex is the SERS spectrum; and the more polarizable is the molecule, the stronger the SERS signal. As for the metal substrates, structures with junctions and sharp tips that can produce large EM enhancement are better suited for SERS.

However, given the issues discussed, the exact location and orientation of molecules with respect to the position and polarization of the EM field can be very challenging to control at the molecule level. It is therefore highly desired that metal substrate with uniform and monodisperse structure be produced and fabricated for SERS. While the EM field effect may not vary substantially on the length scale of a few nm, atomic precision may be necessary when considering the interaction between the molecule and substrate. Resonance absorption of the incident light by the metal substrate is also essential for SERS to be effective.

One application of SERS is DNA detection [57, 58]. The general approach involves using a Raman probe molecule, e.g. a dye molecule, to report the binding between two complementary DNA strands, one strand labeled by the probe dye molecule and conjugated to a gold nanoparticle and another strand immobilized onto a chip surface or glass microbeads. When two DNA strands bind, the Raman probe molecule is expected to report the event of binding based on its Raman or SERS signal. It was found that further deposition of silver is necessary to enhance the SERS signal to an easily detectable level.

Another interesting application of SERS is cellular sensing and analysis. Nanoparticles are generally small enough to fit into cells without impeding their normal function. There is a size limit, too large or too small, where the particles will not be stable inside the cell. If too small (below 20 nm), the particles will be able to slip out of the cell. If too large (above 80 nm), they begin to damage or impede the function of the cell [59]. There is an optimal size range for nanoparticles to be practically useful in SERS study of living cells. The first example of intracellular SERS study involved the SERS spectrum measured at several positions within the cell with greater than 1 μm resolution [60]. A wide variety of different Raman signals were detected, implying a very complex chemical environment within the cell. It was found that the nanoparticles tend to aggregate within the cell and become immobile [55]. The strong signal from the aggregates allowed the observation of transient SERS signals as the cell performed its normal functions.

For example, Fig. 10.2 shows representative SERS spectra of J774 macrophage cells after incubation with gold nanoparticles [61]. The goal of this experiment is to test for possible dependences of the spectra on a specific cell type and to assess differences from the immortalized rat renal proximal tubule (IRPT) epithelial cell type regarding the applicability of gold nanoparticles as SERS nanosensors, and endosomal composition. Similar to IRPT cells, J774 cells also show the maximum signal strengths of the SERS spectra in the time window 120 min after incubation. This indicates that the fate of the gold nanoparticles, and in particular, the formation of nanoaggregates after uptake, must be very similar in both cell lines. The results of an uptake study for gold nanoparticles of a similar size range also suggested the endocytotic mechanism for particle entry and their accumulation in lysosomes of macrophages [62]. Similar behavior with respect to the SERS enhancement factor was observed providing evidence that not only uptake and final accumulation but also the endosomal trafficking of gold structures of 30–50 nm size range must be handled in a similar fashion by epithelial cells and macrophages. However, despite the similar behavior in nanoparticle trafficking, both cell lines were very different regarding the typical SERS signatures observed for their endosomal compartment (Fig. 10.2). Similar to the data obtained from the IRPT cells, the spectral fingerprints contain contributions from proteins,

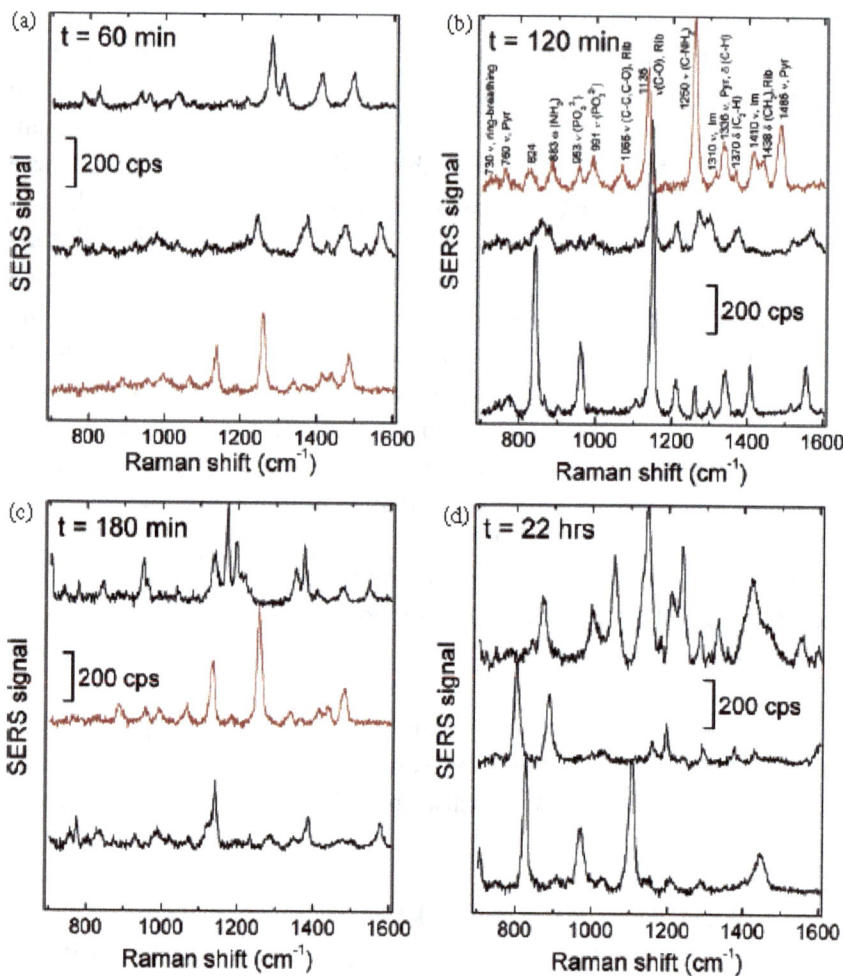

Fig. 10.2. Typical SERS spectra collected from J774 macrophage cells after incubation with gold nanoparticles, excitation $< 3 \times 10^5$ W/cm² at 785 nm, collection time 1 s. The incubation time, including the 60 min nanoparticle pulse, is indicated in each panel. The spectra shown in red in panels (a)–(c) represent an identical SERS fingerprint that was found at all time points and was identified as SERS spectrum of AMP and/or ATP. Abbreviations: cps, counts per second; ν, stretching mode; ω, wagging mode; δ, bending mode; Rib, ribose; Pyr, pyrimidine; Im, imidazol. Reproduced with permission from Ref. 61.

carbohydrates, lipids and nucleotides. On the other hand, unlike the spectra found for IRPT, a specific spectral signature appears at all time points [red spectra in Figs. 10.2(a)–10.2(c)], with the exception of the lysosomal stage (22 h incubation). Almost all bands in this spectrum are characteristic of adenosine phosphate and show contributions from all constituents of this nucleotide (adenine, ribose, and phosphate) [Fig. 10.2(b)].

SERS is potentially useful for monitoring local pH, important for many applications. For instance, SERS has been investigated for mercaptobenzoic acid functionalized silver aggregates [63]. By monitoring the ratio of non-pH sensitive to pH sensitive Raman modes, it is possible to sense, at the nano scale, the pH of the local environment. This type of pH sensing has been expanded by others [64–68]. The general field of using SERS-sensitive particles for intracellular sensing seems to be progressing towards surface functionalization and the use of reactive species to detect various analytes [55]. Many of these SERS sensors utilize large particles or aggregates and it is not clear how these large particles may affect cell function.

10.1.4. *Chemical and biochemical imaging*

Imaging is attractive since it provides spatial information besides spectral information compared to conventional optical detection based on spectroscopy. However, it is understandably more involved in terms of instrumentation as well as signal processing and data analysis. Optical imaging is a 2D display of optical signal, usually emitted or scattered photons, containing 3D information. Usually the vertical axis represents the optical signal intensity while the two axes in the plane represent a combination of space and/or spectral frequency.

QDs have been successfully used for imaging in cell biology and animal biology. In cell biology applications, QDs are used as immunolabels and offer potential advantages over conventional organic dye molecule in terms of stability against photobleaching, tunable and broad absorption, tunable and narrow PL band, possibility for multiplexing, and flexibility for surface modification [1, 69–73]. QDs have been used not only to label

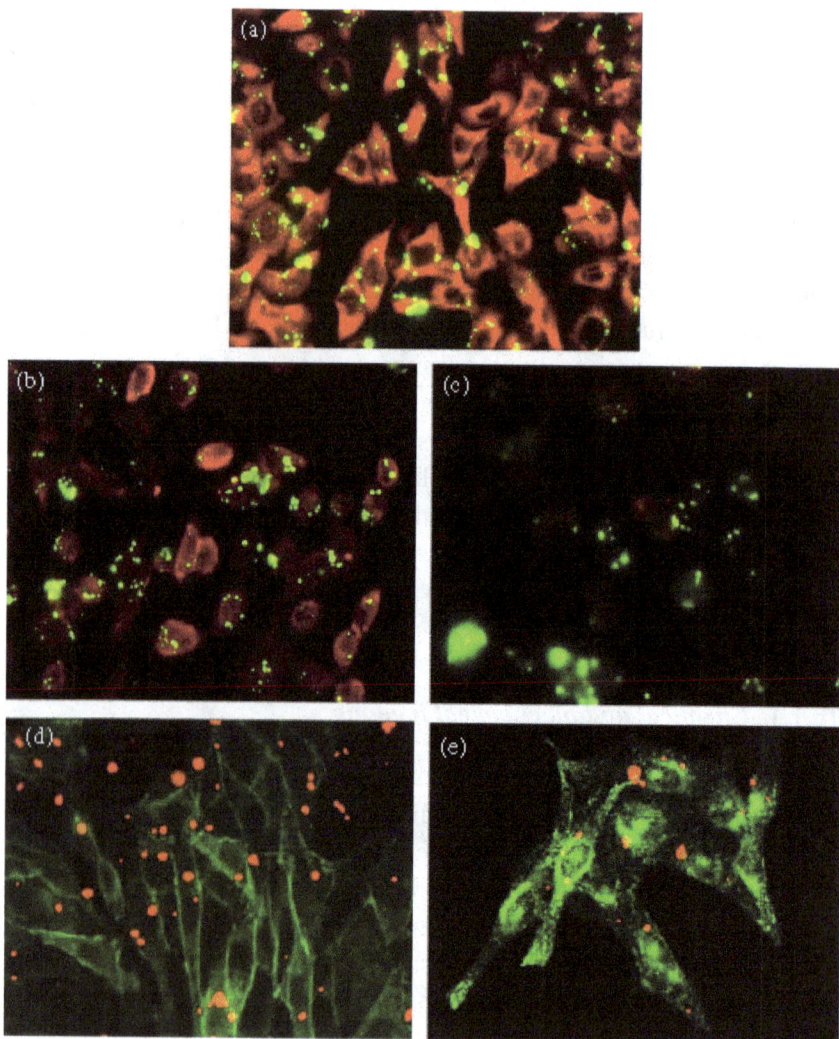

Fig. 10.3. Fluorescent microscopy images of encoded cells in various cellular assays. (a) Immunostaining of tubulin in Chinese hamster ovary (CHO) cells. Cells were encoded with 530-nm QDs and fixed for immunostaining using rabbit antitubulin IgG fraction, biotinylated goat anti-rabbit IgG and streptavidin-conjugated Cy3 (a cyanine dye). Shown is a composite of the Cy3 and QD images. (b and c) Binding of CGP-12177 (a fluorescent ligand) to CHO cells expressing the 2-adrenergic receptor. CHO cells were encoded with 530-nm QDs and incubated with 250 nM BODIPY TMR (±) CGP-12177 (a long-acting fluorescent beta$_2$-adrenoceptor agonist) in the absence (b) or presence

cellular structures and receptors and to incorporate into living cells but also to track the path and fate of individual cells including stems cells [5, 72, 74, 75]. For example, different QDs have been used to label various subpopulations of Chinese hamster ovary (CHO) carcinoma cells on one substrate [75]. Figure 10.3 shows representative fluorescent microscopy images of encoded cells in different cellular assays. This study demonstrates the usefulness of multicolored QDs in cell labeling applications.

In live animal imaging applications, the surface of QDs is very important for toxicity and other considerations such as binding or accumulation in different organs. For example, surface modification of QDs with high molecular weight polyethylene glycol (PEG) molecules has been found to be effective in reducing undesired accumulation in the liver and bone marrow, as shown in Figs. 10.4(a) and 10.4(b) [76]. Compared to visible light, near IR light penetrates deeper into tissues and has less scattering. Therefore, near IR emitting QDs are desired for animal imaging applications, as has been demonstrated using small bandgap CdTe/CdSe core/shell QDs (Fig. 10.4c) [77]. It is clear that both the optical and surface properties of QDs are critical for animal imaging applications. More research is needed to better understand some of the fundamental issues including interaction between QDs and different cells, toxicity and mechanism of biodegradation of the QDs.

SPR discussed in Sec. 10.1 has also been developed into a simple imaging technique for biodetection applications [41, 79, 80]. For instance, a novel surface enzymatic amplification process utilizing RNase H and RNA microarrays has been demonstrated to enhance the sensitivity of SPR imaging in the detection of DNA oligonucleotides down to a concentration of 1 fM, a significant improvement in sensitivity over previous

---

Fig. 10.3. (*continued*) (c) of 1 μM unlabeled CGP-12177 (a beta$_3$-receptor agonist). Shown are composites of the QD and BODIPY images. Binding was measured as total pixel intensity of CGP-12177 Xuorescence. (d and e) Agonist-induced internalization of the β$_2$-adrenergic receptor. CHO cells expressing HA-tagged β$_2$-adrenergic receptor were encoded with 608-nm QDs and incubated in the absence (d) or presence (e) of 10 μM isoproterenol. The cells were assayed for receptor internalization. Shown is a composite of the FITC (Fluorescein isothiocyanate) and QD images. Reproduced with permission from Ref. 75.

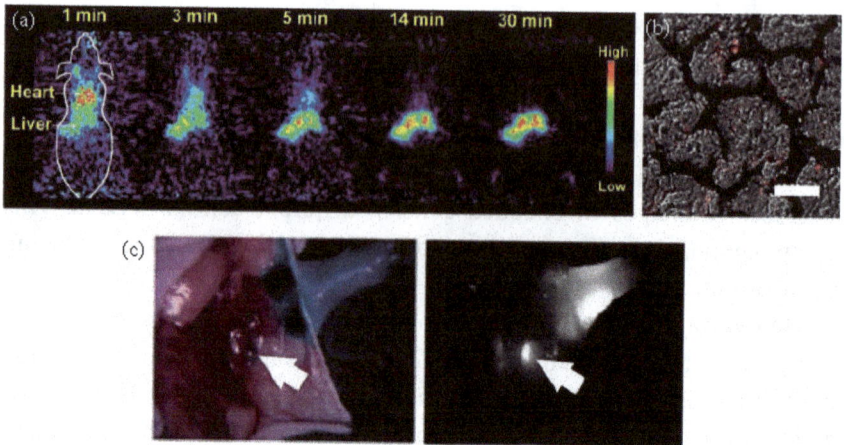

Fig. 10.4. Examples of animal use of quantum dots (qdots or QDs). (a and b) microPET and fluorescence imaging of qdots. QDs having DOTA (a chelator used for radiolabeling) and 600-dalton PEG on their surface were radiolabeled with 64Cu (positron-emitting isotope with half-life of 12.7 hours). These QDs were then injected via the tail vein into nude mice (~80 μCi per animal) and imaged in a small animal scanner. (a) Rapid and marked accumulation of QDs in the liver quickly follows their intravenous injection in normal adult nude mice. This could be avoided by functionalizing QDs with higher molecular weight PEG chains [76]. (b) Overlay of DIC and fluorescence images of hepatocytes from a mouse shows the accumulation of QDs within liver cells. Scale bar, 20 μm. A further step could involve TEM imaging of the precise localization of qdots in cells, illustrating the potential of QDs as probes at the macro-, micro-, and nanoscales. (c) Surgical use of NIR QDs. A mouse was injected intradermally with 10 pmol of NIR qdots in the left paw, 5 min after reinjection with 1% isosulfan blue and exposure of the actual sentinel lymph node. (left) color video; (right) NIR fluorescence image. Isosulfan blue and NIR QDs were localized in the same lymph node (arrows) [77]. Reproduced with permission from Ref. 78.

detection limit of nM [79]. Figure 10.5 shows SPR imaging data obtained on a three-component microarray to demonstrate the selectivity of RNase H for the hydrolysis of surface-bound RNA-DNA heteroduplexes. A three-component array (see Fig. 10.5 inset) composed of two noninteracting RNA probe molecules (with sequences denoted R_A and R_B) and a third DNA probe molecule (D_A) with the same sequence as R_A was created by attaching thiol-modified RNA and DNA oligonucleotides onto an alkanethiol-modified gold thin film. The array was exposed to a 500 nM

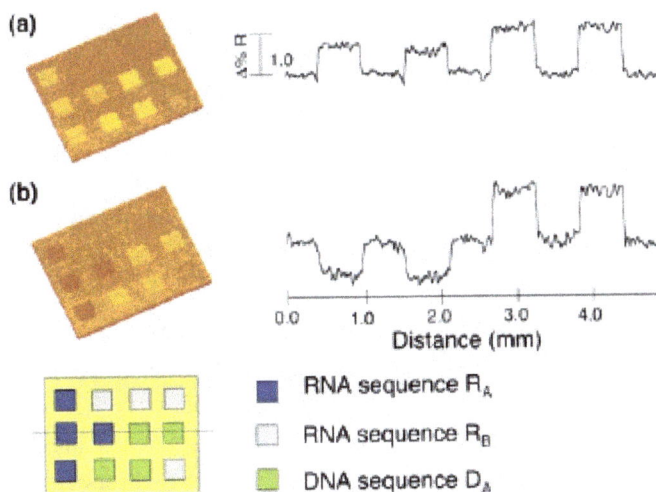

Fig. 10.5. SPR imaging data showing the specificity of RNase H hydrolysis. A three-component array created using two thiol-modified RNA sequences and one thiol-modified DNA sequence (see inset). (a) An SPR difference image for complementary DNA binding to array elements R_A and D_A. (b) An SPR image after the induction of RNase H shows a decrease in percent reflectivity for the RNA–DNA duplex, corresponding to the removal of PNA probes from the surface. Array elements are 500 μm × 500 μm square and contain approximately 4 fmol of oligonucleotide. Reproduced with permission from Ref. 79.

solution of target DNA that was complementary to sequences R_A and D_A; the SPR reflectivity difference image was obtained by subtracting the images taken before and after exposure [see Fig. 10.5(a)]. The DNA target bound specifically to both the R_A and the D_A array elements on the surface, and nonspecific adsorption was not observed to either the mismatched R_B sequence or the array background. This array then contained array elements with heteroduplexed RNA-DNA, dsDNA, and ssRNA. The array was exposed to a solution of RNase H, and Fig. 10.5(b) shows the SPR reflectivity difference image obtained after approximately a 5-min reaction time. A large decrease in percent reflectivity was obtained only for the R_A array elements, which were RNA–DNA heteroduplexes prior to exposure to RNase H. This change in SPR signal is attributed to the selective removal of these RNA probes from the surface. Both the dsDNA and the ssRNA array elements show no change in SPR signal after exposure to RNase H, demonstrating the specificity and selectivity of this enzyme.

Raman scattering, discussed in Chapter 2, has also been developed into an imaging tool from a traditional spectroscopy technique [81–83]. One may expect that it would be a simple and logical extension in going from Raman imaging to SERS imaging (SERS has been discussed in detail in Chapter 7). However, to date, there is very limited development in SERS imaging [84–86], which is somewhat surprising. It should only be a matter of time that SERS imaging will be developed and popularized. The principle for SERS imaging should be the same as Raman imaging with the added advantage of much enhanced signal with SERS. However, SERS is in itself more complicated than Raman since it involves metal nanostructures as substrates that can affect the SERS spectrum, as discussed in Chapter 7.

Figure 10.6 shows an example of SERS imaging of Chinese hamster ovary (CHO) cells incubated with cresyl violet (CV)-labeled silver colloidal particles [86]. A comparison between a bright field image and the image from the total SERS signal of a fixed CHO cell indicates that the total SERS signal is inhomogeneously distributed in the cell as indicated by the arrow in the merged image. This is due to the fact that some regions within the cell do not contain the CV-labeled silver clusters. Variation in the SERS signal intensities that are observed within the cell has been attributed to possible enhancement due to the SERS-active site, resulting from silver nanostructure and/or the location of the laser within the field of view. The large clusters formed by the activated colloid are believed to be the most efficient site for SERS enhancement. Inhomogeneous distribution of the untargeted particles in the presence of Cl^- ions resulted in particle aggregate, thus generating a localized plasmonic effect. However, other factors such as the nonuniform SERS enhancement factor inside the cells may also result in the variation of the scattering signal. Though these results do not show structural characterization of biomolecular components within a cell, they demonstrated system capabilities to localize untargeted CV-labeled nanoparticles within a cell.

Other imaging techniques based on metal nanostructures include photothermal imaging [87–93] and photoacoustic imaging [94–97].

| Brightfield Image | SERS Image @ 596 cm⁻¹ | MERGE |

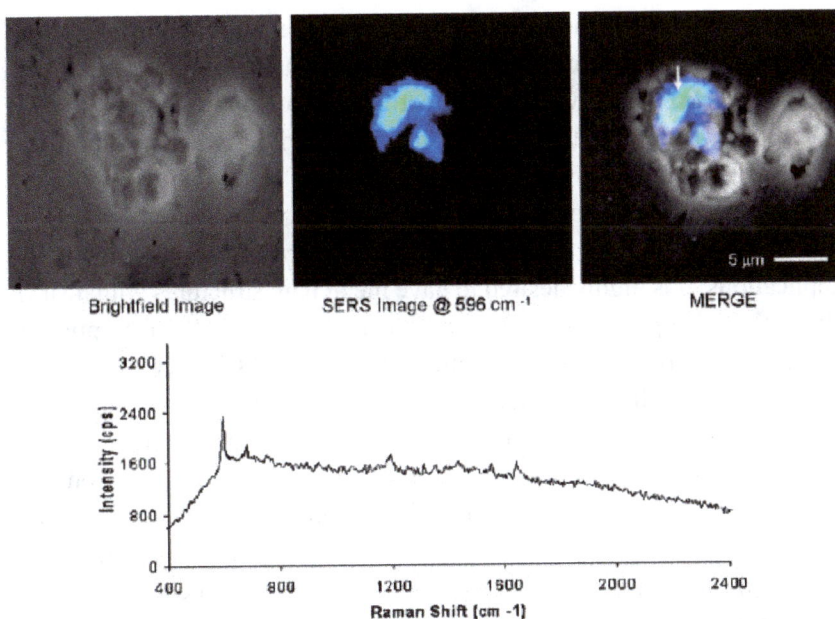

Fig. 10.6. SERS images (top) and spectrum (bottom) of Chinese hamster ovary (CHO) cells incubated with cresyl violet-labeled silver colloidal particles. The bright field image (top left), the total SERS image (top middle) and the merged image (top right) of a fixed CHO cell are shown. The arrow indicates regions within the cell where clusters of silver nanostructures are located. The SERS image was acquired at a SERS intensity signal of 596 cm^{-1} using a 632.8 nm HeNe laser for excitation. Scale bar = 5 μm. Reproduced with permission from Ref. 86.

Photothermal imaging is based on detection of thermal energy or heat generated from photoexcitation, while photoacoustic imaging is based on detection of acoustic wave generated from photoexcitation.

In a typical photothermal imaging system, metal nanostructures are conjugated to biological targets such as cancer cells or tissues and used to absorb light and convert light energy into thermal energy. The thermal energy or heat is detected and used for the purpose of imaging the sample target or therapeutic treatment such as cancer cell destruction or ablation (to be discussed in the next subsection). Specificity or selectivity of the technique relies on mechanisms such as antibody–antigen interaction while the metal nanostructures serve as the light absorber and heat convertor.

Strong light absorption and efficient heat conversion are critical to this process. Metal nanostructures have been demonstrated to be very useful for this purpose. However, to be effective, it is highly desired to have the absorption of the nanostructure matching to that of the wavelength of laser light used for photothermal imaging or therapy.

As discussed in Chapter 7, the surface plasmon absorption can be controlled by varying the shape of metal nanostructures. For biomedical applications, it is highly desired to have metal nanostructures with near IR absorption for optimal tissue penetration and minimal light absorption by the tissue. Many of the nanostructures discussed in Chapter 7 can easily provide the near IR absorption desired. However, most of the nanostructures have sizes that are too large or shapes that are too complex for biomedical applications. Appropriate size and shape combinations are needed for effective delivery to locations of interest for detection, imaging or treatment. In addition, the spectral line shape or bandwidth of the surface plasmon absorption should ideally be narrow for better match with the laser wavelength that is usually very narrow.

Therefore, in the ideal case, nanostructures with strong, narrow and tunable absorption band, small size and spherical shape are preferred. Narrow absorption band usually indicates structural uniformity. Small and uniform nanoshell structures with hollow or dielectric core and spherical shape on the outside are ideal or well suited for meeting these requirements. This has been recently demonstrated using unique hollow gold nanospheres (HAuNS) that turn out to be effective for both *in vivo* and *in vitro* photothermal imaging and therapy [93, 98].

10.1.5. *Biomedical therapy*

Besides detection and imaging applications, some nanomaterials have also been explored as therapeutic agents and, understandably, there are questions related to their toxicity and potential side-effects. In relation to optical properties, an interesting application is the use of nanomaterials as photosensitizer in a process called *photodynamic therapy (PDT)*. PDT is a medical technique for cancer treatment that uses a combination of light, photosensitizers and oxygen [99]. The basic mechanism is that photoexcitation of the photosensitizer leads to the generation of radical and/or singlet oxygen that results in cancer cell necrosis. Most photosensitizers are molecules with strong visible or near IR absorption, e.g. porphyrins [100]. Ideal photosensitizers should have low dark toxicity and strong light absorption, can selectively localize onto cancer tissues, are stable and efficient in generating useful singlet oxygen and/or radicals.

Semiconductor nanoparticles have been exploited and successfully demonstrated to be potentially useful for PDT applications. For example, Si nanocrystals have been shown convincingly to be effective in singlet oxygen generation via an exchange of electrons with mutually opposite spins between the photoexcited Si nanocrystal and ground state (triplet) oxygen [101]. Supporting evidences include PL quenching of Si NC by O_2 and light emission from the singlet oxygen at 0.98 eV [102]. Similarly, CdSe QDs have been explored for PDT application and found to be able to generate singlet oxygen directly via a triplet energy transfer mechanism [103]. Nanoparticles, such as gold and silica, have been used for photosensitizer delivery in PDT applications [104–106].

As discussed in the last subsection, photothermal imaging is a simple and promising technique for biomedical imaging applications [87–92]. It turns out that this same method can be extended beyond imaging to therapeutic treatment such as cancer cell destruction or ablation. This technique is sometimes called *photothermal ablation therapy* (*PTA*) [88, 93, 97, 98, 107–112].

For example, a recent study demonstrates that gold nanorods are useful for optical imaging and photothermal therapy applications due to their strong absorption and scattering of near IR light [88]. By using surface plasmon resonant absorption spectroscopy and light scattering imaging, HSC and HOC (oral epithelial cancer cells) malignant cells are easily distinguished from HaCaT (human keratinocyte *cell* line) nonmalignant cells due to the molecular targeting of overexpressed epidermal growth factor receptor (EGFR) on the malignant cell surface. After exposure of these cells, incubated with anti-EGFR antibody-conjugated Au nanorods, to a near IR CW laser at 800 nm, different laser power energies were observed to cause photothermal destruction among malignant and nonmalignant cells, as shown in Fig. 10.7. Increased uptake of the nanorods by the two malignant cells reduced the energy needed to cause destruction of these

HaCat nonmalignant cells   HSC malignant cells   HOC malignant cells

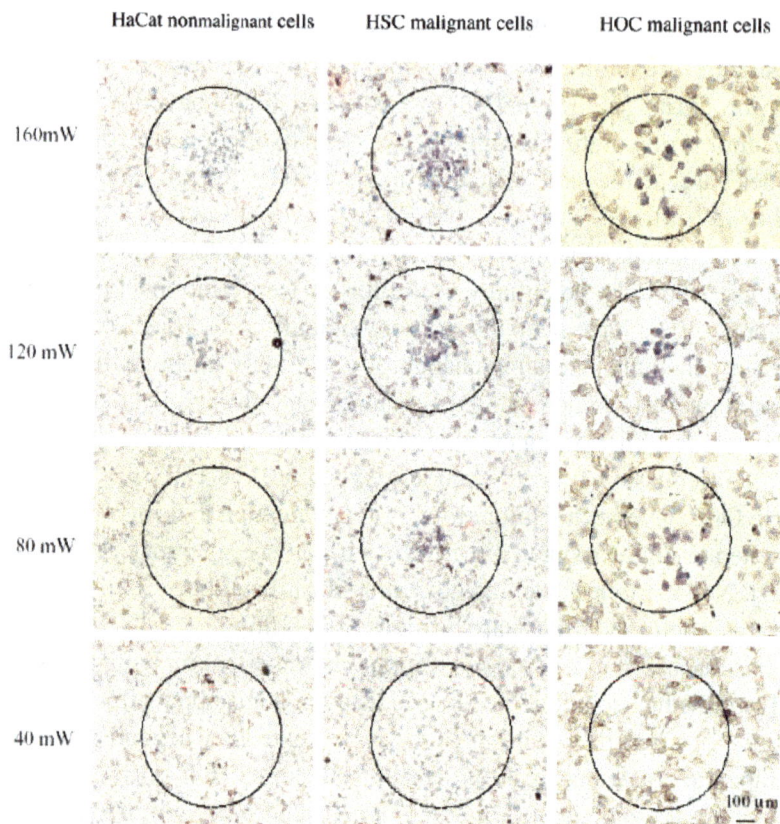

Fig. 10.7. Selective photothermal therapy of cancer cells with anti-EGFR/Au nanorods incubated. The circles show the laser spots on the samples. At 80 mW (10 W/cm^2), the HSC and HOC malignant cells are obviously injured while the HaCat normal cells are not affected. The HaCat normal cells start to be injured at 120 mW (15 W/cm^2) and are obviously injured at 160 mW (20 W/cm^2). Abbreviations: EGFR for epidermal growth factor receptor; HSC and HOC for oral epithelial cancer cells; HaCat for human keratinocyte cell line. Reproduced with permission from Ref. 88.

cells to about half of that required to cause death to the nonmalignant cells. This experiment demonstrates the potential of gold nanorods in cancer diagnostics and therapy [88].

Figure 10.8 shows another example where unique hollow gold nanospheres (HGNs or HAuNS), discussed in Chapter 5, have been successfully used for *in vitro* and *in vivo* photothermal cancer imaging and destruction

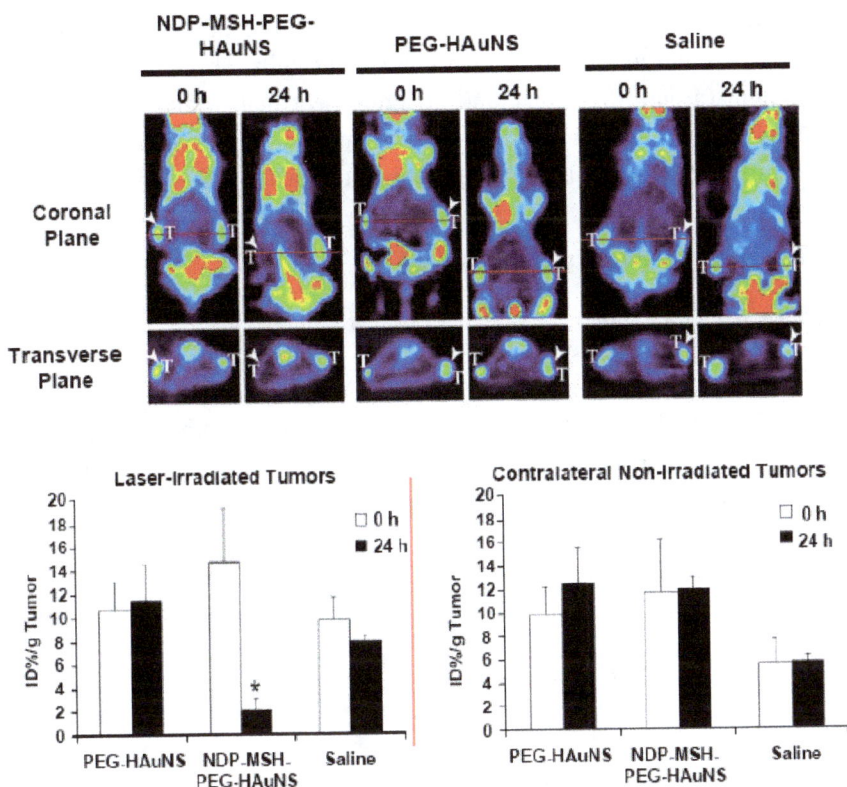

Fig. 10.8. *In vivo* PTA with targeted NDP-MSH-PEG-HAuNS induced selective destruction of B16/F10 melanoma in nude mice. [^{18}F]FDG PET imaging shows significantly reduced metabolic activity in tumors after PTA in mice pretreated with NDP-MSH-PEG-HAuNS but not in mice pretreated with PEG-HAuNS or saline (top). [^{18}F]FDG PET was conducted before (0 h) and 24 h after NIR laser irradiation (0.5 W/cm^2 at 808 nm for 1 min), which was commenced 4 h after intravenous injection of HAuNS or saline. T: tumor. Arrowheads: tumors irradiated with NIR light. [^{18}F]FDG uptakes (%ID/g) before and after laser treatment (bottom). Bars, SD (*n* = 3). *, *P* < 0.01 for %ID/g posttreatment versus %ID/g pretreatment. Abbreviations: MSH for melanocyte-stimulating hormone; NDP-MSH for the peptide H-Ser-Tyr-Ser-Nle-Glu-His-*d*-Phe-Arg-Trp-Gly-Lys(Dde)-Pro-Val-NH$_2$; PEG for polyethylene glycol; HAuNS for hollow Au nanospheres; [^{18}F]FDG-PET for [^{18}F]fluorodeoxyglucose positron emission tomography; SD for standard deviation; ID/g for injected dose per gram of tissue. Reproduced with permission from Ref. 98.

[93, 98]. The HGNs are conjugated to a specific peptide that targets a corresponding antigen of melanoma cancer cells, absorb near IR CW laser light (around 800 nm), and generate heat that destroys the cancer cells by thermal ablation [98]. These preliminary studies show the promise of metal nanostructures for biomedical applications.

10.2. Energy conversion: PV and PEC

10.2.1. *PV solar cells*

Conversion of solar energy into electrical energy using photovoltaic cells (PVC) is arguably one of the most successful and environmentally friendly technologies for energy generation [113]. The energy flux towards the earth surface is about 3×10^{24} J/year. A 1% coverage of the land on earth with solar cells with a 10% conversion efficiency would be able to produce about 26 TW of energy, which is sufficient to cover our current energy needs. Figure 10.9 shows the solar radiation spectrum. It is predicted that in the next three decades large-scale industrial photovoltaics with

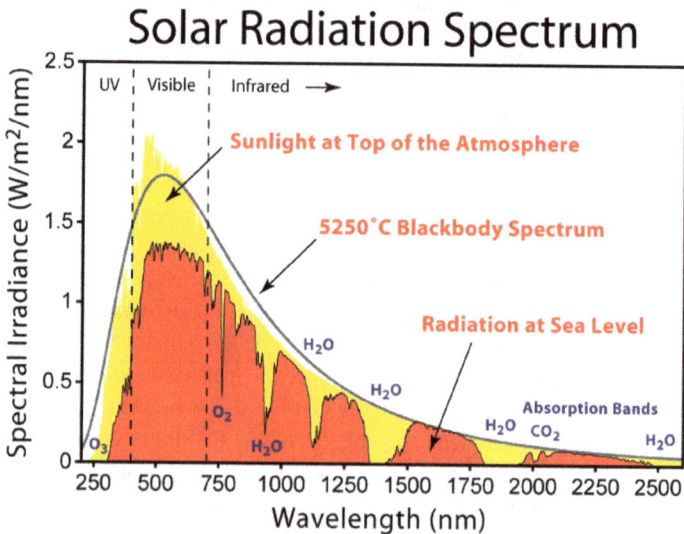

Fig. 10.9. Solar radiation spectrum. Reproduced from http://www.globalwarmingart.com/wiki/Image:Solar_Spectrum_png (figure prepared by Robert A. Rohde).

a throughput on the order of 10 m² of modules per minute (1 GW/year) will be needed to satisfy a global demand for clean energy sources [114]. Increasing solar cell efficiency while maintaining a low production cost has been the primary objective of solar energy conversion. Thick single- and multi-crystalline silicon structures have been the dominant technology in the photovoltaic industry. Commercially available single-junction, single crystalline silicon cells offer 13% conversion efficiency while $GaInP_2$/GaAs-based multiple-junction solar cells have reached 27%, and InGaAsN-based solar cell systems have achieved 30% efficiency [115]. Although these devices exhibit impressive efficiencies, the process of material synthesis is often complex, resulting in low throughput and high production costs.

Conventional silicon solar cells make use of semiconductor diodes based on p-n junctions. Photogenerated electron and holes are driven to separation by a built-in electrical field and then collected as electrical current. Solar cells are usually characterized by current (I)-voltage curves and key parameters include the short-circuit current, I_{sc}, open-circuit voltage, V_{oc}, and fill factor (ff) defined as [116]:

$$ff = P_{max}/(V_{oc} \times I_{sc})$$ (10.3)

where P_{max} is the electrical power delivered by the solar cell at the maximum power point (MPP), as illustrated in Fig. 10.10. From these parameters, one can calculate the overall solar power conversion efficiency, $\eta_{overall}$, using:

$$\eta_{overall} = P_{max}/(I_s \times A_s)$$ (10.4)

where A_s is the area of the solar cell (m²) and I_s is the irradiance of the input light (W/m²), which is usually taken as 1000 W/m² for AM (air mass) at 1.5. AM is defined as AM = 1/cos α with α as the angle between the overhead and the actual position of the sun. The radiation standard assumes an AM of 1.5, which corresponds to $\alpha = 48°$ [117]:

Besides Si, other semiconductor materials, e.g. CdTe, CuInSe and GaAs, have been used for solar cells in bulk crystalline, thin film or poly-crystalline forms. These materials tend to be more costly even though

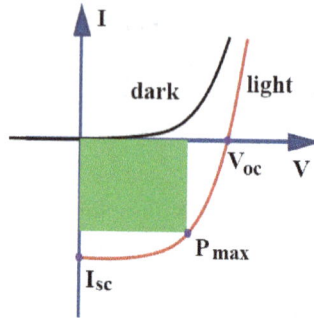

Fig. 10.10. Typical *I-V* characteristics of a solar cell, with the three key characteristic parameters: short circuit current I_{sc}, open-circuit voltage V_{oc}, and fill factor, ff $= P_{max}/(V_{oc} \times I_{sc})$; P_{max} is the electrical power delivered by the cell at the maximum power point (MPP), which is the point at which $V \times I$ is the maximum.

some of them have demonstrated better efficiency than Si solar cells [118–120]. The relatively high cost of manufacturing commercial silicon cells and the use of toxic chemicals in their manufacturing have limited them from widespread use. These aspects have prompted the search for environmentally friendly and low cost solar cell alternatives.

Solar cells based on nanomaterials have been investigated as alternatives for solar cells based on bulk materials such as Si, especially in terms of reducing the cost. Nanostructured materials offer some unique advantages that are ideal for applications in future generation PV devices. Their small size affords a large effective surface area, and thereby potentially low weight cells; quantum confinement effects lead to size-dependent and readily controllable material properties; nanoengineering and synthesis techniques may enable the design of sophisticated nanostructures at low cost and with improved properties. Nanostructured materials also lend themselves conveniently for large area fabrication and on flexible substrates.

Two general types of nanomaterial-based solar cells have been studied, dye-sensitized solar cells [121] and quantum dot solar cells [122]. In the original dye-sensitized solar cell (DSSC) developed in 1991 by O'Regan and Gratzel [121], TiO_2 nanoparticle films were sensitized with the N3 dye [123]. The nanocrystalline nature of the materials provides significantly increased surface area relative to volume as compared to bulk materials. In such a solar cell, dye molecules are adsorbed onto the surface of the nanoparticles. Photoexcitation of the dye molecules results in injection of electrons into the conduction band of TiO_2. The resulting oxidized dye is reduced by electron transfer from a redox couple such as I^-/I^{3-}, for the dye to be regenerated. Charge transport occurs through particle that are interconnected and collected by the electrodes. These types of cells have reached solar power conversion efficiencies of 7–10% under full sun conditions [121, 123]. Recently, there has been increasing research effort to use inorganic semiconductor QDs as sensitizers for large bandgap metal oxide nanostructures for solar cell applications [124–126]. In these cases, the QDs replace organic dye molecule and potential advantages include thermal stability and color tenability by controlling nanoparticle size and shape.

One practical problem is leakage of the electrolyte under working conditions. This has led to the search for replacing the liquid electrolyte with a solid hole conducting material using organic conjugated polymers or inorganic solid compounds [127–132]. Conjugated polymers that have been studied and show promise as a hole conductor include MEH-PPV (poly[2-methoxy-5(2'-ethylhexyloxy)-1,4-phenylene vinylene]) [127, 129, 130, 133] and polythiophene [134–138]. An interesting finding from these studies is that the conjugated polymer can function both as a hole transporter as well as a sensitizer [129, 131, 137, 139]. It was also found that the relative size of the sensitizer molecule and the pores of the nanocrystalline films may be a critical factor to consider in designing photovoltaic devices such as solar cells based on nanoporous materials [131, 132].

Another major limitation with nanoparticle-based solar cells is poor transport properties due to grain boundaries or trap states that significantly limit their conversion efficiency. One possible solution to this problem, while maintaining the advantages of nanomaterials for solar cells, is to use 1D nanostructures such as nanowires and nanorods that are expected to have better transport properties than nanoparticles [140–152]. The 1D nanostructures still have large surface area and are suitable for large area processing. Investigations on solar cells based on 1D MO nanostructures have started recently, e.g. Nd-doped TiO_2 nanorods [153] and ZnO nanorods [145].

In the meantime, efforts have been made to enhance the visible absorption of MO by doping with elements such as N [154–163]. A recent study has also demonstrated the use of combining QD sensitization *and* N-doping to enhance visible absorption and facilitate charge transfer/transport in TiO_2 nanoparticle solar cells [124]. This approach can be extended to 1D or 2D nanostructures, and is currently under investigation.

10.2.2. *Photoelectrochemical cells (PEC)*

Photoelectrochemical cells provide a powerful means for converting light energy into chemical energy that can be stored in molecules such as hydrogen through electrochemical reactions, e.g. water splitting, as first demonstrated in 1972 by Fujishima *et al.* [164]. Besides electrolytes, the key components in a PEC are the electrodes (cathode and anode) on which redox chemical reactions involving electron transfer take place. Figure 10.11 shows a simple schematic of a typical PEC device. Most electrodes are based on bulk materials. However, photoelectrodes based on nanostructured materials have been explored for a number of systems, including CdS [165], Bi_2S_3 [166], Sb_2S_3 [167], SnO_2/ZnO composite [168], WO_3 [169–175], and more commonly TiO_2 [176–182].

Similar to nanomaterials used for PVC, nanomaterials offer some potential advantages for photoelectrodes in PEC, including large surface area, fast diffusion and reaction time, ease of surface modifications, and possibility for manipulating material properties by structural control. It is also possible to use multicomponent composite materials to improve the performance of the PEC over that based on individual components by counteracting the electron-hole recombination process [168]. 1D nanostructures such as nanowires and nanorods have expected to exhibit much improved transport properties than nanoparticles. However, to date, only limited PEC studies based on 1D nanostructures, e.g. TiO_2 nanotubes and nanorods, has been reported [148, 151, 183]. Also similar to PVC, QD sensitization, e.g. with CdSe, and doping, with elements such as C and S, have been exploited for improving the performance of PECs [148, 184, 185].

The PEC performance depends strongly on the properties of the photoanode, especially the bandgap and bandedge positions with

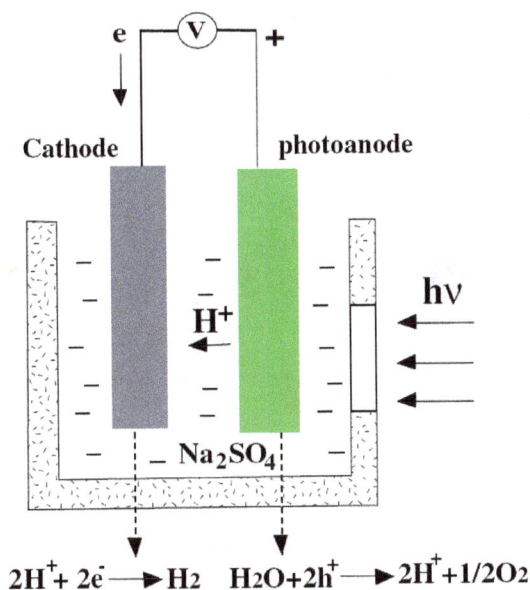

Fig. 10.11. Schematic of a typical PEC device and its basic operation mechanism for hydrogen generation from water splitting.

respect to the vacuum level or normal hydrogen electrode (NHE) level. Figure 10.12 shows the bandgap and bandedge energies of some typical metal oxide materials at pH = 2 [186]. As pH changes, the energy levels can shift significantly with respect to the vacuum level or NHE level [117].

10.3. Environmental protection: photocatalytic and photochemical reactions

Nanomaterials have played a critical role in many important chemical reactions as reactants, catalysts or photocatalysts. In relation to optical properties, nanomaterials have been used in photochemical and photocatalytic reactions. Their reactivities are often altered or enhanced due to size dependent changes in their redox potentials and high density of active surface states associated with a very large surface-to-volume ratio.

Photocatalysis based on semiconductors plays an important role in chemical reactions of small inorganic, large organic and biological molecules. Photocatalytic reactivities are strongly dependent on the nature

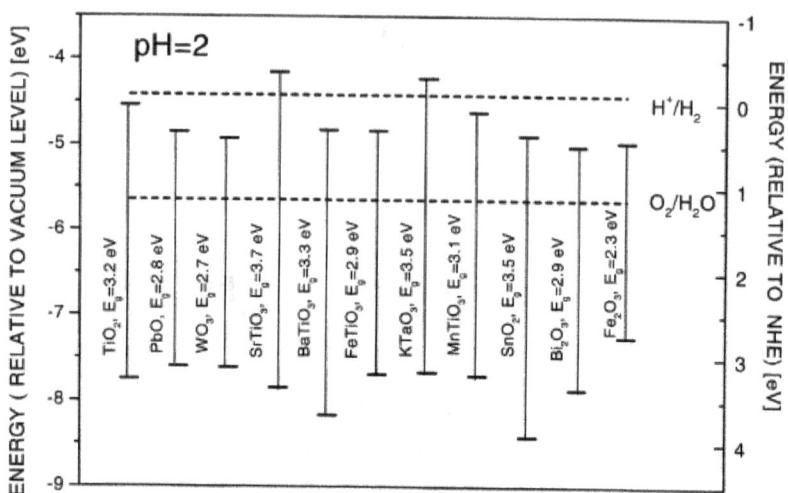

Fig. 10.12. Bandgap and bandedge energy levels of some metal oxide materials with respect to vacuum and normal hydrogen electrode level in electrolyte at pH = 2. Reproduced with permission from Ref. 186.

and properties of the photocatalysts, including pH of the solution, particle size, shape and surface characteristics [187]. These properties are sensitive to preparation methods [187]. Impurities or dopants can significantly alter these properties as well as reactivities. For example, it has been shown that selectively doped nanoparticles have a much greater photoreactivity as measured by their quantum efficiency for oxidation and reduction than their undoped counterparts [188]. A systematic study of the effects of over 20 different metal ion dopants on the photochemical reactivity of TiO_2 colloids with respect to both chloroform oxidation and carbon tetrachloride reduction has been conducted [188, 189]. A maximum enhancement of 18-fold for CCl_4 reduction and 15-fold for $CHCl_3$ oxidation in quantum efficiency for Fe(III)-doped TiO_2 colloids have been observed [190]. Studies have shown that the surface photovoltage spectra (SPS) of TiO_2 and ZnO nanoparticles can be an effective method for evaluating the photocatalytic activity of semiconductor materials since it can provide a rapid, nondestructive monitor of the semiconductor surface properties such as surface band bending, surface and bulk carrier recombination and surface states [191]. It has been demonstrated that for weaker

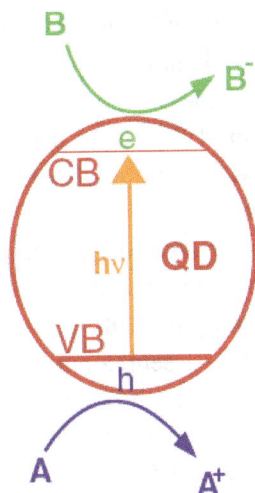

Fig. 10.13.   Illustration of the basic principle of photocatalysis based on semiconductor quantum dots (QDs).

surface photovoltage signal, the higher is the photocatalytic activity in the case of nanosized semiconductor photocatalysts.

Figure 10.13 shows a schematic illustration of the basic principle of photocatalytic reactions involving semiconductor QDs as a photocatalyst. The photogenerated electron in the conduction band (CB) results in reduction of B into B⁻ while the photogenerated hole in the valence band (VB) participates in oxidation of A into A⁺. When both photooxidation and photoreduction are complete, the QD recovers to its original state and thus acts as a photocatalyst. If only one of the two reactions take place, the QD will not recover to its original state and a net photochemical reaction will have taken place.

Photocatalytic oxidation of organic and biological molecules is of great interest for environmental applications, especially in the destruction of hazardous wastes or pollutants. The ideal outcome is complete mineralization of the organic or biological compounds, including aliphatic and aromatic chlorinated hydrocarbons, into small inorganic, non- or less-hazardous molecules, such as CO_2, H_2O, HCl, HBr, SO_4^{2-} and NO_3^-. Photocatalysts include various metal oxide semiconductors, such as TiO_2, in both bulk and particulate forms. Compounds that have been degraded by semiconductor photocatalysis include alkanes, haloalkanes, aliphatic alcohols, carboxylic acids, alkenes, aromatics, haloaromatics, polymers, surfactants, herbicides, pesticides and dyes, as summarized in an excellent review article by Hoffmann [190]. It has been found in many cases that the colloidal particles show new or improved photocatalytic reactivities over their bulk counterparts.

One of the most important areas of application of photocatalytic reactions is removal or destruction of contaminates in water treatment or purification [192–194]. Major pollutants in waste waters are organic compounds. As discussed above, semiconductor nanoparticles offer an attractive system for degrading both organic and inorganic pollutants in water. Water treatment based on photocatalysis provides an important alternative to other advanced oxidation technologies such as UV-H_2O_2 and UV-O_3 designed for environmental remediation by oxidative mineralization. The photocatalytic mineralization of organic compounds in aqueous media typically proceeds through the formation of a series of intermediates of progressively higher oxygen to carbon ratios. For example, photodegradation of phenols yields hydroquinone, catechol and benzoquinone as the major intermediates that are eventually oxidized quantitatively to carbon dioxide and water [195]. The reaction mechanisms have been proposed to involve initial generation of electrons and holes and subsequent generation of protons and radicals such as ·OH that reacts with phenol [196].

In general, the details of the surface morphology, crystal structure and chemical composition critically influence the photochemical and photocatalytic performance of the photocatalysts [197–201]. Therefore, these parameters need to be carefully controlled and evaluated when comparing photocatalytic activities of different materials.

Besides photocatalysis, NPs can also directly participate in photochemical reactions that leave the NPs or QDs in a state different from the state before reaction. It has been demonstrated that photooxidation of some small molecules on semiconductor nanoparticles can lead to the formation of biologically important molecules such as amino acids, peptide oligomers and nucleic acids [202, 203]. In addition to photooxidation, photoreduction based on semiconductor nanoparticles have also been explored for the synthesis of organic molecules [204, 205]. For example, photoinduced reduction of p-dinitrobezene and its derivatives on TiO_2 particles in the presence of a primary alcohol has been found to lead to the formation of benzimidoles with high yields [206].

10.4. Lasers, LEDs, and solid state lighting

10.4.1. *Lasing and lasers*

An area of application of nanomaterials that has attracted considerable interest is lasers. It is in principle possible to build lasers with different wavelengths by simply changing the particle size. There are two practical problems with this idea. First, the spectrum of most nanoparticles is usually quite broad due to homogeneous and inhomogeneous broadening. Second, the high density of trap states leads to fast relaxation of the excited charge carrier, making it difficult to build up population inversion necessary for lasing. When the surface of the particles is clean and has little defects, the idea of lasing can indeed be realized. This has been demonstrated mostly for nanoparticles self-assembled in clean environments based on physical methods, e.g. MBE (molecular beam epitaxy) [207–210] or MOCVD (metal organic chemical vapor deposition) [211]. Examples of quantum dot lasers include InGaAs [207], InAs [209], AlInAs [208, 210], and InP [212]. Stimulated emission has also been observed in GaN quantum dots by optical pumping [211]. The lasing action or stimulated emission has been observed mostly at low temperature [210, 212]. However, room temperature lasing has also been achieved [207, 208].

Recently, lasing action has been observed in colloidal nanoparticles of CdSe based on wet chemistry synthesis and optical pumping [213]. It was found that, despite highly efficient intrinsic nonradiative Auger recombination, large optical gain can be developed at the wavelength of the emitting transition for close-packed solids of CdSe quantum dots. Narrow band stimulated emission with a pronounced gain threshold at wavelengths tunable to the size of the nanocrystal was observed. This work demonstrates the feasibility of nanocrystal quantum dot lasers based on wet chemistry synthesis. Whether real laser devices can be built based on this type of nanoparticles remains to be seen. Also, it is unclear if electrical pumping of such lasers can be realized. Likewise, nanoparticles can be potentially used for laser amplification and such application has yet to be explored. Nanoparticles such as TiO_2 have also been used to enhance stimulated emission for conjugated polymers based on multiple reflection effect [214].

Very recently, room-temperature ultraviolet lasing in ZnO nanowire arrays has been demonstrated [215]. The ZnO nanowires grown on sapphire substrates were synthesized with a simple vapor transport and condensation process. The nanowires form a natural laser cavity with diameters varying from 20 to 150 nm and lengths up to 10 μm. Under optical excitation at 266 nm, surface-emitting lasing action was observed at 385 nm with emission line width less than 0.3 nm (Fig. 10.14). Such miniaturized lasers could have many interesting applications ranging from optical storage to integrated optical communication devices.

10.4.2. *Light emitting diodes (LEDs)*

Nanoparticles have been used for LED applications in two ways. First, they are used to enhance light emission of LED devices with other materials, e.g. conjugated polymers, as the active media. The role of the nanoparticles, such as TiO_2, is not completely clear but thought to enhance either charge injection or transport [216]. In some cases the presence of semiconductor nanocrystals in carrier-transporting polymers has been found to not only enhance the photoinduced charge generation efficiency but also extends the sensitivity range of the polymers, while the polymer matrix is responsible for charge transport [217, 218]. This type of polymer/nanocrystal composite materials can have improved properties over the individual constituent components and may have interesting applications.

Second, the nanoparticles are used as the active materials for light generation directly [122, 210, 219–223]. In this case, the electron and hole are injected directly into the CB and VB, respectively, of the NPs and the recombination of the electron and hole results in light emission (Fig. 5.8). Figure 10.15 shows the absorption, PL and EL spectra of CdSe/ZnS core/shell quantum dots in multilayered LEDs [223].

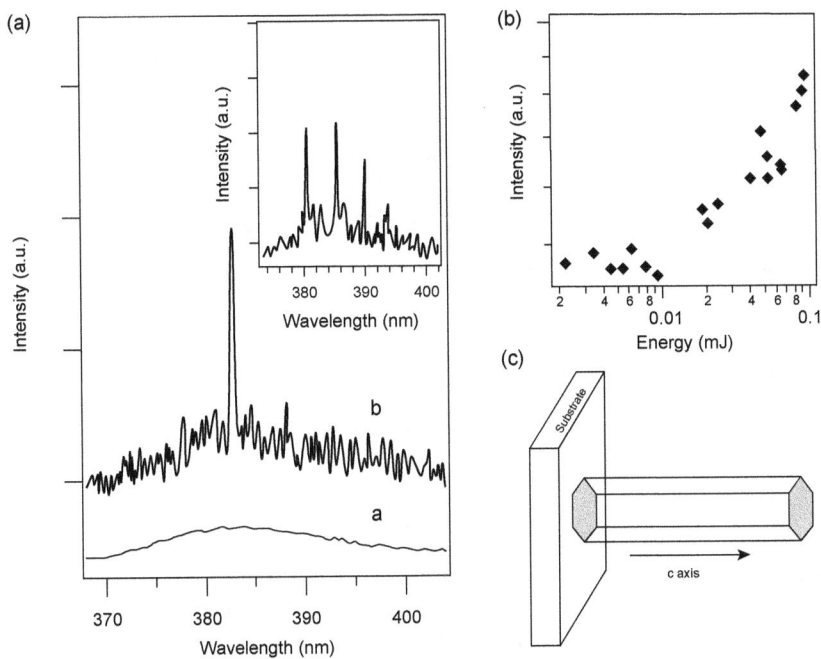

Fig. 10.14. (a) Emission spectra from nanowire arrays below (line a) and above (line b and inset) the lasing threshold. The pump power for these spectra are 20, 100, and 150 kW/cm², respectively. The spectra are offset for easy comparison. (b) Integrated emission intensity from nanowires as a function of optical pumping energy intensity. (c) Schematic illustration of a nanowire as a resonance cavity with two naturally faceted hexagonal end faces acting as reflecting mirrors. Stimulated emission from the nanowires was collected in the direction along the nanowire's end-plane normal (the symmetric axis) with a monochromator combined with a Peltier-cooled charge-coupled device. The 266-nm pump beam was focused on the nanowire array at an angle 10° to the end-plane normal. All experiments were carried out at room temperature. Reproduced with permission from Ref. 215.

Several studies have been reported with the goal to optimize injection and charge transport in such device structures using CdS [221] and CdSe nanoparticles [122, 219]. Since the mobility of the charge carriers is usually much lower than in bulk single crystals, charge transport is one of the major limitations in efficient light generation in such devices. For example, photoconductivity and electric field induced photoluminescence quenching

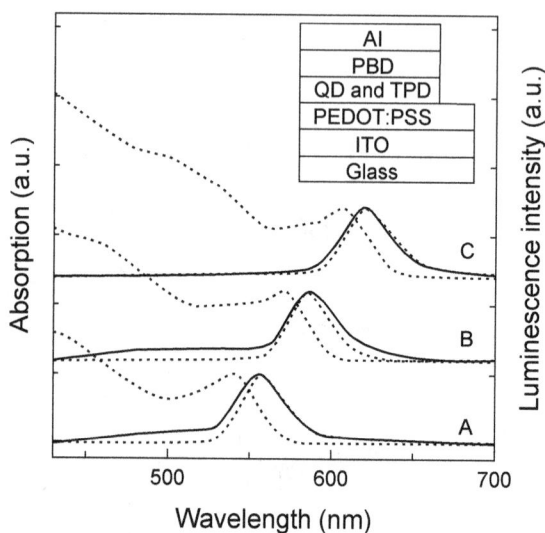

Fig. 10.15. Absorption (dotted lines), photo- (dashed lines), and electroluminescence (solid lines) spectra for 3.2 nm (A), 4.1 nm (B), and 5.4 nm (C) diameter CdSe QDs covered by 2 MLs of ZnS. Voltages and currents are (A) 5.5 V, 0.11 mA; (B) 5.0 V, 0.08 mA; and (C) 4.0 V, 0.06 mA, respectively. Schematic of the structure of the multilayered CdSe/ZnS QD LEDs is shown in the inset. Abbreviations: TPD for N,N′-Diphenyl-N,N′-bis(3-methylphenyl)-1, 1′-biphenyl-4,4′-diamine; PEDOT:PSS for poly(3, 4-ethyl-enedioxythiophene): poly(styrenesulfonate; ITO for indium tin oxide; and PBD for 2-biphenyl-4-yl-5-(4-tert-butylphenyl)-1, 3, 4-oxadiazole. Reproduced with permission from Ref. 223.

studies of close-packed CdSe quantum dot solids suggest that photoexcited, quantum confined excitons are ionized by the applied electric field with a rate dependent on both the size and surface passivation of the quantum dots [224, 225]. Separation of electron-hole pairs confined to the core of the dot requires significantly more energy than separation of carriers trapped at the surface and occurs through tunneling processes. New nanostructures, such as nanowires [215, 226], nanorods [134, 227–229], and nanobelts [230], may provide some interesting alternatives with better transport properties than nanoparticles. Devices such as LEDs and solar cells based on such nanostructures are expected to be developed in the next few years.

10.4.3. *Solid state lighting: ACPEL*

Solid state lighting is an area of fast growing interest. Approximately 30% of the United States electricity is consumed by lighting, an industry that is largely dominated by relatively old technologies such as the incandescent and fluorescent light bulb. New innovations in lower cost and higher efficiency of solid state lighting are expected to significantly reduce our dependence on fossil fuels. A solid state lighting technology that is already compatible with low cost manufacturing is AC powder electroluminescence (ACPEL). Discovered in 1936 [231], powder electroluminescence utilizes emission from ~40–50 micron sized doped-ZnS:Cu,Cl phosphor particles and requires relatively low applied electric fields (10^4 V/cm) compared to DC EL which requires electric fields near 10^6 V/cm. More recently, microencapsulation technology has been successfully applied to ZnS:Cu,Cl powder phosphors so that the emissive particles can be deposited on plastic substrates under open air, nonclean-room conditions using low cost, large area print-based manufacturing. Consequently, ACPEL lights are one of the least expensive large area solid state lighting technologies, and the characteristic blue–green light can now be found in many products. Furthermore, fluorescent energy conversion and doping can readily be used to convert the blue–green light to the white light preferred for normal everyday lighting. Using a variety of dopants (I, Br, Al, Mn, Pr, Tm etc.), other colors can also be obtained [232, 233].

The mechanism for light emission from ZnS:Cu,Cl particles is thought to be due to localized electron and hole injection near Cu_xS inclusions, and this requires an alternating current (AC) to enable the frequency-dependent electron-hole recombination. While this process is highly efficient, the lifetime and power efficiency of ACPEL lights are currently too low to provide a replacement for white lights. The power efficiency is limited by the large ZnS:Cu,Cl particle size (>20 μm) over which the electric field is dropped, resulting in the need for higher voltages, ~120 V, to achieve the brightness needed for solid state lighting. Research on phosphor nanoparticles has attempted to address the voltage issue; however, these systems typically have significantly reduced lifetime and quantum efficiency due to poor charge transport and/or trapping.

For example, an EXAFS study has been conducted to understand how the degradation depends on the local microstructure about Cu in doped-ZnS:Cu,Cl and have found that the degradation process is reversible through modest application of elevated temperatures (~200°C) [234]. Note that although the particle size is large (typically 20–50 µm), the active regions within each particle are near Cu_xS precipitates a few nm in size; hence it is essential to understand the Cu_xS nanoparticle precipitates within the ZnS host and how they interact with other optically active centers. Nanoparticle and nanorod systems exhibiting AC EL have been reported recently [233, 235]. Figure 10.16 shows EL spectra of ZnS:Cu$^+$, Al^{3+} nanodot and nanorods that clearly exhibit different radiative recombination pathways based on their different EL spectral features [235].

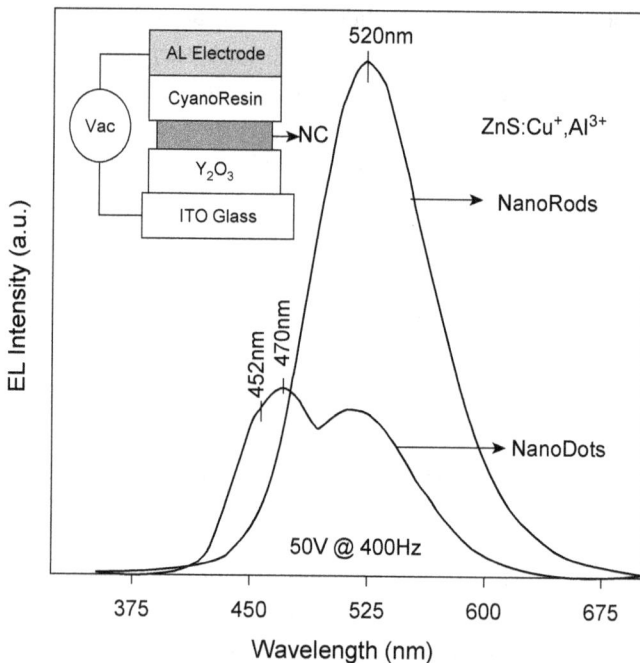

Fig. 10.16. EL spectra of ZnS:Cu$^+$, Al^{3+} nanodots and nanorods. Inset: Schematic structure of AC EL device based on nanodots/nanorods. Reproduced with permission from Ref. 235.

A recent study of ZnS:Cu NPs using combined PL and EXAFS studies has discovered that Cu exists in the +1 oxidation state and Cu dopant ions are in the interior but near the surface of the host ZnS NPs [236]. Further research is needed to better understand the local atomic structure of dopants and its correlation to the PL and EL properties relevant to solid state lighting applications.

10.4.4. *Optical detectors*

Semiconductor nanoparticles have also been investigated for optical detectors. For example, copper iodide nanocrystals have been suggested as potentially economical material for UV detectors [237]. Single photon detectors based on NbN nanowires have been successfully demonstrated [238, 239]. This is an area where further development is expected in the future along with the demand for optical detectors with improved performance including sensitivity, spectral range and speed. Nanomaterials are potentially useful for high energy radiation (e.g. X-ray and γ-ray) detection as well and this is a new area of growing interest.

10.5. Optical filters: photonic bandgap materials or photonic crystals

One of the earliest applications of semiconductor nanoparticles is in glass filters [240, 241]. The nanoparticles are usually produced at high temperature in glass and the size is controlled by the annealing temperature. The different-size particles generate different colored filters. This is one of the simplest and best uses of the quantum confinement effects of nanoparticles. The production of many glass filters or stained glass is based on this arguably oldest nanotechnology.

A special type of filter can be constructed from so-called *photonic bandgap materials* (PBM) or *photonic crystals* (PC) that are composed of nanoparticles with periodically packed structures [242–255]. Mesoscale nanoparticles assembled to form patterned superlattice nanostructures over large areas can be used to function as tunable, narrow-band optical filters [256, 257]. The position (λ_{max}) of the first-order diffraction peak is related to the spacing between planes (d), the refractive index of the crystalline assembly, and the angle (ϕ) between the incident light and the normal to the surface of the filter by the Bragg equation [256, 257]:

$$\lambda_{max} = 2d(n^2 - \sin^2 \phi)^{1/2} \qquad (10.5)$$

The peak wavelength is thus strongly dependent on the incident angle and plane spacing. These materials are also considered as diffractive components for other photonics applications, such as sensors [258, 259], tunable gratings [260], and photonic bandgap structures [261–263].

In some case, semiconductor nanoparticles are introduced in the voids of the photonic bandgap materials to influence the gap and optical properties such as photoemission. It was found that the emission lifetime of Mn^{2+} in Mn-doped ZnS nanoparticles is dependent on the dielectric constant of the photonic material [264]. In another example, 2D and 3D composite PCs of macroporous silicon and nanocrystalline PbS thin films have been fabricated and the influence of these films on both the passive and active optical properties of the PCs has been studied [265]. A consistent red-shift in the photonic stopbands of a composite 3D PC relative to its PC substrate was found and high PL extraction efficiency due to bandedge singularities of a composite 2D PC was demonstrated. Figure 10.17 (top) shows a cross-section SEM image of a 3D composite PC that is characterized by a planar 2D square pattern with a lattice constant of $a_x = 2.5$ μm and vertical periodicity of modulated pores with $a_z = 1.6$ μm. The thin PbS films covering the pore walls are about 60 nm in thickness and consist of nearly spherical NCs, as can be seen in the inset of Fig. 10.17 (top). Figure 10.17 (bottom) shows normal incident IR transmission spectra, measured by FTIR spectroscopy from the original PC substrate and from the composite 3D PCs. The experimental results are shown together with the calculated photonic band structure along the pore direction and the calculated transmission coefficient for a finite slab. It was suggested that for the composite PC, a larger portion of the IR beam propagates in the high index medium of PbS ($n = 4.01$) relative to the case of propagation in the PC substrate. As a result, the effective optical thickness of the PC medium is larger and consequently, a shorter normalized frequency is required to approach the photonic stopbands.

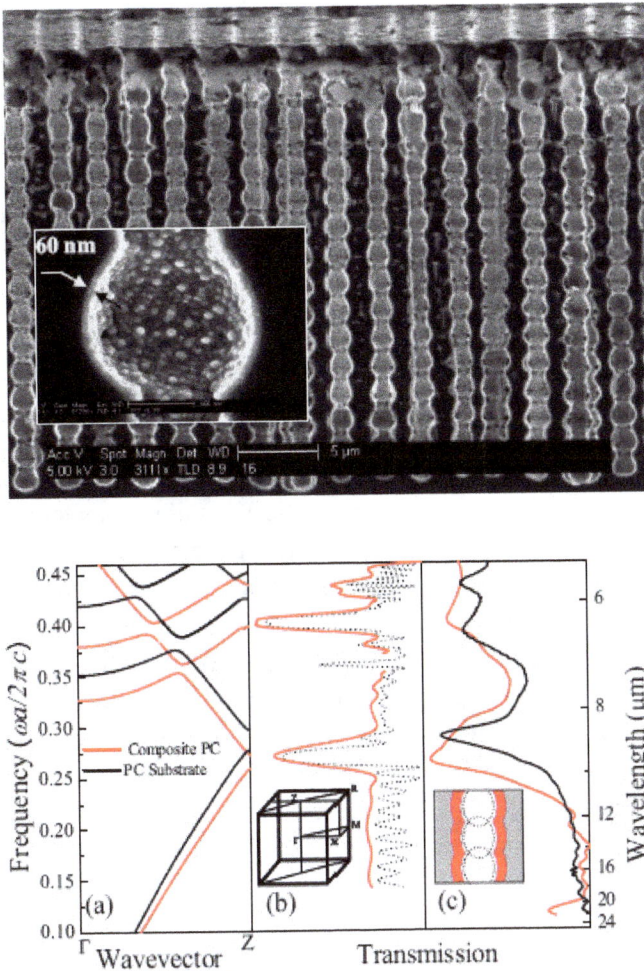

Fig. 10.17. (top) SEM cross-section view of a 3D macroporous silicon composite PC. The inset shows a magnified image of a PC unit cell with a thin film of spherical PbS NCs, about 60 nm in thickness and covering the pore walls. (bottom) (a) Calculated photonic band structure, (b) simulated transmission coefficient, and (c) measured transmission along the Γ-Z direction for the 3D PC substrate (black solid lines) and the 3D composite PC (red solid lines). The inset of (b) shows the high symmetry points of the first Brillouin zone, while the geometrical model of a PC unit cell is schematically illustrated in the inset of (c). The black dotted line in (b) shows the calculated fast oscillations (due to Fabry–Pérot modes) of the transmission coefficient, while the red solid line represents the slowly varying envelope of the transmission that better fits the experimental data. Reproduced with permission from Ref. 265.

Much research has been done to fabricate photonic crystals based on organic, such as liquid crystals and polymers, e.g. polystyrene, PS, and poly(methyl methacrylate), PMMA and inorganic, such as semiconductors and insulators [266]. Opals are examples of this type of material in nature and they are precious gems composed of submicron sized silica spheres packed together and cemented with another slightly different silica. Periodic variation in the dielectric function or index of refraction is key to the unique photonic properties of these materials.

10.6. Summary

This chapter provides a brief overview of some of the applications or emerging applications of various nanomaterials in relation to their optical properties. As can be seen, potential applications of nanomaterials are limited only by our imagination. As one can expect, with advancement in synthesis and fabrication techniques and characterization tools, the fundamental properties of nanomaterials will become better understood, current technologies will be improved, and new applications will be discovered. With the many challenges facing the world today, we can anticipate that nanomaterials and nanotechnology will play an increasingly more important role in the future in areas ranging from medicine to energy and environment. It is our hope that nanoscience and nanotechnology will be properly utilized to improve human life.

References

1. W.C.W. Chan and S.M. Nie, *Science* **281**, 2016 (1998).
2. P. Alivisatos, *Pure Appl. Chem.* **72**, 3 (2000).
3. H. Mattoussi, J.M. Mauro, E.R. Goldman, G.P. Anderson, V.C. Sundar, F.V. Mikulec and M.G. Bawendi, *J. Am. Chem. Soc.* **122**, 12142 (2000).
4. D. Gerion, F. Pinaud, S.C. Williams, W.J. Parak, D. Zanchet, S. Weiss and A.P. Alivisatos, *J. Phys. Chem. B* **105**, 8861 (2001).
5. J.K. Jaiswal, H. Mattoussi, J.M. Mauro and S.M. Simon, *Nat. Biotechnol.* **21**, 47 (2003).
6. W. Jiang, A. Singhal, J.N. Zheng, C. Wang and W.C.W. Chan, *Chem. Mater.* **18**, 4845 (2006).

7. K. Susumu, H.T. Uyeda, I.L. Medintz, T. Pons, J.B. Delehanty and H. Mattoussi, *J. Am. Chem. Soc.* **129**, 13987 (2007).

8. J.R. Lakowicz, *Principles of Fluorescence Spectroscopy*, New York: Kluwer. **371** (1999).

9. D.M. Willard, L.L. Carillo, J. Jung and A. Van Orden, *Nano Lett.* **1**, 469 (2001).

10. I.L. Medintz, A.R. Clapp, H. Mattoussi, E.R. Goldman, B. Fisher and J.M. Mauro, *Nat. Mater.* **2**, 630 (2003).

11. D.J. Zhou, J.D. Piper, C. Abell, D. Klenerman, D.J. Kang and L.M. Ying, *Chem. Commun.* 4807 (2005).

12. J.M. Zhang, Z. Dai, N. Guo, S.C. Xu, Q.X. Dong and B. Sun, *Chem. J. Chinese Universities-Chinese* **28**, 254 (2007).

13. C.R. Kagan, C.B. Murray, M. Nirmal and M.G. Bawendi, *Phys. Rev. Lett.* **76**, 1517 (1996).

14. C.R. Kagan, C.B. Murray and M.G. Bawendi, *Phys. Rev. B-Condensed Matter* **54**, 8633 (1996).

15. X.Y. Huang, L. Li, H.F. Qian, C.Q. Dong and J.C. Ren, *Angew Chem. Int Edit* **45**, 5140 (2006).

16. S.P. Wang, N. Mamedova, N.A. Kotov, W. Chen and J. Studer, *Nano Lett.* **2**, 817 (2002).

17. R. Robelek, L.F. Niu, E.L. Schmid and W. Knoll, *Anal. Chem.* **76**, 6160 (2004).

18. L.A. Bentolila and S. Weiss, *Cell Biochem. Biophys.* **45**, 59 (2006).

19. L. Ma, S.M. Wu, J. Huang, Y. Ding, D.W. Pang and L.J. Li, *Chromosoma* **117**, 181 (2008).

20. D.R. Larson, W.R. Zipfel, R.M. Williams, S.W. Clark, M.P. Bruchez, F.W. Wise and W.W. Webb, *Science* **300**, 1434 (2003).

21. Y.F. Chen and Z. Rosenzweig, *Anal. Chem.* **74**, 5132 (2002).

22. C.Q. Dong, H.F. Qian, N.H. Fang and J.C. Ren, *J. Phys. Chem. B* **110**, 11069 (2006).

23. V. Maurel, M. Laferriere, P. Billone, R. Godin and J.C. Scaiano, *J. Phys. Chem. B* **110**, 16353 (2006).

24. R.J. Heaton, P.I. Haris, J.C. Russell and D. Chapman, *Biochem. Soc. Trans.* **23**, S502 (1995).

25. R. Rella, J. Spadavecchia, M.G. Manera, P. Siciliano, A. Santino and G. Mita, *Biosens. Bioelectron.* **20**, 1140 (2004).

26. H. Li, D.F. Cui, J.Q. Liang, H.Y. Cai and Y.J. Wang, *Chinese Chem. Lett.* **17**, 1481 (2006).

27. I.C. Stancu, A. Fernandez-Gonzalez and R. Salzer, *J. Optoelectr. Adv. Mater.* **9**, 1883 (2007).
28. J.B. Beusink, A.M.C. Lokate, G.A.J. Besselink, G.J.M. Pruijn and R.B.M. Schasfoort, *Biosens. Bioelectron.* **23**, 839 (2008).
29. J.S. Yuk, J.W. Jung, Y.M. Mm and K.S. Ha, *Sensor Actuat B-Chem.* **129**, 113 (2008).
30. R. Elghanian, J.J. Storhoff, R.C. Mucic, R.L. Letsinger and C.A. Mirkin, *Science* **277**, 1078 (1997).
31. J.J. Storhofff, R. Elghanian, C.A. Mirkin and R.L. Letsinger, *Langmuir* **18**, 6666 (2002).
32. J.H. Li, X. Chu, Y.L. Liu, J.H. Jiang, Z.M. He, Z.W. Zhang, G.L. Shen and R.Q. Yu, *Nucleic Acids Res.* **33** (2005).
33. H.X. Li and L. Rothberg, *Proc. Nat. Acad. Sci. USA* **101**, 14036 (2004).
34. F. Abiko, K. Tomoo, A. Mizuno, S. Morino, H. Imataka and T. Ishida, *Biochem. Bioph. Res. Co.* **355**, 667 (2007).
35. L.L. Chen, L. Deng, L.L. Liu and Z.H. Peng, *Biosens. Bioelectron.* **22**, 1487 (2007).
36. X.Z. Du and Y.C. Wang, *J. Phys. Chem. B* **111**, 2347 (2007).
37. H. Dong, X.D. Cao, C.M. Li and W.H. Hu, *Biosens. Bioelectron.* **23**, 1055 (2008).
38. H. Petrova, C.H. Lin, S. de Liejer, M. Hu, J.M. McLellan, A.R. Siekkinen, B.J. Wiley, M. Marquez, Y.N. Xia, J.E. Sader and G.V. Hartland, *J. Chem. Phys.* **126**, 094709 (2007).
39. T. Wilkop, Z.Z. Wang and Q. Cheng, *Langmuir* **20**, 11141 (2004).
40. J.S. Yuk, D.G. Hong, H.I. Jung and K.S. Ha, *Sensor Actuat B-Chem.* **119**, 673 (2006).
41. Y. Li, H.J. Lee and R.M. Corn, *Nucleic Acids Res.* **34**, 6416 (2006).
42. H. Petrova, C.H. Lin, M. Hu, J.Y. Chen, A.R. Siekkinen, Y.N. Xia, J.E. Sader and G.V. Hartland, *Nano Lett.* **7**, 1059 (2007).
43. K.A. Willets and R.P. Van Duyne, *Ann. Rev. Phys. Chem.* **58**, 267 (2007).
44. D.A. Stuart, C.R. Yonzon, X.Y. Zhang, O. Lyandres, N.C. Shah, M.R. Glucksberg, J.T. Walsh and R.P. Van Duyne, *Anal. Chem.* **77**, 4013 (2005).
45. S. Haemers, G.J.M. Koper, M.C. van der Leeden and G. Frens, *Langmuir* **18**, 2069 (2002).
46. A.J. Haes and R.P. Van Duyne, *J. Am. Chem. Soc.* **124**, 10596 (2002).
47. A.J. Haes, L. Chang, W.L. Klein and R.P. Van Duyne, *J. Am. Chem. Soc.* **127**, 2264 (2005).

48. L.A. Dick, A.D. McFarland, C.L. Haynes and R.P. Van Duyne, *J. Phys. Chem. B* **106**, 853 (2002).

49. C.R. Yonzon, D.A. Stuart, X.Y. Zhang, A.D. McFarland, C.L. Haynes and R.P. Van Duyne, *Talanta* **67**, 438 (2005).

50. E. Fu, T. Chinowsky, K. Nelson, K. Johnston, T. Edwards, K. Helton, M. Grow, J.W. Miller and P. Yager, in *Oral-Based Diagnostics* **335** (2007).

51. K. Kneipp, H. Kneipp, I. Itzkan, R.R. Dasari and M.S. Feld, *J. Phys. Condens Matter* **14**, R597 (2002).

52. T.A. Alexander, P.M. Pellegrino and J.B. Gillespie, in *Smart Medical and Biomedical Sensor Technology.* 2004. Bellingham, WA: SPIE.

53. W.R. Premasiri, D.T. Moir, M.S. Klempner, N. Krieger, G. Jones and L.D. Ziegler, *J. Phys. Chem. B* **109**, 312 (2005).

54. X.H. Ji, S.P. Xu, L.Y. Wang, M. Liu, K. Pan, H. Yuan, L. Ma, W.Q. Xu, J.H. Li, Y.B. Bai and T.J. Li, *Colloid Surface A* **257–58**, 171 (2005).

55. J. Kneipp, H. Kneipp and K. Kneipp, *Proc. Nat. Acad. Sci. USA* **103**, 17149 (2006).

56. S. Shanmukh, L. Jones, J. Driskell, Y.P. Zhao, R. Dluhy and R.A. Tripp, *Nano Lett.* **6**, 2630 (2006).

57. Y.W.C. Cao, R.C. Jin and C.A. Mirkin, *Science* **297**, 1536 (2002).

58. R.C. Jin, Y.C. Cao, C.S. Thaxton and C.A. Mirkin, *Small* **2**, 375 (2006).

59. B.D. Chithrani and W.C.W. Chan, *Nano Lett.* **7**, 1542 (2007).

60. K. Kneipp, A.S. Haka, H. Kneipp, K. Badizadegan, N. Yoshizawa, C. Boone, K.E. Shafer-Peltier, J.T. Motz, R.R. Dasari and M.S. Feld, *App. Spectroscopy* **56**, 150 (2002).

61. J. Kneipp, H. Kneipp, M. McLaughlin, D. Brown and K. Kneipp, *Nano Lett.* **6**, 2225 (2006).

62. R. Shukla, V. Bansal, M. Chaudhary, A. Basu, R.R. Bhonde and M. Sastry, *Langmuir* **21**, 10644 (2005).

63. C.E. Talley, L. Jusinski, C.W. Hollars, S.M. Lane and T. Huser, *Anal. Chem.* **76**, 7064 (2004).

64. S.W. Bishnoi, C.J. Rozell, C.S. Levin, M.K. Gheith, B.R. Johnson, D.H. Johnson and N.J. Halas, *Nano Lett.* **6**, 1687 (2006).

65. A.M. Schwartzberg, T.Y. Oshiro, J.Z. Zhang, T. Huser and C.E. Talley, *Anal. Chem.* **78**, 4732 (2006).

66. R.A. Jensen, J. Sherin and S.R. Emory, *Appl. Spectroscopy* **61**, 832 (2007).

67. J. Kneipp, H. Kneipp, B. Wittig and K. Kneipp, *Nano Lett.* **7**, 2819 (2007).

68. A. Shamsaie, M. Jonczyk, J. Sturgis, J.P. Robinson and J. Irudayaraj, *J. Biomed. Opt.* **12**, 205021 (2007).

69. M. Bruchez, M. Moronne, P. Gin, S. Weiss and A.P. Alivisatos, *Science* 281, 2013 (1998).

70. A. Sukhanova, L. Venteo, J. Devy, M. Artemyev, V. Oleinikov, M. Pluot and I. Nabiev, *Laboratory Investigation* 82, 1259 (2002).

71. X.Y. Wu, H.J. Liu, J.Q. Liu, K.N. Haley, J.A. Treadway, J.P. Larson, N.F. Ge, F. Peale and M.P. Bruchez, *Nat. Biotechnol.* 21, 41 (2003).

72. A.M. Derfus, W.C.W. Chan and S.N. Bhatia, *Advan. Mater.* 16, 961 (2004).

73. A. Sukhanova, M. Devy, L. Venteo, H. Kaplan, M. Artemyev, V. Oleinikov, D. Klinov, M. Pluot, J.H.M. Cohen and I. Nabiev, *Anal. Biochem.* 324, 60 (2004).

74. J.A. Kloepfer, R.E. Mielke, M.S. Wong, K.H. Nealson, G. Stucky and J.L. Nadeau, *Appl. Environ. Microb.* 69, 4205 (2003).

75. L.C. Mattheakis, J.M. Dias, Y.J. Choi, J. Gong, M.P. Bruchez, J.Q. Liu and E. Wang, *Anal. Biochem.* 327, 200 (2004).

76. B. Ballou, B.C. Lagerholm, L.A. Ernst, M.P. Bruchez and A.S. Waggoner, *Bioconjugate Chem.* 15, 79 (2004).

77. S. Kim, Y.T. Lim, E.G. Soltesz, A.M. De Grand, J. Lee, A. Nakayama, J.A. Parker, T. Mihaljevic, R.G. Laurence, D.M. Dor, L.H. Cohn, M.G. Bawendi and J.V. Frangioni, *Nat. Biotechnol.* 22, 93 (2004).

78. X. Michalet, F.F. Pinaud, L.A. Bentolila, J.M. Tsay, S. Doose, J.J. Li, G. Sundaresan, A.M. Wu, S.S. Gambhir and S. Weiss, *Science* 307, 538 (2005).

79. T.T. Goodrich, H.J. Lee and R.M. Corn, *J. Am. Chem. Soc.* 126, 4086 (2004).

80. L.K. Wolf, D.E. Fullenkamp and R.M. Georgiadis, *J. Am. Chem. Soc.* 127, 17453 (2005).

81. C.L. Jahncke, H.D. Hallen and M.A. Paesler, *J. Raman Spectroscopy* 27, 579 (1996).

82. C.J.L. Constantino, R.F. Aroca, C.R. Mendonca, S.V. Mello, D.T. Balogh and O.N. Oliveira, *Spectrochim Acta A* 57, 281 (2001).

83. J.C. Carter, W.A. Scrivens, M.L. Myrick and S.M. Angel, *Appl. Spectroscopy* 57, 761 (2003).

84. V. Deckert, D. Zeisel, R. Zenobi and T. Vo-Dinh, *Anal. Chem.* 70, 2646 (1998).

85. T. Vo-Dinh, F. Yan and M.B. Wabuyele, *J. Raman Spectroscopy* 36, 640 (2005).

86. M.B. Wabuyele, F. Yan, G.D. Griffin and T. Vo-Dinh, *Rev. Sci. Instr.* 76, 063710 (2005).

87. C. Loo, A. Lin, L. Hirsch, M.H. Lee, J. Barton, N. Halas, J. West and R. Drezek, *Technol. Cancer Res. Treatment* 3, 33 (2004).

88. X.H. Huang, I.H. El-Sayed, W. Qian and M.A. El-Sayed, *J. Am. Chem. Soc.* **128**, 2115 (2006).

89. I.H. El-Sayed, X.H. Huang and M.A. El-Sayed, *Cancer Lett.* **239**, 129 (2006).

90. X.H. Huang, P.K. Jain, I.H. El-Sayed and M.A. El-Sayed, *Nanomed.* **2**, 681 (2007).

91. A.K. Oyelere, P.C. Chen, X.H. Huang, I.H. El-Sayed and M.A. El-Sayed, *Bioconjugate Chem.* **18**, 1490 (2007).

92. P.K. Jain, X. Huang, I.H. El-Sayed and M.A. El-Sayad, *Plasmonics* **2**, 107 (2007).

93. M.P. Melancon, W. Lu, Z. Yang, R. Zhang, Z. Cheng, A.M. Elliot, J. Stafford, T. Olsen, J.Z. Zhang and C. Li, Mole. *Cancer Therapeutics* **7**, 1730 (2008).

94. J.A. Copland, M. Eghtedari, V.L. Popov, N. Kotov, N. Mamedova, M. Motamedi and A.A. Oraevsky, *Mol. Imag. Biol.* **6**, 341 (2004).

95. X.M. Yang, S.E. Skrabalak, Z.Y. Li, Y.N. Xia and L.H.V. Wang, *Nano Lett.* **7**, 3798 (2007).

96. M. Eghtedari, A. Oraevsky, J.A. Copland, N.A. Kotov, A. Conjusteau and M. Motamedi, *Nano Lett.* **7**, 1914 (2007).

97. J. Shah, S. Park, S. Aglyamov, T. Larson, L. Ma, K. Sokolov, K. Johnston, T. Milner and S.Y. Emelianov, *J. Biomed. Opt.* **13**, 034024 (2008).

98. W. Lu, C. Xiong, G. Zhang, Q. Huang, R. Zhang, J.Z. Zhang and C. Li, *Clinical Cancer Res.* **15**, 876 (2008).

99. T.J. Dougherty, *J. Opt. Soc. Am. B* **1**, 555 (1984).

100. T.G. Truscott, A.J. McLean, A. Phillips and W.S. Foulds, *Molecular Aspects of Med.* **11**, 106 (1990).

101. D. Kovalev and M. Fuji, in *Annual Review of Nano Research*, G. Cao and C.J. Brinker, Editors., World Scientific: Singapore. **159** (2008).

102. D. Kovalev, E. Gross, N. Kunzner, F. Koch, V.Y. Timoshenko and M. Fujii, *Phys. Rev. Lett.* **89** (2002).

103. A.C.S. Samia, X.B. Chen and C. Burda, *J. Am. Chem. Soc.* **125**, 15736 (2003).

104. M.E. Wieder, D.C. Hone, M.J. Cook, M.M. Handsley, J. Gavrilovic and D.A. Russell, *Photochem. Photobiol. Sci.* **5**, 727 (2006).

105. T.Y. Hulchanskyy, I. Roy, L.N. Goswami, Y. Chen, E.J. Bergey, R.K. Pandey, A.R. Oseroff and P.N. Prasad, *Nano Lett.* **7**, 2835 (2007).

106. Y. Cheng, A.C. Samia, J.D. Meyers, I. Panagopoulos, B.W. Fei and C. Burda, *J. Am. Chem. Soc.* **130**, 10643 (2008).

107. X.H. Huang, P.K. Jain, I.H. El-Sayed and M.A. El-Sayed, *Photochemistry and Photobiology*, **82**, 412 (2006).

108. B. Khlebtsov, V. Zharov, A. Melnikov, V. Tuchin and N. Khlebtsov, *Nanotechnology* **17**, 5167 (2006).

109. J.Y. Chen, D.L. Wang, J.F. Xi, L. Au, A. Siekkinen, A. Warsen, Z.Y. Li, H. Zhang, Y.N. Xia and X.D. Li, *Nano Lett.* **7**, 1318 (2007).

110. A.M. Gobin, M.H. Lee, N.J. Halas, W.D. James, R.A. Drezek and J.L. West, *Nano Lett.* **7**, 1929 (2007).

111. X. Huang, W. Qian, I.H. El-Sayed and M.A. El-Sayed, *Lasers in Surgery and Medicine* **39**, 747 (2007).

112. T.A. Larson, J. Bankson, J. Aaron and K. Sokolov, *Nanotechnology* **18**, 325101 (2007).

113. M.A. Green, *Third Generation Photovoltaics: Advanced Solar Energy Conversion*, Berlin; New York: Springer. xi (2003).

114. H.A. Atwater, B. Sopori, T. Ciszek, L.C. Feldman, J. Gee and A. Rohatgi, Snadia National Labs: Albuquerque. 111 (1999).

115. L.L. Kazmerski, in *The 29th IEEE PV Specialists Conference*. 2002. New Orleans, LA: NREL.

116. A. Shah, P. Torres, R. Tscharner, N. Wyrsch and H. Keppner, *Science* **285**, 692 (1999).

117. T. Bak, J. Nowotny, M. Rekas and C.C. Sorrell, *Int. J. Hydrogen Energy* **27**, 991 (2002).

118. M.S. Tomar and F.J. Garcia, *Thin Solid Films* **90**, 419 (1982).

119. M. Alonealaluf, J. Appelbaum and N. Croitoru, *Thin Solid Films* **320**, 159 (1998).

120. K.D. Dobson, I. Visoly-Fisher, G. Hodes and D. Cahen, *Sol. Energ. Mat. Sol. C* **62**, 295 (2000).

121. B. Oregan and M. Gratzel, *Nature* **353**, 737 (1991).

122. N.C. Greenham, X.G. Peng and A.P. Alivisatos, *Phys. Rev. B-Condensed Matter* **54**, 17628 (1996).

123. M.K. Nazeeruddin, A. Kay, I. Rodicio, R. Humphrybaker, E. Muller, P. Liska, N. Vlachopoulos and M. Gratzel, *J. Am. Chem. Soc.* **115**, 6382 (1993).

124. T. Lopez-Luke, A. Wolcott, L.-P. Xu, S. Chen, Z. Wen, J.H. Li, E. De La Rosa and J.Z. Zhang, *J. Phys. Chem. C* **112**, 1282 (2008).

125. A. Kongkanand, K. Tvrdy, K. Takechi, M. Kuno and P.V. Kamat, *J. Am. Chem. Soc.* **130**, 4007 (2008).

126. P.V. Kamat, *J. Phys. Chem. C* **112**, 18737 (2008).

127. G. Yu, G. Pakbaz and A.J. Heeger, *Appl. Phys. Lett.* **64**, 3422 (1994).
128. G. Yu and A.J. Heeger, *J. Appl. Phys.* **78**, 4510 (1995).
129. T.J. Savenije, J.M. Warman and A. Goossens, *Chem. Phys. Lett.* **287**, 148 (1998).
130. A.C. Arango, S.A. Carter and P.J. Brock, *Appl. Phys. Lett.* **74**, 1698 (1999).
131. C.D. Grant, A.M. Schwartzberg, G.P. Smestad, J. Kowalik, L.M. Tolbert and J.Z. Zhang, *J. Electroanal. Chem.* **522**, 40 (2002).
132. C.D. Grant, A.M. Schwartzberg, G.P. Smestad, J. Kowalik, L.M. Tolbert and J.Z. Zhang, *Synthet. Metal.* **132**, 197 (2003).
133. A.C. Arango, L.R. Johnson, V.N. Bliznyuk, Z. Schlesinger, S.A. Carter and H.H. Horhold, *Advan. Mater.* **12**, 1689 (2000).
134. W.U. Huynh, X.G. Peng and A.P. Alivisatos, *Advan. Mater.* **11**, 923 (1999).
135. D. Gebeyehu, C.J. Brabec, F. Padinger, T. Fromherz, S. Spiekermann, N. Vlachopoulos, F. Kienberger, H. Schindler and N.S. Sariciftci, *Synthet. Metal* **121**, 1549 (2001).
136. T. Fromherz, F. Padinger, D. Gebeyehu, C. Brabec, J.C. Hummelen and N.S. Sariciftci, *Sol. Energ. Mat. Sol. C* **63**, 61 (2000).
137. S. Spiekermann, G. Smestad, J. Kowalik, L.M. Tolbert and M. Gratzel, *Synthet. Metal* **121**, 1603 (2001).
138. S. Gan, A. El-azab and Y. Liang, *Surface Sci.* **479**, L369 (2001).
139. D. Godovsky, L.C. Chen, L. Pettersson, O. Inganas, M.R. Andersson and J.C. Hummelen, *Adv. Mater. Opt. Electron.* **10**, 47 (2000).
140. N. Beermann, L. Vayssieres, S.E. Lindquist and A. Hagfeldt, *J. Electrochem. Soc.* **147**, 2456 (2000).
141. T. Lindgren, H.L. Wang, N. Beermann, L. Vayssieres, A. Hagfeldt and S.E. Lindquist, *Sol. Energ. Mat. Sol. C* **71**, 231 (2002).
142. W.U. Huynh, J.J. Dittmer and A.P. Alivisatos, *Science* **295**, 2425 (2002).
143. S. Uchida, R. Chiba, M. Tomiha, N. Masaki and M. Shirai, in *Nanotechnology in Mesostructured Materials.* 791(2003).
144. S. Ngamsinlapasathian, S. Sakulkhaemaruethai, S. Pavasupree, A. Kitiyanan, T. Sreethawong, Y. Suzuki and S. Yoshikawa, *J. Photoch Photobio A* **164**, 145 (2004).
145. M. Law, L.E. Greene, J.C. Johnson, R. Saykally and P.D. Yang, *Nat. Mater.* **4**, 455 (2005).
146. B.R. Mehta and F.E. Kruis, *Sol. Energ. Mat. Sol. C* **85**, 107 (2005).
147. S. Pavasupree, S. Ngamsinlapasathian, M. Nakajima, Y. Suzuki and S. Yoshikawa, *J. Photoch Photobio A* **184**, 163 (2006).
148. J.H. Park, S. Kim and A.J. Bard, *Nano Lett.* **6**, 24 (2006).

149. H.M. Jia, H. Xu, Y. Hu, Y.W. Tang and L.Z. Zhang, *Electrochem. Commun.* **9**, 354 (2007).

150. K.S. Kim, Y.S. Kang, J.H. Lee, Y.J. Shin, N.G. Park, K.S. Ryu and S.H. Chang, *B Kor. Chem. Soc* **27**, 295 (2006).

151. K. Pan, Q.L. Zhang, Q. Wang, Z.Y. Liu, D.J. Wang, J.H. Li and Y.B. Bai, *Thin Solid Films* **515**, 4085 (2007).

152. Y.B. Liu, B.X. Zhou, B.T. Xiong, J. Bai and L.H. Li, *Chinese Sci. Bull.* **52**, 1585 (2007).

153. Q.H. Yao, J.F. Liu, Q. Peng, X. Wang and Y.D. Li, *Chemistry-an Asian J.* **1**, 737 (2006).

154. C. Burda, Y.B. Lou, X.B. Chen, A.C.S. Samia, J. Stout and J.L. Gole, *Nano Lett.* **3**, 1049 (2003).

155. J.L. Gole, J.D. Stout, C. Burda, Y.B. Lou and X.B. Chen, *J. Phys. Chem. B* **108**, 1230 (2004).

156. X.B. Chen and C. Burda, *J. Phys. Chem. B* **108**, 15446 (2004).

157. K. Kobayakawa, Y. Murakami and Y. Sato, *J. Photoch. Photobio. A* **170**, 177 (2005).

158. M. Sathish, B. Viswanathan, R.P. Viswanath and C.S. Gopinath, *Chem. Mater.* **17**, 6349 (2005).

159. Y. Nakano, T. Morikawa, T. Ohwaki and Y. Taga, *Appl. Phys. Lett.* **86**, 132104 (2005).

160. S.W. Yang and L. Gao, *J. Inorg. Mater.* **20**, 785 (2005).

161. A.R. Gandhe, S.P. Naik and J.B. Fernandes, *Micropor. Mesopor. Mat.* **87**, 103 (2005).

162. K. Yamada, H. Nakamura, S. Matsushima, H. Yamane, T. Haishi, K. Ohira and K. Kumada, *Cr. Chim.* **9**, 788 (2006).

163. P. Xu, L. Mi and P.N. Wang, *J. Cryst. Growth* **289**, 433 (2006).

164. A. Fujishima and K. Honda, *Nature* **238**, 37 (1972).

165. B.H. Wang, D.J. Wang, Y. Cui, J. Zhang, X.D. Chai and T.J. Li, *Synthet. Metal* **71**, 2239 (1995).

166. R.S. Mane, B.R. Sankapal and C.D. Lokhande, *Mater. Chem. Phys.* **60**, 196 (1999).

167. R.S. Mane and C.D. Lokhande, *Mater. Chem. Phys.* **78**, 385 (2002).

168. G. Kumara, K. Tennakone, I.R.M. Kottegoda, P.K.M. Bandaranayake, A. Konno, M. Okuya, S. Kaneko and K. Murakami, *Semiconductor Science & Technology* **18**, 312 (2003).

169. I. Bedja, S. Hotchandani and P.V. Kamat, *J. Phys. Chem.* **97**, 11064 (1993).

170. I. Bedja, S. Hotchandani, R. Carpentier, K. Vinodgopal and P.V. Kamat, *Thin Solid Films* **247**, 195 (1994).

171. I. Saeki, N. Okushi, H. Konno and R. Furuichi, *J. Electrochem. Soc.* **143**, 2226 (1996).

172. H.L. Wang, T. Lindgren, J.J. He, A. Hagfeldt and S.E. Lindquist, *J. Phys. Chem. B* **104**, 5686 (2000).

173. U.O. Krasovec, M. Topic, A. Georg, A. Georg and G. Drazic, *J. Sol. Gel. Sci. Techn.* **36**, 45 (2005).

174. C.V. Ramana, S. Utsunomiya, R.C. Ewing, C.M. Julien and U. Becker, *J. Phys. Chem. B* **110**, 10430 (2006).

175. A. Wolcott, T.R. Kuykendall, W. Chen, S.W. Chen and J.Z. Zhang, *J. Phys. Chem. B* **110**, 25288 (2006).

176. F. Cao, G. Oskam, G.J. Meyer and P.C. Searson, *J. Phys. Chem.* **100**, 17021 (1996).

177. E. Stathatos and P. Lianos, *Int. J. Photoenergy* **4**, 11 (2002).

178. T. Stergiopoulos, I.M. Arabatzis, G. Katsaros and P. Falaras, *Nano Lett.* **2**, 1259 (2002).

179. H. Saitoh, K. Takayama, H. Sugata, S. Ohshio, H. Takada, Y. Yamazaki, Y. Yamaguchi and Y. Ono, *Jpn. J. Appl. Phys.* **2**, 41, L1250 (2002).

180. P.R. Mishra, P.K. Shukla and O.N. Srivastava, *Int. J. Hydrogen Energ.* **32**, 1680 (2007).

181. H.L. Zhao, D.L. Jiang, S.L. Zhang and W. Wen, *J. Catal.* **250**, 102 (2007).

182. D. Chen, Y.F. Gao, G. Wang, H. Zhang, W. Lu and J.H. Li, *J. Phys. Chem. C* **111**, 13163 (2007).

183. A. Wolcott, W. Smith, T.R. Kuykendall, Y. Zhao and J.Z. Zhang, *Small* **5**, 104 (2008).

184. Q. Shen, K. Katayama, T. Sawada, M. Yamaguchi and T. Toyoda, *Jpn. J. Appl. Phys.* **1**, 45, 5569 (2006).

185. Y. Murakami, B. Kasahara and Y. Nosaka, *Chem. Lett.* **36**, 330 (2007).

186. S. Chandra, *Photoelectrochemical Solar Cells*, New York: Gorden and Breach. 270 (1985).

187. M.A. Fox and M.T. Dulay, *Chem. Rev.* **93**, 341 (1993).

188. W.Y. Choi, A. Termin and M.R. Hoffmann, *J. Phys. Chem.* **98**, 13669 (1994).

189. W.Y. Choi, A. Termin and M.R. Hoffmann, *Angew. Chem., Int. Ed. Engl.* **33**, 1091 (1994).

190. M.R. Hoffmann, S.T. Martin, W.Y. Choi and D.W. Bahnemann, *Chem. Rev.* **95**, 69 (1995).

191. L.Q. Jing, X.J. Sun, J. Shang, W.M. Cai, Z.L. Xu, Y.G. Du and H.G. Fu, *Sol. Energ. Mat. Sol. C* **79**, 133 (2003).

192. D.F. Ollis and H. Al-Ekabi, *Photocatalytic Purification and Treatment of Water and Air: Proc. 1st Int. Conf. TiO* Photocatalytic Purification and*

Treatment of Water and Air, London, Ontario, Canada, 8–13 November, 1992. Trace metals in the environment; 3, Amsterdam; New York: Elsevier. xiii (1993).

193. A. Mills, R.H. Davies and D. Worsley, *Chem. Soc. Rev.* **22**, 417 (1993).
194. N. Serpone and R.F. Khairutdinov, in *Semiconductor Nanoclusters-Physical, Chemical, and Catalytic Aspects,* P.V. Kamat and D. Meisel, Editors, Elsevier: New York. **417** (1997).
195. K. Okamoto, Y. Yamamoto, H. Tanaka, M. Tanaka and A. Itaya, *Bull. Chem. Soc. Japan* **58**, 2015 (1985).
196. M. Andersson, L. Osterlund, S. Ljungstrom and A. Palmqvist, *J. Phys. Chem. B* **106**, 10674 (2002).
197. C.H. Kwon, H.M. Shin, J.H. Kim, W.S. Choi and K.H. Yoon, *Mater. Chem. Phys.* **86**, 78 (2004).
198. Y. Nemoto and T. Hirai, *B Chem. Soc. Jpn.* **77**, 1033 (2004).
199. Y. Jiang, P. Zhang, Z.W. Liu and F. Xu, *Mater. Chem. Phys.* **99**, 498 (2006).
200. Z.X. Wang, S.W. Ding and M.H. Zhang, *Chinese J. Inorg. Chem.* **21**, 437 (2005).
201. Y.C. Lee and S. Cheng, *J. Chinese Chem. Soc.* **53**, 1355 (2006).
202. W.W. Dunn, Y. Aikawa and A.J. Bardm, *J. Am. Chem. Soc.* **6893** (1981).
203. H. Harada, T. Ueda and T. Sakata, *J. Phys. Chem.* **93**, 1542 (1989).
204. L.F. Lin and R.R. Kuntz, *Langmuir* **8**, 870 (1992).
205. P. Zuman and Z. Fijalek, *J. Electroanal. Chem. Interf. Electrochem.* **296**, 583 (1990).
206. H.Y. Wang, R.E. Partch and Y.Z. Li, *J. Organic Chem.* **62**, 5222 (1997).
207. R. Mirin, A. Gossard and J. Bowers, *Electron. Lett.* **32**, 1732 (1996).
208. S. Fafard, K. Hinzer, A.J. Springthorpe, Y. Feng, J. McCaffrey, S. Charbonneau and E.M. Griswold, *Mat. Sci. Eng. B-Solid* **51**, 114 (1998).
209. K. Hinzer, C.N. Allen, J. Lapointe, D. Picard, Z.R. Wasilewski, S. Fafard and A.J.S. Thorpe, *J. Vac. Sci. Technol. A* **18**, 578 (2000).
210. K. Hinzer, J. Lapointe, Y. Feng, A. Delage, S. Fafard, A.J. SpringThorpe and E.M. Griswold, *J. Appl. Phys.* **87**, 1496 (2000).
211. S. Tanaka, H. Hirayama, Y. Aoyagi, Y. Narukawa, Y. Kawakami and S. Fujita, *Appl. Phys. Lett.* **71**, 1299 (1997).
212. M.K. Zundel, K. Eberl, N.Y. Jin-Phillipp, F. Phillipp, T. Riedl, E. Fehrenbacher and A. Hangleiter, *J. Cryst. Growth* **202**, 1121 (1999).
213. V.I. Klimov, A.A. Mikhailovsky, S. Xu, A. Malko, J.A. Hollingsworth, C.A. Leatherdale, H.J. Eisler and M.G. Bawendi, *Science* **290**, 314 (2000).

214. F. Hide, B.J. Schwartz, M.A. Diazgarcia and A.J. Heeger, *Chem. Phys. Lett.* **256**, 424 (1996).

215. M.H. Huang, S. Mao, H. Feick, H.Q. Yan, Y.Y. Wu, H. Kind, E. Weber, R. Russo and P.D. Yang, *Science* **292**, 1897 (2001).

216. S.A. Carter, J.C. Scott and P.J. Brock, *Appl. Phys. Lett.* **71**, 1145 (1997).

217. Y. Wang and N. Herron, *J. Luminescence* **70**, 48 (1996).

218. S. Chaudhary, M. Ozkan and W.C.W. Chan, *Appl. Phys. Lett.* **84**, 2925 (2004).

219. V.L. Colvin, M.C. Schlamp and A.P. Alivisatos, *Nature* **370**, 354 (1994).

220. M.C. Schlamp, X.G. Peng and A.P. Alivisatos, *J. Appl. Phys.* **82**, 5837 (1997).

221. S. Nakamura, K. Kitamura, H. Umeya, A. Jia, M. Kobayashi, A. Yoshikawa, M. Shimotomai and K. Takahashi, *Electron. Lett.* **34**, 2435 (1998).

222. V.V. Khorenko, S. Malzer, C. Bock, K.H. Schmidt and G.H. Dohler, *Phys. Status Solidi B* **224**, 129 (2001).

223. J.L. Zhao, J.Y. Zhang, C.Y. Jiang, J. Bohnenberger, T. Basche and A. Mews, *J. Appl. Phys.* **96**, 3206 (2004).

224. C.A. Leatherdale, C.R. Kagan, N.Y. Morgan, S.A. Empedocles, M.A. Kastner and M.G. Bawendi, *Phys. Rev. B* **62**, 2669 (2000).

225. H. Mattoussi, A.W. Cumming, C.B. Murray, M.G. Bawendi and R. Ober, *Phys. Rev. B-Condensed Matter* **58**, 7850 (1998).

226. L. Brus, *J. Phys. Chem.* **98**, 3575 (1994).

227. Z.L. Wang, M.B. Mohamed, S. Link and M.A. El-Sayed, *Surface Sci.* **440**, L809 (1999).

228. L. Manna, E.C. Scher and A.P. Alivisatos, *J. Am. Chem. Soc.* **122**, 12700 (2000).

229. Z. Adam and X. Peng, *J. Amer. Chem. Soc.* **123**, 183 (2001).

230. Z.W. Pan, Z.R. Dai and Z.L. Wang, *Science* **291**, 1947 (2001).

231. Y.A. Ono, Annual Review of Materials. *Science* **27**, 283 (1997).

232. S. Tanaka, H. Kobayashi and H. Sasakura, in *Phosphor Handbook*, S. Shionoya and W.M. Yen, Editors., CRC Press: New York. 601 (1999).

233. K. Manzoor, S.R. Vadera, N. Kumar and T.R.N. Kutty, *Appl. Phys. Lett.* **84**, 284 (2004).

234. M. Warkentin, F. Bridges, S.A. Carter and M. Anderson, *Phys. Rev. B* **75**, 075301 (2007).

235. K. Manzoor, V. Aditya, S.R. Vadera, N. Kumar and T.R.N. Kutty, *Solid State Commun.* **135**, 16 (2005).

236. C. Corrado, Y. Jiang, F. Oba, M. Kozina, F. Bridges and J.Z. Zhang, *J. Phys. Chem. C*, in press (2009).

237. R. DeMeis, *Laser Focus World* **33**, 46 (1997).
238. B.S. Robinson, A.J. Kerman, E.A. Dauler, R.O. Barron, D.O. Caplan, M.L. Stevens, J.J. Carney, S.A. Hamilton, J.K.W. Yang and K.K. Berggren, *Opt. Lett.* **31**, 444 (2006).
239. K.M. Rosfjord, J.K.W. Yang, E.A. Dauler, A.J. Kerman, V. Anant, B.M. Voronov, G.N. Gol'tsman and K.K. Berggren, *Opt. Express* **14**, 527 (2006).
240. N.F. Borrelli, D.W. Hall, H.J. Holland and D.W. Smith, *J. Appl. Phys.* **61**, 5399 (1987).
241. I. Okur and P.D. Townsend, *Nuclear Instruments & Methods in Physics Research Section B-Beam Interactions With Materials and Atoms* **124**, 76 (1997).
242. S.H. Park and Y.N. Xia, *Langmuir* **15**, 266 (1999).
243. Y.N. Xia, J.A. Rogers, K.E. Paul and G.M. Whitesides, *Chem. Rev.* **99**, 1823 (1999).
244. A. Rogach, A. Susha, F. Caruso, G. Sukhorukov, A. Kornowski, S. Kershaw, H. Mohwald, A. Eychmuller and H. Weller, *Advan. Mater.* **12**, 333 (2000).
245. X.H. Hu and C.T. Chan, *Appl. Phys. Lett.* **85**, 1520 (2004).
246. X.D. Wang, C. Neff, E. Graugnard, Y. Ding, J.S. King, L.A. Pranger, R. Tannenbaum, Z.L. Wang and C.J. Summers, *Advan. Mater.* **17**, 2103 (2005).
247. M. Matsuu, S. Shimada, K. Masuya, S. Hirano and M. Kuwabara, *Advan. Mater.* **18**, 1617 (2006).
248. S. Noda, M. Fujita and T. Asano, *Nature Photon.* **1**, 449 (2007).
249. V. Reboud, N. Kehagias, M. Zelsmann, C. Schuster, M. Fink, F. Reuther, G. Gruetzner and C.M.S. Torres, *Opt. Express*, **15**, 7190 (2007).
250. Y. Yang, S.H. Zhang and G.P. Wang, *Appl. Opt.* **46**, 84 (2007).
251. R.B. Wehrspohn, H.S. Kitzerow and K. Busch, *Phys. Status Solidi a — Applications and Materials Science* **204**, 3583 (2007).
252. C.K. Lee, R. Radhakrishnan, C.C. Chen, J. Li, J. Thillaigovindan and N. Balasubramanian, *J. Lightwave Technol.* **26**, 839 (2008).
253. D.A.S. Razo, L. Pallavidino, E. Garrone, F. Geobaldo, E. Descrovi, A. Chiodoni and F. Giorgis, *J. Nanoparticle Res.* **10**, 1225 (2008).
254. M.R. Singh, *Appl. Phys. B-Lasers O* **93**, 91 (2008).
255. M.J. Ventura and M. Gu, *Advan. Mater.* **20**, 1329 (2008).
256. P.L. Flaugh, S.E. O'Donnel and S.A. Asher, *Appl. Spectroscopy* **38**, 847 (1984).
257. R.J. Spry and D.J. Kosan, *Appl. Spectroscopy* **40**, 782 (1986).
258. J.H. Holtz and S.A. Asher, *Nature* **389**, 829 (1997).

259. J.H. Holtz, J.S.W. Holtz, C.H. Munro and S.A. Asher, *Anal. Chem.* **70**, 780 (1998).

260. J.M. Weissman, H.B. Sunkara, A.S. Tse and S.A. Asher, *Science* **274**, 959 (1996).

261. Tarhan, II and G.H. Watson, *Phys. Rev. Lett.* **76**, 315 (1996).

262. W.L. Vos, M. Megens, C.M. Vankats and P. Bosecke, *J. Phys-Condens Matter* **8**, 9503 (1996).

263. H. Miguez, F. Meseguer, C. Lopez, A. Blanco, J.S. Moya, J. Requena, A. Mifsud and V. Fornes, *Advan. Mater.* **10**, 480+ (1998).

264. J. Zhou, Y. Zhou, S. Buddhudu, S.L. Ng, Y.L. Lam and C.H. Kam, *Appl. Phys. Lett.* **76**, 3513 (2000).

265. N. Gutman, A. Armon, A. Sa'ar, A. Osherov and Y. Golan, *Appl. Phys. Lett.* **93**, 073111 (2008).

266. A. Blanco and C. Lopez, in *Annual Review of Nano Research*, G. Cao and C.J. Brinker, Editors., World Scientific: Singapore. **81** (2007).

Index

www.ingramcontent.com/pod-product-compliance
Lightning Source LLC
Chambersburg PA
CBHW061616220326
41598CB00026BA/3781